Blueprints for Future Living

Dimitrios Sargiotis

Blueprints for Future Living

AI-Driven Urban Innovation and Resilient Cities

Springer

Dimitrios Sargiotis
School of Civil Engineering
National Technical University of Athens
Zografou Campus, Athens, Greece

ISBN 978-3-032-24177-1 ISBN 978-3-032-24178-8 (eBook)
https://doi.org/10.1007/978-3-032-24178-8

This Springer imprint is published by the registered company Springer Nature Switzerland AG
The registered company address is: Gewerbestrasse 11, 6330 Cham, Switzerland

Preface

We no longer merely inhabit cities—we coevolve with them. Cities today are no longer inert containers of human activity; they are becoming algorithmically sentient entities, co-constituted by sensors, data flows, ethical algorithms, and civic desire. This book, Blueprints for Future Living: AI-Driven Urban Innovation and Resilient Cities, is not a technocratic manual. It is a scholarly cartography and philosophical invocation—mapping how artificial intelligence (AI), when deliberately aligned with justice, ecology, and participation, can metamorphose urban life into a more intelligent and more humane condition.

This volume refuses the simplistic optimism of the "smart city" and instead assembles an architecture of urban intelligence that thinks with and for the public good. It chronicles a global gallery of deeply technical, rigorously modeled case studies—from Barcelona's proximal policy optimization (PPO)-driven traffic governance to Singapore's LSTM-powered transit modulation; from Amsterdam's predictive DRL–LSTM energy grid orchestration to New York's AI-augmented waste logistics; and from Dubai's ethically contested algorithmic policing to Ålesund's participatory digital twins. Each of these systems is not simply a marvel of computational precision—they are ethical propositions. They reveal how code becomes governance, how architecture becomes algorithm, and how data becomes destiny.

At its core, this book interrogates a profound epistemic inversion: AI is no longer a tool deployed within cities. It is the epistemology of the city itself. The infrastructures of mobility, energy, waste, and public health do not merely host AI—they are increasingly defined by it. With each algorithm trained, each reinforcement policy deployed, and each feedback loop closed, the city becomes less a fixed terrain and more a learning system: self-aware, feedback-driven, and continuously becoming.

But this transformation is not ethically neutral. The very same AI that enables anticipatory bus scheduling or grid-stabilizing renewable energy dispatch can also be repurposed toward surveillance, extractive efficiency, or unjust exclusion. This book embraces that ambivalence, confronting it not with fear but with design. Its mathematical models are paired with civic ethics. Its architectures are animated by governance constraints. Its simulations are not endgames, but provocations for participatory dialogue and responsible intervention.

Indeed, what is at stake in this work is not merely the optimization of urban systems, but the optimization of urban conscience. Can we design AI to learn not only traffic patterns, but also democratic values? Can we embed explainability, fairness, and human dignity directly into the decision matrices that shape lived urban experience? This book asserts: yes, but only if we reframe intelligence itself—not as performance, but as care.

To that end, this volume advances a theoretical and practical framework for algorithmic empathy. It proposes cities where machine learning is not merely fast, but fair; where neural networks do not just compute, but collaborate; and where infrastructure becomes the visible form of an invisible ethics.

This book is both a report and a declaration. It documents the leading edge of AI-infused urbanism while inviting readers—engineers, policymakers, scholars, and citizens alike—to participate in shaping a future where intelligence is not just artificial, but just. In that sense, this work is not complete. It is an evolving infrastructure of ideas, a generative prototype for reimagining how cities can think, feel, and act in the age of AI—not for capital or control, but for climate resilience, social equity, and collective flourishing.

Let this volume be read not as the end of an inquiry, but as the beginning of a new contract between machine logic and human values—one in which intelligence is not merely deployed in the city, but co-authored by it.

Zografou Campus, Athens, Greece Dimitrios Sargiotis

Introduction

The contemporary city is being asked to do more under conditions of less certainty. Densification, climate volatility, strained infrastructure, and rising expectations of public service have converged to place urban systems under a form of pressure that incremental reform seems unlikely to relieve. It is within this predicament that Artificial Intelligence (AI) has moved from the periphery of urban management toward its center. What was until recently an analytical aid, useful for forecasting or classification, now increasingly shapes how mobility is scheduled, how energy is dispatched, how waste is routed, and how risk is anticipated. The argument advanced across this volume is that AI is no longer merely deployed within the city; it is becoming part of the city's operating logic, an infrastructure in its own right rather than an instrument laid over the ones that already exist. Much of the existing literature on the smart city tends to settle at one of two poles. One celebrates technological capability and reports efficiency gains with limited attention to the normative stakes; the other offers valuable critique of surveillance and datafication yet rarely engages the technical substance of the systems it examines. This book attempts to occupy the more demanding middle ground. It takes the engineering seriously, presenting explicit mathematical formulations and layered computational architectures, while insisting that every such architecture also encodes a set of choices about justice, participation, and ecological responsibility. Each case study is therefore read twice over: once as a technical artifact, and once as an ethical proposition.

Methodologically, the volume proceeds through grounded, comparative analysis rather than abstraction. It draws on documented implementations across a range of urban contexts and problem domains, allowing the reader to see not only what a given model optimizes but also what it assumes, what it excludes, and what it makes possible. The comparative frame is deliberate. Setting cities with different governance traditions, resources, and political cultures alongside one another helps to distinguish what may be genuinely transferable in AI-driven urbanism from what remains contingent on local conditions.

The organization of the book follows this logic. The opening chapter establishes the core domains of urban life, mobility, energy, waste, and public safety, through comparative case studies that introduce the principal AI methods used throughout. The chapters that follow extend the inquiry outward. One turns to integrated governance frameworks in two Polish voivodships, examining how layered AI architectures can be aligned with regional resilience and social inclusion. Another investigates AI-empowered digital twins across five cities, treating governance itself as a design problem of roles, processes, and controls. A further chapter draws lessons on evaluation from five United Kingdom cities, arguing that measurement belongs at the heart of decision-making rather than at its margins. The volume then considers AI-enabled personalization and digital mental health care, before closing with the role of AI in disaster resilience and climate adaptation, where flood forecasting serves as a particularly demanding test case.

Taken together, these chapters support a single, if carefully qualified, contention: that to design the cities of the coming decades is, in large part, to design the AI systems that will govern them, and that such design cannot responsibly be separated from questions of value. Intelligence embedded in urban infrastructure may extend efficiency, yet it may just as readily extend exclusion; the difference lies less in the models themselves than in how they are governed. The book is addressed, accordingly, to a mixed readership of engineers, planners, policymakers, and scholars, in the hope that the technical and the civic might be held together rather than pursued apart.

Acknowledgments

This book stands as the culmination of a solitary intellectual odyssey—one anchored in the belief that cities, at their most profound, are not built solely of steel, stone, or silicon, but of ethics, inference, and shared aspiration. Blueprints for Future Living was conceived not as a technical monograph, but as a philosophical investigation into how artificial intelligence may be architected to resonate with the civic soul of the urban condition. Every algorithmic framework herein is, at its core, a gesture toward a more dignified future.

Though authored alone, this work is indebted to a constellation of minds and movements. It draws on decades of pioneering scholarship at the intersection of machine learning, governance, and urban ecology—scholarship that continues to illuminate the contested terrain between intelligence and responsibility. I owe much to the interdisciplinary visionaries whose insights have seeded the intellectual soil from which this book emerged. Their influence is neither quoted nor footnoted in full—but it is everywhere, breathing through every framework, every case study, and every page.

My profound thanks extend to the editorial and production teams at Springer, whose rigor, precision, and intellectual generosity transformed a complex manuscript into a volume worthy of global readership. Their stewardship upheld the highest standards of academic publishing, ensuring that this book enters the world as both artifact and invitation: a provocation for thoughtful engagement, and a benchmark of scholarly excellence.

To the international institutions, research consortia, and civic organizations who persistently advocate for ethical, inclusive, and sustainable technological futures—you are the invisible infrastructure beneath this book. Your unwavering commitment to digital justice, participatory governance, and climate-aware urbanism informs the deepest strata of this work's moral architecture.

And to those closest to me—my family, my quietest interlocutors, and my enduring sources of patience and grace—this work owes its resilience. In moments of conceptual difficulty, philosophical fatigue, or sleepless scrutiny of a single equation, your presence restored balance. You offered silence when thought demanded solitude, and solidarity when vision required courage.

This book is not merely research. It is a reflection of conviction. A declaration that intelligence, if it is to matter, must first serve. That cities, if they are to thrive, must be not only smart, but also just. And that AI, if it is to endure, must be designed not to dominate, but to dignify.

May this volume stand as a testament to those who believe that the future is not only a technical problem, but also a moral horizon—and that our responsibility is not only to build systems, but also to build futures worth inheriting.

Competing Interests
The author affirms that this book is the result of wholly independent scholarly labor. Blueprints for Future Living: AI-Driven Urban Innovation and Resilient Cities was neither commissioned, financed, nor shaped by any external entity—corporate, governmental, or institutional. At no point during its conceptualization, composition, or production was any form of sponsorship, partnership, or financial inducement sought or accepted.

Every theoretical construct, AI architecture, and urban case study herein is the product of autonomous inquiry, grounded in interdisciplinary research and animated by an unwavering commitment to methodological integrity. This work was not conceived as a vehicle for commercial promotion, technological endorsement, or proprietary advocacy. No algorithmic model, dataset, or implementation example included in this volume serves the interests of any vendor, platform, or political actor.

Rather, this monograph is dedicated to the unfettered advancement of knowledge at the intersection of artificial intelligence, civic governance, and ethical urbanism. Its purpose is scholarly—not strategic; educational—not instrumental. The author's sole allegiance is to the principles of scientific transparency, epistemic clarity, and the public good.

This declaration stands as a testament to the intellectual independence and ethical orientation of the work. Readers may engage with this volume in full confidence that it reflects only the author's academic judgment, unmediated by private interest or institutional bias.

About the Book

In an era demanding unprecedented urban adaptability, sustainability, and intelligence, this book "Blueprints for Future Living: AI-Driven Urban Innovation and Resilient Cities," part of the book series "AI and the Intelligent City: Data-Driven Innovations for Urban Transformation," reconceives the city as an evolving computational organism—one that senses, predicts, and autonomously optimizes its functions through artificial intelligence (AI). Anchored in rigorously documented real-world implementations, it synthesizes state-of-the-art AI systems across transportation, energy, waste management, and public safety.

Barcelona's deep reinforcement learning (DRL) traffic control framework, employing proximal policy optimization (PPO), exemplifies emission-aware, real-time adaptation of urban mobility infrastructure. Singapore's LSTM-augmented transit networks dynamically forecast and recalibrate commuter flows, operationalizing AI as a proactive mobility governor. Amsterdam's dual-layer long short-term memory (LSTM)–deep reinforcement learning (DRL) model orchestrates energy consumption and dispatch aligned with renewable inflows, while Stockholm's extreme gradient boosting (XGBoost) optimizes smart grid sustainability through data-driven source allocation.

In waste management, AI accelerates circular economy ambitions: New York City utilizes deep reinforcement learning (DRL)-enhanced route optimization, while San Francisco achieves real-time convolutional neural network (CNN)-based material sorting accuracy. Urban safety systems in Dubai and Barcelona integrate long short-term memory (LSTM), support vector machine (SVM), and proximal policy optimization (PPO) engines to preemptively coordinate law enforcement and emergency responses.

These interconnected frameworks collectively herald a paradigm shift—where AI transcends tool status to become the very infrastructure of urban life, intrinsically imbued with ethical values and anticipatory governance. By uniting algorithmic precision with civic purpose, this book challenges readers to envision cities, where designing AI systems is inseparable from architecting just, sustainable, and sentient urban futures.

Contents

About the Author

Dimitrios Sargiotis is a Faculty Dr. and Deputy Director of the PC-Laboratories within the Digital Technology Laboratory for Technical Projects at the School of Civil Engineering, National Technical University of Athens. With more than three decades of academic and professional experience, he stands at the intersection of Artificial Intelligence, Computer Science, Data Governance, and Civil Engineering, pioneering the digital transformation of infrastructures, cities, and societies.

He has authored seventeen scholarly books, one hundred thirty-seven book chapters, and over fifty scientific publications, articles, research reports, and other scholarly contributions disseminated through Springer and other prestigious international outlets, establishing himself as one of the leading voices in AI for Smart Cities, Digital Twins, Intelligent Transportation Systems, Green and Sustainable Infrastructure, and Human-Centric AI. His frameworks emphasize ethics, explainability, and governance, aligning technological innovation with public value and societal well-being.

Dr. Sargiotis provides leadership in European Commission–funded research projects, serves as a reviewer and program committee member for international journals and conferences, and collaborates with esteemed organizations, including Springer, the Programming Research Group of Paris, and the Union of Computer Engineers and Communications of Greece.

Beyond academia, he is an advocate of life-long learning, physical well-being, and cultural engagement, embodying a vision where scientific rigor, ethical responsibility, and human creativity converge. His career demonstrates how Artificial Intelligence can act as a catalyst for sustainable development, social justice, and the flourishing of digitally governed societies worldwide.

Abbreviations

AI	Artificial intelligence—a computational paradigm wherein machines simulate cognitive functions such as learning, reasoning, and perception, applied here to optimize urban infrastructure, resilience, and governance.
BERT	Bidirectional encoder representations from transformers—a deep learning model for language understanding, employed in NLP tasks such as policy analysis and civic discourse evaluation.
CNN	Convolutional neural network—a class of deep neural networks used primarily in visual processing tasks, including waste classification and real-time monitoring.
DRL	Deep reinforcement learning—a learning paradigm combining deep neural networks with reinforcement learning to enable adaptive decision-making in complex urban environments.
FC	Fully connected—a type of neural network layer where each node is connected to every node in the previous layer, foundational to deep learning architectures.
GAT	Graph attention network—a form of GNN that uses attention mechanisms to weigh the influence of neighboring nodes dynamically, improving representational power in urban graph data.
GIS	Geographic information system—a framework for gathering, managing, and analyzing spatial and geographic data, supporting AI-driven urban planning and environmental monitoring.
GNN	Graph neural networks—models designed to operate on graph-structured data, used in the book to analyze interconnected urban systems and infrastructure networks.
GPR	Gaussian process regression—a nonparametric Bayesian approach for regression tasks, used for uncertainty-aware predictions in urban data modeling.

GRU Gated recurrent unit—a simplified variant of LSTM networks, capturing temporal patterns with fewer parameters, suitable for sequential modeling in urban AI systems.
ICT Information and communication technology—an umbrella term for technology infrastructure enabling digital communication, foundational for smart city AI deployment.
LSTM Long short-term memory—a recurrent neural network architecture designed to learn and retain temporal dependencies in sequential data, used in this volume for predictive modeling in transport and energy systems.
MVI Multivariate index—a composite indicator aggregating multiple variables, used in AI evaluation metrics for smart city performance.
NLP Natural language processing—the field of AI focused on enabling machines to interpret, generate, and learn from human language, applied in participatory governance and feedback systems.
PPO Proximal policy optimization—a policy gradient reinforcement learning algorithm that balances performance and stability, utilized in dynamic traffic control and energy dispatch optimization.
RL Reinforcement learning—a machine learning paradigm in which agents learn optimal behavior by interacting with an environment to maximize cumulative reward.
RNN Recurrent neural network—a class of neural networks that process sequences by maintaining hidden states, used in forecasting and temporal analysis.
SHAP SHapley Additive exPlanations—a game theoretic approach to interpreting machine learning models used in the book to ensure transparency and accountability in AI decisions.

Annotations

Citizen engagement The involvement of residents in the design, feedback, and oversight of AI-enabled services to ensure transparency, legitimacy, and civic alignment.
Climate change A global environmental phenomenon addressed through AI models that forecast impacts, optimize resource allocation, and support mitigation strategies in cities.
Climate resilience The capacity of urban systems to adapt to, withstand, and recover from climate-related stresses through AI-supported modeling, prediction, and responsive adaptation strategies.

Deep reinforcement learning	An advanced AI technique combining deep neural networks with reinforcement learning, enabling complex decision-making in high-dimensional and dynamic urban environments.
Digital twin technology	The technological infrastructure enabling digital twins, including data integration, modeling platforms, and simulation engines that mirror physical urban dynamics.
Digital twin	A real-time digital replica of a physical urban system or asset, continuously updated with sensor data to enable monitoring, simulation, and intelligent decision-making.
Federated learning	A privacy-preserving machine learning approach where models are trained collaboratively across distributed edge devices or nodes without exchanging raw data.
Layered AI architecture	A multilevel computational framework organizing AI modules hierarchically to handle data ingestion, learning, control, and feedback across urban infrastructure systems.
Long short-term memory	A type of recurrent neural network capable of capturing long-range temporal dependencies in sequential data, used for forecasting demand in transportation and energy systems.
Mental health	A critical domain in urban AI systems involving the deployment of technologies such as chatbots or diagnostic models to support emotional well-being and psychiatric care.
Predictive maintenance	An AI-driven strategy that anticipates equipment failures or service degradation in infrastructure by analyzing patterns in operational data.
Proximal policy optimization	A stable and efficient reinforcement learning algorithm that updates policies using clipped surrogate objectives, widely applied in adaptive traffic and energy systems.
Public health	The application of AI in the management, monitoring, and forecasting of health outcomes and risks within urban populations.
Reinforcement learning	A machine learning paradigm where autonomous agents learn to make sequential decisions by receiving feedback through reward signals, used for adaptive urban control.

Smart city	An urban environment that leverages digital technology, data analytics, and interconnected systems to optimize public services, infrastructure, and sustainability outcomes.

Chapter 1
AI-Driven Urban Sustainability: Comparative Case Studies of Smart Cities in Mobility, Energy, and Waste Systems

The twenty-first century city stands at a decisive intersection between intelligence and sustainability, where artificial intelligence (AI) no longer serves as an auxiliary analytical tool but as the central nervous system of urban transformation. This chapter examines how data-centric intelligence, machine learning architectures, and reinforcement learning control systems have become the structural drivers of sustainable urban governance. It explores how pioneering cities such as Barcelona, Singapore, Amsterdam, Stockholm, New York City, and San Francisco operationalize AI to harmonize environmental performance with social efficiency through real-time adaptive management, predictive optimization, and circular resource design.

The discussion begins with urban mobility, where Deep Reinforcement Learning (DRL) guided by Proximal Policy Optimization (PPO) enables Barcelona's adaptive traffic signal control to minimize congestion and emissions, while Singapore's Long Short-Term Memory (LSTM) neural networks forecast passenger demand to optimize public transport schedules through autonomous, demand-responsive bus allocation. The analysis then advances into urban energy systems, where Amsterdam integrates LSTM-based demand forecasting with DRL-driven grid control to balance consumption, stabilize renewable integration, and reduce waste, and Stockholm enhances this paradigm through Extreme Gradient Boosting (XGBoost) and reinforcement learning frameworks to manage supply-demand equilibrium, cost-efficiency, and emission mitigation. The narrative culminates with the deployment of AI in waste management, where sensor-based smart bins and DRL-optimized routing transform traditional collection logistics into anticipatory, low-emission systems aligned with circular-economy principles.

By uniting predictive analytics, neural forecasting, and optimization under a single methodological umbrella, this chapter demonstrates that AI is not merely a digital extension of existing infrastructures—it is the generative intelligence that learns, corrects, and redesigns the metabolic flows of cities. Through explicit mathematical formulations, layered architectural blueprints, and empirically grounded case studies, it offers a reproducible and exportable framework for

D. Sargiotis, *Blueprints for Future Living*,
https://doi.org/10.1007/978-3-032-24178-8_1

municipal leaders, data engineers, and sustainability strategists seeking to embed intelligence directly into the urban fabric. The result is a vision of cities as evolving cognitive ecosystems—responsive, self-regulating, and ethically adaptive—where technology becomes a means of sustaining life rather than consuming it.

1.1 Transportation and Mobility

1.1.1 AI-Driven Traffic Management in Barcelona

Barcelona has emerged as a paradigm of AI-integrated traffic intelligence. The city's real-time traffic management platform consolidates multi-source data streams—surveillance feeds, inductive loop sensors, and mobile app-derived crowd signals—to orchestrate dynamic traffic regulation strategies (Yigitcanlar & Cugurullo, 2020). Through advanced machine learning architectures, these heterogeneous inputs are transformed into actionable insights that predict traffic congestion points and activate real-time signal reconfigurations.

A foundational feature of this system lies in its capacity to adaptively manage intersection behavior under varying vehicular densities. During high-congestion intervals, the platform recalibrates green-light durations to alleviate directional bottlenecks. Inversely, it minimizes green time in underloaded approaches, a function that introduces both flow optimization and environmental benefits. Where traditional traffic cycles remain rigidly timed and often oblivious to real-world variability, Barcelona's adaptive model leverages continuous data inputs to orchestrate smoother vehicular transitions through the urban grid (Yigitcanlar & Cugurullo, 2020).

Perhaps most critical is the system's mitigation of vehicular idle times—a common by-product of static scheduling—by synchronizing intersection performance with real-time spatial flow data. Reduced idling not only diminishes localized emissions but also contributes to the city's wider decarbonization targets. Recent deployments of similar AI-powered systems report emission reductions of up to 30%, supporting broader urban sustainability objectives through traffic flow optimization (Yigitcanlar & Cugurullo, 2020).

To mathematically model this dynamic signal control regime, we adopt a Deep Reinforcement Learning (DRL) framework governed by Proximal Policy Optimization (PPO)—a policy-gradient algorithm recognized for its convergence reliability and adaptability to non-stationary urban environments. The agent continuously refines intersection-level policies based on real-time feedback, aiming to optimize a composite objective: maximizing traffic throughput, stabilizing signal transitions, and minimizing emissions.

At the foundation of this architecture lies a high-resolution sensing layer that captures the evolving physical state of urban intersections. This interface translates raw traffic data—such as vehicle counts, queue lengths, emission indices, and signal phase timers—into temporally indexed state vectors s_t, forming the perceptual substrate for all downstream decision-making. By establishing this empirical-to-symbolic mapping, the agent maintains continuous awareness of spatial-temporal conditions and environmental fluctuations across the control horizon.

Before the agent can engage in reward-based learning and policy optimization, it must first perceive the environment in a form that is computationally meaningful. Raw urban traffic inputs—such as vehicle counts, queue lengths, emission levels, and light phase timers—are inherently noisy, heterogeneous, and high-dimensional. These must be mapped into a structured format suitable for decision-making within the reinforcement learning framework. To achieve this, the model invokes a transformation function that performs perceptual abstraction and dimensional coherence. This encoding mechanism formally defines the state vector s_t, which becomes the temporal reference for all subsequent control actions and value evaluations.

$$s_t = \varphi(x_t)$$

where s_t, structured traffic state vector at time t, $\varphi(\cdot)$, deterministic feature extraction or encoding function, x_t, raw observation vector containing real-time traffic data. This function transforms raw environmental inputs into a state-space representation compatible with the learning policy $\pi(a_t | s_t)$, enabling informed signal control under uncertainty.

Mathematically, the transformation $\varphi(x_t)$ can assume various forms depending on system complexity. For low-dimensional cases, it may be expressed as a linear projection:

$$\varphi(x_t) = Wx_t + b$$

where $\varphi(x_t)$encoded state vector, W, learnable weight matrix, x_t, raw observation vector, and b, learnable bias term. This function performs affine transformation of the input space into a latent subspace where policy evaluation and optimization are computationally tractable.

For environments requiring hierarchical or nonlinear feature abstraction—such as those involving spatial congestion distributions, temporal patterns, or emission surges—$\varphi(x_t)$ is often implemented via multilayer neural networks. A minimal feedforward realization can be written as:

$$\varphi(x_t) = \sigma(W_2 \cdot \sigma(W_1 x_t + b_1) + b_2)$$

where $\varphi(x_t)$, encoded state vector, x_t, raw observation vector, W_1, W_2, learnable weight matrices, b_1, b_2, learnable biases, $\sigma(\cdot)$, nonlinear activation function (e.g., ReLU: ($\sigma(z) = max(0, z)$), tanh: ($\sigma(z) = \frac{e^z - e^{-z}}{e^z + e^{-z}}$)). This function applies a two-layer neural encoding that maps the observation space into a high-level latent feature space, supporting expressive policy learning in non-stationary traffic contexts.

The resulting policy $\pi\pi$ is optimized via a multi-objective reward function that jointly encodes traffic efficiency and sustainability constraints:

$$\pi^* = \arg\max_{\pi} \mathbb{E}\left[\sum_{t=0}^{T} \gamma^t (R(s_t, a_t) - \lambda_1 \cdot c(a_t) - \lambda_2 \cdot e(a_t))\right]$$

where, π^*, optimal signal control policy, π, stochastic policy mapping state-action pairs, $\mathbb{E}[\cdot]$, expectation operator over time horizon T, γ, discount factor weighting future versus immediate performance, $R(s_t, a_t)$, reward function measuring vehicle throughput efficiency, $c(a_t)$, transition cost penalizing frequent signal changes, $e(a_t)$, emission estimate linked to idling time and stop-start cycles, λ_1, penalty coefficient for transition stability, and λ_2, penalty coefficient for environmental sustainability. This function directs the reinforcement learning agent to maximize operational efficiency while adhering to physical and environmental stability constraints.

To isolate the component that drives traffic flow optimization, we formalize the reward function $R(s_t, a_t)$, which incentivizes maximal vehicle clearance and minimum delay at each intersection. This reward reflects the immediate performance of signal action a_t under state s_t, typically based on empirical measures like queue reduction or throughput rate.

$$R(s_t, a_t) = \eta_1 \cdot q_{\text{cleared}}(t) - \eta_2 \cdot w_{\text{avg}}(t)$$

where η_1, weight for queue clearance benefit, $q_{\text{cleared}}(t)$, number of vehicles cleared from intersection queues, η_2, weight for penalizing average wait time, and $w_{\text{avg}}(t)$, average vehicle waiting time. This function promotes signal decisions that enhance real-time traffic throughput while minimizing congestion latency.

In tandem, the stability of signal transitions is enforced through a regularization function $c(a_t)$, designed to penalize frequent or abrupt switching between signal phases. This control-theoretic term suppresses oscillatory behaviors that can emerge from over-reactive policies in dynamic urban environments.

$$c(a_{\text{t}}) = 1_{\{a_t \neq a_{t-1}\}} \cdot \delta$$

where $1_{\{a_t \neq a_{t-1}\}}$, indicator function for signal phase change and δ, fixed transition cost scalar. This function penalizes policy trajectories that frequently alter intersection phases, fostering signal consistency and reducing behavioral noise.

The final component in the learning signal addresses environmental sustainability by quantifying the emissions generated due to traffic idling and acceleration cycles, both of which are exacerbated by poor signal timing. This is captured through the emission penalty function $e(a_t)$, which links traffic signal actions to carbon and particulate output.

$$e(a_t) = \rho \cdot \text{IdleTime}(a_t) + \kappa \cdot \text{StopGoCycles}(a_t)$$

where ρ, weight assigned to total idling time, IdleTime(a_t), cumulative idling duration induced by action a_t, κ, weight assigned to start-stop emission cost, and StopGoCycles(a_t), count of abrupt deceleration-acceleration transitions. This function incorporates real-world emission factors into the learning objective, enabling the model to internalize environmental externalities during policy training.

Following the multi-objective training established in the prior layer, the policy π^* now operates within a real-time execution environment, mapping dynamically evolving traffic states to immediate signal control actions. At each discrete time step, the model samples from its optimized policy distribution to determine the most contextually appropriate intervention. This interaction closes the decision loop between learned policy inference and the evolving urban environment.

$$a_t \sim \pi_\theta(a_t \mid s_t)$$

where a_t, selected traffic signal control action at time t, $\pi_\theta(a_t \mid s_t)$, stochastic policy parameterized by θ, and s_t, current intersection-level traffic state. This function defines how the trained policy generates probabilistic signal control decisions conditioned on the present system state, enabling real-time responsiveness and adaptability under uncertainty.

Upon executing the action a_t, the system transitions into a new environmental state s_{t+1}, which becomes observable through sensor input. The cumulative emission levels, queue metrics, and transition frequencies induced by a_t are also logged. This real-time feedback is instrumental not only for traffic adaptation but also for policy evaluation and future refinement cycles.

Having established the real-time mapping from traffic states to signal actions, the model enters its policy refinement phase—an essential learning loop designed to enhance the long-term quality of decision-making without destabilizing already effective strategies. This is accomplished through the Proximal Policy Optimization (PPO) surrogate loss, which implements a clipped objective ensuring that updates to the control policy remain within a trust region. This formulation balances improvement incentives with bounded risk, yielding reliable convergence behavior across diverse traffic regimes.

$$L^{\text{PPO}}(\theta) = \mathbb{E}_t[\min(r_t(\theta)A_t, \text{clip}(r_t(\theta), 1-\epsilon, 1+\epsilon)A_t)]$$

where $L^{\text{PPO}}(\theta)$, clipped surrogate objective for PPO policy update, $\mathbb{E}_t$, expectation over sampled trajectories, $r_t(\theta)$, probability ratio between new and old policies, A_t, advantage estimate at time t, and ϵ, trust region threshold enforcing stability. This function optimizes the policy by balancing exploitation of advantageous actions against the risk of overfitting to stochastic feedback, preserving robustness in volatile traffic environments.

To operationalize this refinement mechanism, the core sub-function is the probability ratio $r_t(\theta)$, which quantifies how the updated policy diverges from the previous one under the same action-state context. This term dynamically scales the advantage, determining whether an update should proceed unmodified or be constrained.

$$r_t(\theta) = \frac{\pi_\theta(a_t \mid s_t)}{\pi_{\theta_{\text{old}}}(a_t \mid s_t)}$$

where $r_t(\theta)$, policy probability ratio at time t, $\pi_\theta(a_t \mid s_t)$, current policy evaluated at action-state pair, and $\pi_{\theta_{\text{old}}}(a_t \mid s_t)$, previous policy distribution before update. This function dynamically adjusts the learning signal based on relative policy improvement, ensuring that only substantial and reliable gains influence the next iteration of control strategies.

The clipping operator further regulates the refinement process. It applies soft bounds to the policy ratio to prevent excessive parameter changes that could destabilize learned behavior. This results in the formation of a piecewise surrogate objective, favoring moderate, reliable steps.

$$\text{clip}(r_t(\theta), 1 - \epsilon, 1 + \epsilon)$$

where clip(·), bounded linear operator, $r_t(\theta)$, policy ratio, and ϵ, allowed deviation threshold for policy updates. This function enforces conservative learning by truncating policy change magnitude when it exceeds a predefined confidence interval.

The advantage function A_t derives from the contrast between the value of executing a specific action and the baseline expectation for the current state. This contrast hinges on the precise formulation of both the action-value function and the state-value function, each serving a distinct evaluative role in the policy's update logic.

$$A_t = Q(s_t, a_t) - V(s_t)$$

where A_t, advantage function at time t, $Q(s_t, a_t)$, action-value function representing the expected cumulative reward when taking action a_t in state s_t, and $V(s_t)$, state-value function estimating the expected return from state s_t. This function quantifies how much better or worse a specific action performs relative to the policy's expected behavior at that state, providing the principal update signal in PPO training.

The action-value function $Q(s_t, a_t)$ evaluates the expected long-term reward when executing action a_t from state s_t and following the current policy thereafter. It integrates environmental feedback and emission penalties accumulated over the temporal horizon.

$$Q(s_t, a_t) = \mathbb{E}_\pi \left[\sum_{k=0}^{\infty} \gamma^k (R(s_{t+k}, a_{t+k}) - \lambda_1 c(a_{t+k}) - \lambda_2 e(a_{t+k})) \right]$$

where $Q(s_t, a_t)$, expected return from executing action a_t at state s_t, $\mathbb{E}_\pi[\cdot]$, expectation over trajectories induced by policy π, γ, discount factor for future rewards, $R(s_{t+k},$

a_{t+k}), throughput reward function, $c(a_{t+k})$, signal transition penalty, $e(a_{t+k})$, emission cost function, λ_1, penalty coefficient for instability, and λ_2, penalty coefficient for emissions. This function evaluates the long-term return of taking action a_t, capturing both efficiency and environmental costs under the learned policy trajectory.

The state-value function $V(s_t)$, by contrast, aggregates the expected return assuming the agent acts optimally according to the policy ππ from state s_t onward, without committing to a particular initial action. It serves as a dynamic baseline for policy comparison.

$$V(s_t) = \mathbb{E}_{a_t \sim \pi}[Q(s_t, a_t)]$$

where $V(s_t)$, expected return from state s_t under policy π, $\mathbb{E}_{a_t \sim \pi}[\cdot]$, expectation over actions sampled from the policy at state s_t, and $Q(s_t, a_t)$, action-value function for each action a_t. This function serves as a baseline in advantage estimation, providing a policy-conditioned valuation of the current state irrespective of a specific action choice.

Together, the tightly integrated sub-functions $Q(s_t, a_t)$, $V(s_t)$, and A_t form the evaluative backbone of the PPO refinement mechanism. They encode performance differentials that allow the agent to identify and prioritize behavior changes that yield substantive system-wide gains in both traffic efficiency and environmental alignment.

This DRL-based policy framework continuously evaluates and refines intersection-level signal strategies based on cumulative system rewards, promoting fluid traffic throughput while explicitly penalizing erratic phase transitions and emission-intensive control states. The integration of the cost function $c(a_t)$ enforces temporal stability by suppressing disruptive oscillations in signal patterns, whereas the emission penalty term $e(a_t)$ internalizes environmental externalities—such as CO2 and PM2.5 emissions—arising from prolonged idling and stop-go cycles.

Grounded in rigorous policy-gradient optimization, this DRL formulation enables Barcelona's AI infrastructure to function as a closed-loop, real-time adaptive control system. It unifies policy learning, environmental feedback, and signal decision-making into a single cohesive architecture. The use of Proximal Policy Optimization (PPO) ensures stable policy convergence within a trust region framework, while the multi-objective reward structure embeds sustainability directly into the optimization logic of traffic engineering. This model exemplifies how reinforcement learning can transcend reactive, rule-based paradigms—delivering anticipatory, self-optimizing infrastructure that dynamically adapts to urban complexity while advancing both mobility efficiency and environmental resilience.

Framework: DRL-Based Emission-Conscious Urban Signal Optimization (DRL-EUSO)

Layered AI Architecture for Real-Time Urban Traffic Flow and Sustainability Integration

Layer	Main functionality	AI module(s) and components	Operational scope (Input → Processing → Output)
Layer 1	Temporal traffic state observation	Temporal-Spatial Data Fusion Engine	Intersection state, vehicle density, emission metrics → time-series state encoding → structured input to PPO agent
Layer 2	Policy learning and emission-constrained optimization	Proximal Policy Optimization (PPO)	Structured traffic states → DRL policy training loop with emission/stability penalization → optimized traffic signal policies
Layer 3	Real-time policy execution and feedback loop	Proximal Policy Optimization (PPO)	Current state + learned policy → signal control action → intersection update, emission feedback
Layer 4	Policy refinement via trust region surrogate loss	Proximal Policy Optimization (PPO)	Past vs. current policy comparison → clipped objective evaluation → stable update of policy weights
Layer 5	Environmental feedback integration	Environmental impact assessment module	Emission outcome and transition frequency → deviation from sustainability goals → adjusted emission penalty parameters
Layer 6	Human-in-the-loop oversight and exception monitoring	Human oversight and anomaly detection interface	Operator evaluation of signal policy behavior → feedback annotation and override → logged policy adjustments for retraining
Layer 7	Continual policy improvement via historical evaluation	Historical policy analytics and continuous learning module	Logged emission and traffic data → longitudinal policy performance evaluation → retraining trigger based on policy degradation

1.1.2 Optimization of Traffic Flow

The AI system deployed in Barcelona operates as a real-time adaptive control mechanism, continuously monitoring vehicular density across intersections and arterial corridors. Rather than relying on pre-programmed signal cycles, the system dynamically responds to the current volume of vehicles, recalibrating signal phases to maximize throughput and reduce congestion. This decentralized yet harmonized control strategy enables traffic signals to self-adjust based on input from a network of sensors embedded in the road infrastructure, supplemented by real-time user data. For example, in scenarios where a directional surge is detected—such as the buildup of vehicles during peak hours—the system instantaneously extends green light intervals to alleviate congestion, thus preserving flow continuity (Yigitcanlar & Cugurullo, 2020). The importance of this real-time responsiveness cannot be overstated; it transforms the traditionally rigid temporal programming of intersections into an intelligent network capable of minimizing travel delays, enhancing mobility, and sustaining operational resilience under fluctuating demand conditions.

To formalize this dynamic decision-making process, the model adopts a deep reinforcement learning approach in which the AI agent learns an optimal policy for traffic signal selection. The agent observes the system's current state—defined by factors such as queue length, vehicle arrival rates, and current signal phase—and selects an action that modifies the signal configuration to improve flow conditions while minimizing operational disruptions. The central objective of the learning process is to maximize the cumulative long-term benefit of each signal decision. This objective is represented as follows:

$$\pi^* = \arg\max_{\pi} \mathbb{E}\left[\sum_{t=0}^{T} \gamma^t (F(s_t, a_t) - \lambda \cdot T(a_t))\right]$$

where π^*, optimal traffic signal policy, π, current adaptive traffic policy, $\mathbb{E}[\cdot]$, expectation across time horizon T, γ, discount factor balancing short-term versus long-term optimization, $F(s_t, a_t)$, traffic flow reward capturing throughput per time step, $T(a_t)$, penalty function quantifying disruption caused by signal transition, and λ, regularization parameter ensuring signal stability. This formulation captures the essential balance between fluidity and control: the agent is encouraged to favor signal configurations that improve vehicular discharge rates yet is disincentivized from making excessively frequent or abrupt transitions, which can create delays due to start-up loss, all-red intervals, and driver reaction time.

As the policy evolves over time, the agent requires a robust mechanism to determine whether new actions are genuinely superior to previous decisions. This is achieved through a stable learning update grounded in the Proximal Policy Optimization (PPO) algorithm, which refines the policy using a clipped surrogate objective. This function allows the agent to gradually improve its decisions without overcorrecting in response to transient traffic anomalies:

$$L^{\mathrm{PPO}}(\theta) = \mathbb{E}_t[\min(r_t(\theta)A_t, \mathrm{clip}(r_t(\theta), 1-\epsilon, 1+\epsilon)A_t)]$$

where $L^{\mathrm{PPO}}(\theta)$ is the PPO objective function to be maximized during training, $r_t(\theta) = \dfrac{\pi_\theta(a_t \mid s_t)}{\pi_{\theta^{\mathrm{old}}}(a_t \mid s_t)}$ is the probability ratio comparing the likelihood of action a_t under the new and old policies, A_t is the advantage function estimating the relative benefit of the action compared to the average expected return, and ϵ is a clipping parameter constraining the policy update to a trust region to avoid instability. This clipped objective ensures that policy refinement remains conservative, allowing the system to respond to evolving traffic conditions without diverging erratically. The ratio $r_t(\theta)$ serves to assess the extent to which the new policy deviates from the old, and the use of the minimum function ensures that updates only proceed when there is a demonstrable advantage. The advantage function A_t plays a crucial role here, as it quantifies the incremental value of a specific action within its given context.

To estimate this advantage with fidelity to temporal dynamics, the model leverages a sequence of temporal-difference errors that accumulate over multiple steps, capturing how outcomes deviate from expectations as time unfolds. This cumulative estimate is defined as:

$$A_t = \sum_{k=0}^{n-1} \gamma^k \delta_{t+k}, \quad \text{with} \quad \delta_t = r_t + \gamma V(s_{t+1}) - V(s_t)$$

where A_t is the total advantage accumulated over n steps, δ_t is the temporal-difference error at time t, r_t is the immediate reward received (as defined by $F(s_t, a_t) - \lambda T(a_t)$, and $V(s_t)$ is the estimated value of being in state stst under the current policy. This estimate reinforces actions that yield better-than-expected traffic outcomes while discouraging choices that lead to congestion or inefficiency. The dynamic refinement of policy through $L^{\text{PPO}}(\theta)$, guided by this advantage structure, ensures that the AI system converges toward signal strategies that are resilient, fluid, and environmentally aware.

By anchoring the learning process in these principles, the model empowers each intersection to function not as an isolated controller, but as an adaptive agent capable of intelligent, self-correcting behavior. Through repeated exposure to diverse traffic conditions—ranging from routine rush hours to unexpected surges—the system generalizes its decisions across scenarios, continually improving its performance through direct feedback. In the context of Barcelona's urban infrastructure, this framework exemplifies how real-time AI can operationalize sustainable mobility, not merely by reacting to traffic, but by anticipating it, shaping it, and optimizing it for the collective benefit of the city.

Framework: Real-Time Adaptive Traffic Signal Control Framework (RATSCF)

Layered AI Architecture for Urban Traffic Signal Optimization

Layer	Main functionality	AI module(s) and components	Operational scope (Input → Processing → Output)
Layer 1	Infrastructure and sensor input collection	Sensor network integration and data stream normalization	Traffic density data, vehicle counts, intersection signal states → Sensor input acquisition and preprocessing → State variables for AI system
Layer 2	Adaptive policy learning for signal optimization	Deep Reinforcement Learning (DRL)	Traffic state representation → Policy learning via cumulative reward maximization → Adaptive signal phase selection
Layer 3	Stable policy refinement with trust region	Proximal Policy Optimization (PPO)	Policy updates, observed outcomes → Clipped surrogate loss

(continued)

Layer	Main functionality	AI module(s) and components	Operational scope (Input → Processing → Output)
			function → Improved stable adaptive policy
Layer 4	Temporal advantage estimation for traffic states	Temporal-difference learning	Observed reward, next-state value → Multi-step TD advantage accumulation → Refined learning signals for PPO
Layer 5	Continuous policy feedback and real-time adaptation	Real-time performance monitoring and adaptive feedback module	Traffic outcomes, congestion metrics → Feedback assimilation into policy evolution → Improved generalization and responsiveness

1.1.3 Reduction of Idle Times and Emissions

A core advantage of Barcelona's AI-enabled traffic control architecture lies in its capacity to reduce vehicle idle durations at intersections. Traditional systems employ pre-timed, non-adaptive cycles that fail to reflect real-world variations in traffic volumes. This rigidity results in vehicles waiting at red signals unnecessarily—often when no cross traffic is present—thereby wasting fuel, increasing engine idling, and generating avoidable emissions. Barcelona's AI framework disrupts this inefficiency by dynamically adapting signal phases in response to real-time traffic states. In doing so, it minimizes static dwell times and accelerates vehicular discharge during low-conflict periods. The implications are twofold: operational efficiency and substantial environmental benefits. Empirical studies have confirmed that such AI-based adaptations can reduce greenhouse gas emissions by up to 30%, situating traffic intelligence as a critical tool in urban climate policy (Yigitcanlar & Cugurullo, 2020). To mathematically encode the relationship between signal control actions and emission outcomes, the following emission-penalized reward function is integrated into the reinforcement learning framework. This model builds upon the previously defined policy π^*, explicitly incorporating idle time and stop-start transition costs into the reward formulation. The structure captures the environmental impact of each decision, ensuring that traffic efficiency does not come at the cost of sustainability:

$$\pi^* = \arg\max_{\pi} \mathbb{E}\left[\sum_{t=0}^{T} \gamma^t (F(s_t, a_t) - \lambda_1 \cdot T(a_t) - \lambda_2 \cdot E(s_t, a_t))\right]$$

where π^*, optimal signal timing policy, π, policy governing intersection actions, $\mathbb{E}[\cdot]$, expectation across time horizon T, γ, discount factor for future performance, $F(s_t, a_t)$, traffic throughput reward at time t, $T(a_t)$, transition cost penalizing phase switching disruptions, $E(s_t, a_t)$, emission cost function based on idle time and vehicle stops, λ_1, regularization for signal change frequency, and λ_2, environmental penalty coefficient calibrated to emission goals.

The new sub-function $E(s_t, a_t)$ is defined below to model vehicular idle emission cost as a function of current traffic state and signal decision:

$$E(s_t, a_t) = \sum_{i=1}^{n} (\alpha_i \cdot I_i(s_t, a_t) + \beta_i \cdot S_i(s_t, a_t))$$

where $E(s_t, a_t)$, total estimated emissions from idle and stop-start conditions, n, number of approaches or vehicle types, α_i, emission rate per unit idle time for vehicle type i, $I_i(s_t, a_t)$, idle time in seconds for vehicle type i at time t, β_i, emission rate per stop-start event for vehicle type i, and $S_i(s_t, a_t)$, number of stops triggered by action a_t at time t.

This formulation disaggregates emissions into two key contributors: prolonged idling and repeated stopping due to poorly timed signals. The term $I_i(s_t, a_t)$ quantifies how long vehicles remain idle due to current signal decisions, while $S_i(s_t, a_t)$ captures the number of stop-start transitions caused by suboptimal phase sequencing. By introducing α_i and β_i as empirically derived emission coefficients for each vehicle class (e.g., diesel buses, electric cars), the model captures both immediate and cumulative ecological impact at high granularity.

Incorporating this emission function into the broader reinforcement learning structure ensures that the AI agent not only learns to reduce travel delay but also to actively minimize environmental cost. The dual-objective optimization enforced by the λ_2 coefficient allows policymakers to calibrate the model toward aggressive emission targets or more balanced efficiency goals, depending on real-world priorities.

In essence, this emission-integrated reward architecture converts Barcelona's AI traffic control into an eco-aware agent, one that internalizes the cost of inaction at red lights and acts to balance throughput against carbon output. The model evolves continuously based on traffic-environment feedback, enabling urban infrastructure to learn, correct, and align itself with climate resilience strategies—transforming traffic signals into intelligent actuators of sustainability.

Framework: Emission-Aware Adaptive Traffic Control Framework (EAATCF)

Layered AI Architecture for Sustainable Urban Signal Optimization

Layer	Main functionality	AI module(s) and components	Operational scope (Input → Processing → Output)
Layer 1	Traffic emission baseline and signal timing data collection	Sensor-based emission data acquisition and signal log integration	Traffic signal timings, vehicle idle durations → Sensor data aggregation → Environmental state variables
Layer 2			

(continued)

Layer	Main functionality	AI module(s) and components	Operational scope (Input → Processing → Output)
	Emission-aware policy optimization for signal control	Deep Reinforcement Learning (DRL)	Traffic and emission state → Dual-objective reward maximization → Environmentally-optimized signal decisions
Layer 3	Emission cost estimation for idle and stop events	Reward shaping via emission penalty function	Vehicle class, idle time, stop counts → Weighted emission function calculation → Emission penalty for RL reward
Layer 4	Policy calibration toward environmental goals	Environmental policy adjustment and compliance calibration engine	Emission targets, regulatory thresholds → Adjustment of $\lambda 2$ coefficient → Policy alignment with climate strategy
Layer 5	Continuous learning via traffic-emission feedback loop	Adaptive feedback and sustainability optimization module	Emission outcomes, traffic flow data → Feedback to reward model → Updated policy for sustainable signal control

Figure 1.1 depicts a structured cluster diagram illustrating the components of Barcelona's AI-driven traffic management system. At the center, the principal node, "AI-Driven Traffic Management System in Barcelona," connects to three primary nodes: "Data Integration and Sources," "Machine Learning Algorithms," and "Optimization of Traffic Flow." Each primary node further branches into key sub-nodes detailing specific elements. "Data Integration and Sources" includes traffic cameras, road infrastructure sensors, and user-generated data from mobile apps. "Machine Learning Algorithms" involve traffic pattern analysis, congestion prediction, and real-time traffic signal adjustments. "Optimization of Traffic Flow" comprises dynamic adjustments of traffic signals based on real-time traffic volume assessments, green light adjustments for congested directions, and peak hour adaptability. Additionally, this section emphasizes emission reduction through minimized idle times, contributing to decreased greenhouse gas emissions and alignment with sustainability goals.

1.1.4 Enhancing Public Transport Efficiency in Singapore

Singapore is recognized as a leader in applying AI to optimize public transportation, especially through its Land Transport Authority (LTA). The LTA utilizes advanced predictive analytics powered by AI to improve the efficiency and reliability of its public transport system, particularly for bus services.

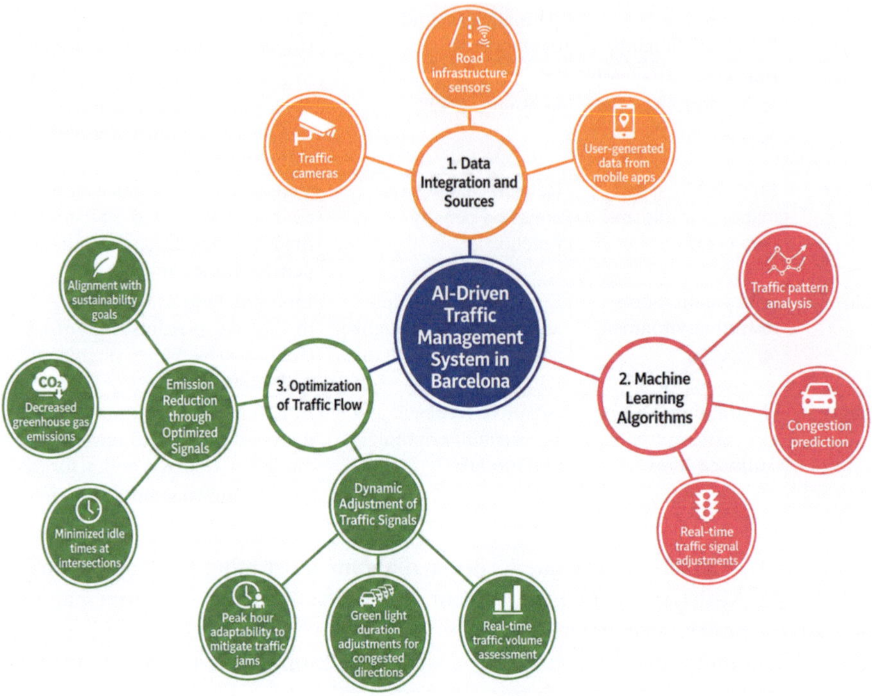

Fig. 1.1 AI-driven traffic management system in Barcelona

1.1.4.1 Predictive Analytics for Demand Forecasting

To ensure optimal public transport allocation, the Land Transport Authority (LTA) leverages predictive analytics powered by artificial intelligence to forecast passenger demand across Singapore's bus network. The forecasting model integrates historical ridership data with real-time indicators, including time of day, day of week, past usage patterns, and spatial route distributions. This enables the system to anticipate demand surges before they occur—particularly during peak periods or near high-traffic nodes—and to proactively adjust service deployment accordingly. Such anticipation prevents overcrowding, minimizes passenger waiting times, and contributes to a more balanced and reliable urban mobility experience (Yigitcanlar & Cugurullo, 2020).

To capture the temporal dynamics of transit behavior, the LTA employs a Long Short-Term Memory (LSTM) neural network. This architecture is designed to model sequential dependencies, incorporating both recent events and longer-term trends into its forecasts. At the heart of this framework lies the output function, which translates the internal state of the model into a demand prediction:

$$\hat{y}_t = \sigma(W_o \cdot \tanh(C_t) + b_o)$$

where $\hat{y}_t$, predicted demand at time t, W_o, output weight matrix, C_t, memory state at time t, b_o, bias vector, tanh, nonlinear activation, σ, output transformation function.

This function generates the ridership forecast by transforming the internal memory state of the model into a numerical estimate. The weights and bias calibrate the contribution of remembered information, while the nonlinearities allow the model to represent complex, fluctuating patterns in real-world data. We use this function to translate accumulated passenger behavior into an actionable demand value for transit scheduling.

The memory state C_t, essential to the forecast above, evolves through the following update function:

$$C_t = f_t \cdot C_{t-1} + i_t \cdot \tilde{C}_t$$

where, C_t, updated memory, f_t, forget gate, C_{t-1}, memory from previous step, i_t, input gate, and $\tilde{C}_t$, candidate memory content.

This formulation allows the LSTM network to balance historical knowledge with current observations. The forget gate f_t selectively attenuates outdated patterns, while the input gate i_t promotes the assimilation of new transit data. The interaction between these gates ensures that long-standing travel trends—such as morning rush-hour cycles—are preserved, while allowing the system to adapt to irregular events such as public holidays, traffic incidents, or large-scale gatherings.

By integrating these mechanisms, the LSTM model continuously learns from time-series transit data and refines its predictions over time. The network's recursive structure enables it to generalize across routine and anomalous travel behaviors, producing high-fidelity demand forecasts at the route and time-slot level. These forecasts serve as actionable inputs for dynamic scheduling systems, enabling transport authorities to allocate buses more effectively, reduce underutilization during off-peak periods, and preemptively address capacity shortfalls in critical segments.

In this way, predictive analytics supported by LSTM architectures transform transit planning from a reactive endeavor into a forward-looking, data-driven strategy. This model exemplifies how artificial intelligence can enhance urban service delivery by converting historical usage data into intelligent operational foresight—ensuring that public transportation remains responsive, efficient, and aligned with the evolving rhythms of city life.

Framework: Predictive Public Transport Demand Forecasting Framework (PPTDFF)

Layered AI Architecture for Passenger Demand Forecasting and Transit Allocation

Layer	Main functionality	AI module(s) and components	Operational scope (Input → Processing → Output)
Layer 1	Historical and real-time ridership data aggregation	Transit data integration and preprocessing engine	Time, date, location, route usage data → Structured time-series dataset → Input features for AI forecasting model

(continued)

Layer	Main functionality	AI module(s) and components	Operational scope (Input → Processing → Output)
Layer 2	Sequential demand prediction via temporal modeling	Long Short-Term Memory (LSTM)	Historical and real-time passenger flows → LSTM memory and output activation → Predicted route-level demand forecasts
Layer 3	Memory cell update for sequential pattern retention	LSTM internal state mechanism	Current and prior passenger trends → LSTM memory gate update → Refined internal state for demand prediction
Layer 4	Forecast-informed dynamic bus allocation	Dynamic scheduling and resource optimization module	Demand forecasts, route saturation levels → Scheduling system adjustment → Optimal bus deployment strategy

1.1.4.2 Dynamic Scheduling and Resource Allocation

In response to the demand forecasts, the AI system dynamically adjusts bus schedules and reallocates resources as needed. For example, if a particular route is predicted to experience higher demand due to a local event or changes in commuter patterns, the system can increase the frequency of buses on that route, improving overall service reliability (Yigitcanlar & Cugurullo, 2020). This capability not only enhances user satisfaction but also helps optimize operational costs by efficiently managing the fleet and reducing unnecessary expenditures on underutilized services. To model this decision-making process, a Deep Reinforcement Learning framework is applied. The AI agent evaluates the predicted demand values from the LSTM model and determines optimal scheduling actions to maximize service efficiency. The policy guiding this adaptive resource management is given by:

$$\pi^* = \arg\max_{\pi} \mathbb{E}\left[\sum_{t=0}^{T} \gamma^t \left(U(s_t, a_t) - \lambda_1 \cdot C(a_t) - \lambda_2 \cdot O(a_t)\right)\right]$$

where π^*, optimal dynamic scheduling policy, π, current policy controlling resource actions, $\mathbb{E}[\cdot]$, expected value over time horizon T, γ, discount factor, $U(s_t, a_t)$, utility of meeting forecasted demand with action a_t, $C(a_t)$, cost of schedule adjustment, $O(a_t)$, penalty for over-deployment of fleet, λ_1, weight for operational efficiency, and λ_2, weight for budget optimization.

This function captures the balance between three elements: the benefit of matching predicted demand, the cost of modifying schedules, and the penalty for inefficient resource use. The utility function $U(s_t, a_t)$ increases when demand is effectively met, while $C(a_t)$ accounts for the effort and disruption caused by rescheduling. $O(a_t)$ prevents the system from deploying excess capacity, which would lead to underutilized services and wasted fuel or labor. We use this formulation to ensure that AI learns how to distribute resources optimally in a changing environment.

This AI-driven model demonstrates how predictive insights and autonomous decision-making can be combined to build a public transport system that is efficient, scalable, and capable of adapting in real time to urban demand.

Framework: AI-Driven Dynamic Bus Scheduling Optimization Framework (AIDBSOF)

Layered AI Architecture for Public Transport Resource Allocation

Layer	Main functionality	AI module(s) and components	Operational scope (Input → Processing → Output)
Layer 1	Integration of demand forecasts with network status	Forecast–network state integration and decision preprocessing engine	Predicted passenger demand, route metadata → Forecast-to-decision integration → State inputs for scheduling model
Layer 2	Adaptive bus scheduling policy optimization	Deep Reinforcement Learning (DRL)	Forecasted demand and traffic state → Reward-based policy learning → Optimal dynamic scheduling policy
Layer 3	Multi-objective reward evaluation for deployment	DRL with utility-cost-penalty objective function	Service utility, cost, over-deployment penalty → Composite reward computation → Balanced deployment policy update
Layer 4	Fleet allocation and scheduling execution	Dispatch orchestration and real-time fleet control module	Optimized bus schedule policy → Dispatch and resource reallocation → Real-time route-level service adjustment

1.1.4.3 Improved Service Reliability and User Experience

In order to operationalize the forecasts produced by LSTM-based models, the AI system dynamically adjusts bus schedules and resource allocations in response to predicted ridership fluctuations. When a spike in demand is anticipated—such as those triggered by localized events or temporal commuting shifts—the system can deploy additional buses along affected routes to maintain service reliability and minimize passenger wait times (Yigitcanlar & Cugurullo, 2020). Conversely, in scenarios of anticipated low demand, resources can be strategically withheld or redirected. This bidirectional responsiveness optimizes both user satisfaction and operational cost-efficiency by aligning transit capacity with real-time demand profiles.

To formalize this decision-making layer, the AI system employs a Deep Reinforcement Learning (DRL) framework, where the agent receives input from the LSTM-based demand predictor and learns to issue optimal scheduling actions. At each decision point, the agent evaluates the expected utility of its action based on the predicted demand and system conditions, while also accounting for adjustment overhead and resource constraints. The objective is to maximize long-term operational efficiency while avoiding unnecessary fleet overcommitment. The resulting policy optimization problem is expressed as:

$$\pi^* = \arg\max_{\pi} \mathbb{E}\left[\sum_{t=0}^{T} \gamma^t (U(s_t, a_t) - \lambda_1 C(a_t) - \lambda_2 O(a_t))\right]$$

where π^* is the optimal dynamic scheduling policy, π is the stochastic policy mapping system states stst to actions a_t, $\mathbb{E}$ denotes the expected cumulative reward, γ is the discount factor prioritizing near-term versus long-term returns, $U(s_t, a_t)$ is the utility derived from matching supply to forecasted demand, $C(a_t)$ represents the cost of modifying schedules, $O(a_t)$ penalizes over-deployment of fleet resources, and λ_1, λ_2 are regularization parameters tuning the trade-off between flexibility and efficiency.

The utility function $U(s_t, a_t)$ captures the degree of demand satisfaction at time t based on LSTM forecasts. To quantify how effectively supply matches demand, it is expressed as:

$$U(s_t, a_t) = -\sum_{j=1}^{n} \left(d_j^t - q_j^t(a_t)\right)^2$$

where d_j^t is the predicted demand for route j at time t, obtained from the LSTM model, $q_j^t(a_t)$is the supplied capacity on that route as determined by action a_t, and n is the total number of monitored routes or service segments. This formulation ensures that the agent is rewarded for minimizing the squared error between forecasted and supplied capacity. When demand and service are closely aligned, the utility increases; when under- or over-provisioning occurs, the penalty grows. This enables the system to internalize both passenger experience and operational accuracy in a single objective.

To train this policy stably and ensure convergence, the model adopts the Proximal Policy Optimization (PPO) algorithm. This method updates the policy network using a clipped surrogate objective, which restricts excessive deviation from previous strategies and enhances robustness under noisy or unpredictable traffic patterns:

$$L^{\text{PPO}}(\theta) = \mathbb{E}_t[\min(r_t(\theta)A_t, \text{clip}(r_t(\theta), 1 - \epsilon, 1 + \epsilon)A_t)]$$

where $L^{\text{PPO}}(\theta)$ is the PPO loss function used to update the parameters θ of the policy network, $r_t(\theta) = \pi_\theta\left(a_{t\theta^{\text{old}}}(a_t s_t)\right.$ is the policy ratio comparing new and old action probabilities, A_t is the advantage function estimating the benefit of action a_t over the expected baseline, and ϵ is a clipping threshold ensuring that updates remain within a trust region. This clipped objective guards against destabilizing policy shifts by allowing improvements only when the advantage is statistically significant. The advantage term A_t is computed through temporal-difference estimation, reflecting the long-term benefit of each scheduling decision relative to the system's overall performance trajectory.

Using predictive demand modeling with this reinforcement-based scheduling architecture, the system learns to balance flexibility, efficiency, and environmental awareness. It continuously adapts to fluctuating demand while avoiding resource overcommitment or service shortfalls. As such, the framework offers a high-fidelity, closed-loop model of anticipatory transport planning—demonstrating how predictive insight, autonomous control, and sustainability objectives can be synthesized within a unified AI-driven system for intelligent public transit management.

Framework: Forecast-Guided DRL Scheduling Framework for Urban Transit (FGDSF-UT)

Layered AI Architecture for Demand-Aligned Scheduling Optimization

Layer	Main functionality	AI module(s) and components	Operational scope (Input → Processing → Output)
Layer 1	Forecast integration with operational transit data	Forecast data fusion and operational state encoding module	LSTM-based demand forecasts, current transit data → Input consolidation → State representation for DRL agent
Layer 2	Dynamic scheduling via deep reinforcement learning	Deep Reinforcement Learning (DRL)	Forecast-informed states → Policy learning to match demand → Optimized scheduling actions
Layer 3	Forecast-supply utility evaluation	Squared error utility function	Predicted vs. scheduled route capacities → Demand-supply utility calculation → Reward signal for DRL
Layer 4	Policy optimization with stability constraints	Proximal Policy Optimization (PPO)	Policy ratio and advantage function → Clipped objective optimization → Updated DRL policy parameters
Layer 5	Fleet scheduling execution and feedback loop	Real-time fleet deployment and adaptive feedback engine	Optimized schedule → Fleet allocation and monitoring → Feedback for continuous DRL training

Figure 1.2 illustrates a structure of Singapore's AI-driven public transport efficiency system. The central node, "AI-Driven Public Transport Efficiency in Singapore," connects to three main nodes: "Predictive Analytics for Demand Forecasting," "Dynamic Scheduling and Resource Allocation," and "Improved Service Reliability and User Experience." "Predictive Analytics for Demand Forecasting" includes the use of historical and real-time data, prediction of passenger demand by route and time, and proactive adjustments to meet demand. "Dynamic Scheduling and Resource Allocation" details adjustments to bus schedules based on demand, increased frequency on high-demand routes, and optimized operational costs through efficient resource management. "Improved Service Reliability and User Experience" highlights enhanced service predictability, reduced wait times for passengers, and increased public transport usage supporting sustainability goals.

Fig. 1.2 AI-driven public transport efficiency in Singapore

The experiences of Barcelona and Singapore demonstrate the transformative potential of AI in urban transportation systems. Through the implementation of advanced traffic management systems and predictive analytics, both cities have achieved remarkable improvements in traffic flow efficiency, public transport reliability, and environmental sustainability. The integration of AI technologies not only addresses the immediate challenges of congestion and operational inefficiencies but also aligns with the broader goals of creating more sustainable and livable urban environments. As urbanization continues to accelerate globally, the strategies employed by these pioneering cities serve as valuable models for other urban centers aiming to enhance their transportation systems through innovative technological solutions.

1.2 Energy Management

The integration of artificial intelligence (AI) in energy management systems is transforming urban environments by optimizing energy consumption, enhancing the efficiency of energy grids, and promoting the use of renewable energy sources. This analysis explores the innovative approaches adopted by Amsterdam and Stockholm in leveraging AI technologies for sustainable energy management.

1.2.1 Case Study 1: AI-Driven Energy Management in Amsterdam

Amsterdam has positioned itself as a frontrunner in implementing AI to optimize energy consumption across its urban infrastructure. The city's AI-driven smart energy management system employs sophisticated algorithms to analyze data from various sources, including energy usage patterns, grid conditions, and renewable energy generation forecasts.

1.2.1.1 Streamlining Energy Consumption

Amsterdam's energy management infrastructure exemplifies how artificial intelligence transforms traditional monitoring systems into predictive, self-optimizing agents. Rather than reacting to observed consumption patterns, the city's AI framework anticipates fluctuations in energy demand and proactively adjusts distribution to reduce system inefficiencies (Badidi, 2022). This predictive capability mitigates grid stress and minimizes energy waste, particularly during demand troughs or critical peak periods. The architecture relies on a dual-layer AI model that combines sequential prediction with dynamic control: a Long Short-Term Memory (LSTM) network forecasts upcoming energy demand, and a Deep Reinforcement Learning (DRL) agent schedules energy dispatch actions in response to those predictions.

The forecasting layer uses LSTM to process historical energy usage alongside exogenous factors—such as temperature, day type, and occupancy levels—to anticipate short- and long-term consumption trends. The hidden state and memory cell of the LSTM evolve through time to capture persistent dependencies. The predicted energy demand is obtained via:

$$\hat{d}_t = \sigma(W_o \cdot \tanh(C_t) + b_o)$$

where $\hat{d}_t$ is the forecasted energy demand at time t, C_t is the internal memory state of the LSTM, C_t is the output weight matrix, bobo is the bias term, tanh introduces non-real-time linearity to the cell state, and σ is an output activation function that scales the result to match demand magnitude.

The memory state C_t evolves dynamically based on the gating mechanisms of the LSTM network. The update rule is given by:

$$C_t = f_t \cdot C_{t-1} + i_t \cdot \tilde{C}_t$$

where f_t is the forget gate, C_{t-1} is the memory from the previous time step, i_t is the input gate, and $\tilde{C}_t$ is the candidate content based on the current input and hidden state. This formulation enables the model to adapt to evolving conditions while retaining essential energy consumption patterns.

The output $\hat{d}_t$ serves as an input to the control layer, which employs a deep reinforcement learning agent to optimize energy distribution. The agent observes the current system state s_t—including the predicted demand and grid status—and selects a dispatch action a_t that minimizes both energy imbalance and grid strain. The optimal policy is expressed as:

$$\pi^* = \arg\max_{\pi} \mathbb{E}\left[\sum_{t=0}^{T} \gamma^t (S(s_t, a_t) - \lambda \cdot L(a_t))\right]$$

where π^*, optimal energy dispatching policy, π, current scheduling policy, $\mathbb{E}[\cdot]$, expected total reward over time horizon T, γ, discount factor for long-term reward prioritization, $S(s_t, a_t)$, supply-demand matching score at time t, $L(a_t)$, load imbalance penalty for action a_t, and λ, trade-off coefficient for stability. This control policy balances predicted demand against real-time grid constraints by encouraging actions that keep energy supply aligned with usage forecasts while penalizing significant mismatches. The term $S(s_t, a_t)$ quantifies how well the energy allocation meets projected demand, and $L(a_t)$ captures the operational strain associated with under- or over-supplying the grid. The coefficient λ adjusts the penalty's importance, enabling policy tuning for either performance optimization or grid stability.

The supply-demand matching function $S(s_t, a_t)$ is defined as:

$$S(s_t, a_t) = -\left(\hat{d}_t - q_t(a_t)\right)^2$$

where $\hat{d}_t$ is the forecasted energy demand, and $q_t(a_t)$ is the supplied energy determined by action a_t. This squared error formulation penalizes mismatches between demand and delivery, rewarding actions that closely align the two.

To ensure the agent learns this policy stably in the face of environmental variability, the PPO algorithm is employed. The clipped surrogate loss function governing the policy update is defined as:

$$L^{\text{PPO}}(\theta) = \mathbb{E}_t[\min(r_t(\theta)A_t, \text{clip}(r_t(\theta), 1-\epsilon, 1+\epsilon)A_t)]$$

where $L^{\text{PPO}}(\theta)$ is the objective optimized during training, $r_t(\theta) = \pi_\theta\left(a_{t\theta^{\text{old}}}(a_t s_t)\right.$ is the probability ratio, A_t is the advantage estimate indicating how much better action a_t performed relative to expected value, and ϵ constrains update magnitude.

By coupling high-resolution demand forecasting with stable policy learning, Amsterdam's energy management system transitions from reactive load balancing to anticipatory optimization. The LSTM model enables the system to forecast periods of heightened or diminished usage with precision, while the DRL policy dynamically schedules energy flows in response, thereby minimizing waste and protecting grid health. This two-layer framework exemplifies how artificial intelligence can support urban sustainability objectives—not only by managing existing conditions more efficiently but by anticipating systemic needs before they arise.

Framework: Predictive Energy Distribution and Grid Optimization Framework (PEDGOF)

Layered AI Architecture for Urban Demand Forecasting and Grid Stabilization

Layer	Main functionality	AI module(s) and components	Operational scope (Input → Processing → Output)
Layer 1	Energy data aggregation and exogenous input encoding	Energy data integration and contextual feature engineering service	Historical energy usage, temperature, occupancy → Time-series and contextual input structuring → Inputs for demand forecasting model
Layer 2	Forecasting energy demand with temporal dependencies	Long Short-Term Memory (LSTM)	Sequential usage and contextual data → LSTM memory update and activation → Predicted short-term energy demand
Layer 3	Supply-demand matching and control optimization	Deep Reinforcement Learning (DRL)	Predicted demand and grid status → Policy learning to minimize mismatch and grid strain → Optimal dispatching actions
Layer 4	Policy stabilization via PPO optimization	Proximal Policy Optimization (PPO)	Action performance and policy ratios → Clipped surrogate loss evaluation → Updated DRL policy network parameters
Layer 5	Grid response execution and feedback assimilation	Grid orchestration and telemetry feedback service	Optimized dispatch policy → Energy allocation, system monitoring → Feedback for policy refinement and retraining

1.2.1.2 Integration of Renewable Energy Sources

One of the standout features of Amsterdam's AI energy management system is its capability to integrate renewable energy sources, such as solar and wind, into the existing energy grid. The system utilizes predictive analytics to forecast energy production from these renewable sources based on weather patterns and historical data (Badidi, 2022). By accurately predicting the availability of renewable energy, the city can adjust its energy mix accordingly, prioritizing clean energy sources when they are most abundant. This not only promotes the use of renewable energy but also contributes to the city's overall sustainability goals by reducing reliance on fossil fuels.

To support this integration, the existing LSTM-based architecture is expanded to include a dedicated forecasting module for renewable energy generation. This layer processes meteorological time-series data to estimate the availability of renewable power at each future time step. The prediction of renewable generation is defined by:

$$\hat{r}_t = \sigma\left(W_o^{(r)} \cdot \tanh\left(C_t^{(r)}\right) + b_o^{(r)}\right)$$

where $\hat{r}_t$ is the predicted renewable energy output at time t, $C_t^{(r)}$ is the renewable-specific memory state capturing historical weather-dependent generation patterns, $W_o^{(r)}$ and $b_o^{(r)}$ are the output weight matrix and bias term for the renewable LSTM head, tanh is a nonlinear activation applied to the memory state, and σ transforms the output into a feasible energy scale.

The memory state $C_t^{(r)}$ evolves through standard LSTM dynamics, selectively retaining and updating key temporal patterns in solar irradiance, wind velocity, humidity, and temperature. These features are encoded via a recurrent update rule:

$$C_t^{(r)} = f_t^{(r)} \cdot C_{t-1}^{(r)} + i_t^{(r)} \cdot \tilde{C}_t^{(r)}$$

where $f_t^{(r)}$, $i_t^{(r)}$, and $\tilde{C}_t^{(r)}$ represent the forget gate, input gate, and candidate memory content for renewable forecasting, respectively. This recursive process allows the model to learn both seasonal patterns and high-resolution short-term fluctuations in renewable availability.

The predicted renewable supply $\hat{r}_t$ is then integrated into the DRL-based control layer, which manages the real-time allocation of energy across the grid. The goal is to minimize deviation between total supply—composed of predicted renewables and scheduled grid allocations—and the forecasted demand. The reward function for this optimization is adapted accordingly:

$$S(s_t, a_t) = -\left|\hat{d}_t - \hat{r}_t - g(a_t)\right|$$

where $S(s_t, a_t)$ quantifies the negative absolute mismatch between total supply and demand, $\hat{d}_t$ is the LSTM-predicted energy demand, $\hat{r}_t$ is the forecasted renewable supply, and $g(a_t)$ is the non-renewable energy allocation governed by dispatch action a_t. This structure ensures that the agent is penalized when total delivered energy exceeds or undershoots consumption forecasts, thus promoting supply-demand equilibrium.

To ensure that this balance is achieved through stable and scalable learning, the control policy π is updated using the Proximal Policy Optimization (PPO) algorithm. The clipped surrogate loss governing policy refinement remains:

$$L^{\text{PPO}}(\theta) = \mathbb{E}_t[\min(r_t(\theta)A_t, \text{clip}(r_t(\theta), 1-\epsilon, 1+\epsilon)A_t)]$$

where $r_t(\theta) = \frac{\pi_\theta(a_t s_t)}{\pi_{\theta \text{old}}(a_t s_t)}$ is the policy ratio, A_t is the advantage function, and ϵ controls the allowable range for safe policy updates. This mechanism ensures that the scheduling strategy evolves incrementally, learning to prioritize renewable integration without destabilizing the broader energy network.

By embedding renewable forecasting directly into both the predictive and control components of the energy system, Amsterdam's AI model supports a real-time shift in energy mix composition. When the renewable forecast peaks, the model reduces reliance on non-renewables, contributing to lower emissions. When renewable availability wanes, grid resources are smoothly activated to maintain delivery continuity. This two-layered architecture exemplifies how artificial intelligence can serve as both a predictive and prescriptive engine for sustainable urban infrastructure—one that learns, adapts, and aligns city-scale energy operations with ecological priorities.

Framework: Renewable Energy Integration and Grid Optimization Framework (REIGOF)

Layered AI Architecture for Dynamic Energy Management with Renewable Integration

Layer	Main functionality	AI module(s) and components	Operational scope (Input → Processing → Output)
Layer 1	Data aggregation for renewable energy forecasting	Renewable data integration and meteorological feature processing module	Historical energy data, weather patterns, solar irradiance, wind speed → Meteorological input processing → Input features for renewable energy forecasting
Layer 2	Forecasting renewable energy generation with LSTM	Long Short-Term Memory (LSTM)	Weather data and energy generation trends → LSTM memory cell update → Predicted renewable energy output
Layer 3	Supply-demand matching and control optimization for energy distribution	Deep Reinforcement Learning (DRL)	Predicted renewable supply and energy demand → Policy learning to optimize grid dispatching → Optimized energy distribution actions
Layer 4	Policy stability and energy dispatch control	Proximal Policy Optimization (PPO)	Action performance, policy ratio → Clipped loss optimization → Updated DRL policy for stable grid control
Layer 5	Energy grid response and feedback for continuous learning	Grid feedback analytics and adaptive policy retraining module	Optimized energy dispatch policy → Grid reallocation and monitoring → Feedback for policy adaptation and retraining

1.2.1.3 Reduction of Energy Waste and Promotion of Clean Energy

The AI algorithms implemented in Amsterdam are instrumental in identifying areas where energy consumption can be reduced. For instance, by analyzing usage patterns, the system can recommend energy-saving measures to residents and businesses, such as adjusting heating and cooling settings based on occupancy patterns. The result is a significant reduction in energy waste and an overall increase in the efficiency of the city's energy consumption (Badidi, 2022). This initiative not only benefits the environment but also translates into financial savings for both the city and its residents. To formalize this functionality, an occupancy-aware consumption optimization function is introduced within the LSTM–DRL architecture. This component identifies discrepancies between forecasted demand and actual occupancy-adjusted usage, serving as a corrective layer for downstream policy learning. The instantaneous efficiency deviation is modeled as:

$$\mathcal{E}(t) = \hat{d}_t - \eta \cdot u_t$$

where $\mathcal{E}(t)$, energy waste at time t, $\hat{d}_t$, predicted energy demand from LSTM model, η, occupancy-efficiency factor, and u_t, occupancy-adjusted usable energy at time t. This function calculates the difference between what the system predicts will be used ($\hat{d}_t$) and what is actually needed based on building occupancy data (u_t). The scalar η accounts for insulation, appliance efficiency, and HVAC behavior. We use this function to guide the system toward minimizing the energy gap between actual needs and expected allocation.

To penalize waste within the energy allocation policy, the DRL reward function is updated by embedding $\mathcal{E}(t)$ directly into the optimization objective as follows:

$$\pi^* = \arg \underset{\pi}{max} \mathbb{E}\left[\sum_{t=0}^{T} \gamma^t (S(s_t, a_t) - \lambda_1 \cdot L(a_t) - \lambda_2 \cdot \mathcal{E}(t))\right]$$

where π^*, optimal policy minimizing waste, π, current energy control policy, $\mathbb{E}[\cdot]$, expected return over time T, γ, discount factor, $S(s_t, a_t)$, supply-demand alignment reward, $L(a_t)$, load stress penalty, $\mathcal{E}(t)$, energy waste as defined above, λ_1, penalty coefficient for grid imbalance, and λ_2, penalty coefficient for consumption inefficiency. This equation links predicted usage, occupancy patterns, and control actions, ensuring that AI decisions reduce both overconsumption and operational stress. By explicitly minimizing $\mathcal{E}(t)$, the system incentivizes actions that suppress unnecessary energy usage without compromising comfort or service availability.

The policy $\pi\pi$ is refined using Proximal Policy Optimization (PPO), ensuring that updates respond proportionally to observed improvements while avoiding destabilizing changes. The PPO loss remains:

$$L^{\mathrm{PPO}}(\theta) = \mathbb{E}_t[\min(r_t(\theta)A_t, \mathrm{clip}(r_t(\theta), 1-\epsilon, 1+\epsilon)A_t)]$$

where $r_t(\theta) = \frac{\pi_\theta(a_t s_t)}{\pi_{\theta \mathrm{old}}(a_t s_t)}$ is the policy ratio, A_t is the advantage function capturing the marginal gain of action a_t, and $\epsilon\epsilon$ regulates the trust region for policy stability. As the agent observes occupancy-adjusted deviations $\mathbb{E}_t$, it learns to select actions that close the gap between allocated and required energy, particularly during low-use periods.

This AI-driven formulation transforms Amsterdam's energy system into a proactive efficiency engine. By continuously learning and minimizing the discrepancy between actual occupancy-adjusted energy needs and projected distribution, the city is able to reduce waste, save costs, and accelerate the promotion of clean energy. The integration of dynamic efficiency optimization into the existing prediction-control architecture provides a scalable and intelligent foundation for future-ready energy governance.

Framework: Occupancy-Aware Energy Consumption Optimization Framework (OAECOF)

Layered AI Architecture for Efficient Energy Allocation and Waste Minimization

Layer	Main functionality	AI module(s) and components	Operational Scope (Input → Processing → Output)
Layer 1	Energy data aggregation and occupancy information integration	Energy data integration and occupancy feature processing module	Historical energy consumption, occupancy data, HVAC settings → Time-series data preprocessing → Inputs for demand forecasting and efficiency optimization
Layer 2	Energy demand forecasting and occupancy-aware deviation calculation	Long Short-Term Memory (LSTM)	Energy usage data, occupancy information → LSTM-based demand prediction and $E(t)$ calculation → Energy waste estimation for corrective action
Layer 3	Energy allocation optimization with efficiency penalty	Deep Reinforcement Learning (DRL)	Predicted energy demand, occupancy-adjusted waste, grid status → Policy learning to optimize energy dispatch and minimize waste → Optimized energy distribution actions
Layer 4	Policy stability and energy control update	Proximal Policy Optimization (PPO)	Action ratios, advantage function → Clipped objective optimization → Updated DRL policy for stable, efficient grid control
Layer 5	Energy grid response and feedback loop for policy refinement	Energy Telemetry Analytics and Continuous Policy Retraining Module	Optimized energy dispatch policy → Grid response, operational adjustments → Feedback for retraining and continuous learning

Fig. 1.3 AI-driven smart energy management system in Amsterdam

Figure 1.3 represents the structure of Amsterdam's AI-driven smart energy management system. The central node, "AI-Driven Smart Energy Management System in Amsterdam," is connected to three main components: "Streamlining Energy Consumption," "Integration of Renewable Energy Sources," and "Reduction of Energy Waste and Promotion of Clean Energy." "Streamlining Energy Consumption" focuses on continuous monitoring, predicting peak usage periods, and dynamic load balancing to reduce waste. "Integration of Renewable Energy Sources" includes forecasting renewable production, adjusting the energy mix, and prioritizing clean energy to reduce reliance on fossil fuels. "Reduction of Energy Waste and Promotion of Clean Energy" highlights identifying energy-saving opportunities, recommending efficient practices to residents and businesses, and achieving financial and environmental benefits through reduced consumption.

1.2.2 *Case Study 2: Enhancing Energy Supply and Demand Balancing in Stockholm*

Stockholm is also at the forefront of using AI technologies in its energy management strategies. The city has integrated AI into its smart grid systems to effectively balance energy supply and demand, thus enhancing overall energy efficiency.

1.2.2.1 Machine Learning for Predictive Analysis

Stockholm's energy management system employs machine learning techniques to analyze historical energy usage data, enabling the city to predict future energy demand accurately. By identifying patterns in energy consumption, the AI system can forecast demand spikes and adjust energy distribution in real time (Badidi, 2022). This predictive capability is crucial for managing the city's energy resources effectively, particularly during peak usage periods when the demand for energy can exceed supply.

To support the intelligent forecasting and adaptive control outlined in the original narrative, we now introduce a mathematically rigorous AI-driven model based on LSTM for temporal prediction and DRL for control optimization, forming a tightly coupled framework that addresses both demand prediction and responsive supply-side regulation within the smart grid infrastructure. The first step involves constructing a temporal prediction model capable of capturing sequential energy usage dependencies. The LSTM-based energy demand prediction function is defined as:

$$h_t = \sigma\left(W_h \cdot \begin{bmatrix} h_{t-1} \\ x_t \end{bmatrix} + b_h\right)$$

where h_t, hidden state representing encoded temporal energy usage features, x_t, real-time input vector (e.g., weather, time, appliance load), h_{t-1}, previous hidden state carrying memory of past usage trends, W_h, learned weight matrix regulating influence of inputs, b_h, bias vector for numerical stability, and σ, activation function introducing nonlinearity for modeling complex patterns. This function dynamically updates the hidden state based on past and present energy consumption data, enabling the AI model to anticipate near-future energy needs across Stockholm's grid. The sequential learning component ensures that seasonal variations, peak hour loads, and temperature-influenced demands are incorporated into the model's understanding of energy patterns.

To retain long-term demand dependencies and decide which energy features to preserve or discard, we define the LSTM memory update function:

$$C_t = f_t \odot C_{t-1} + i_t \odot \tilde{C}_t$$

where C_t, current cell state storing long-term memory, f_t, forget gate filtering outdated demand patterns, i_t, input gate deciding what new information to retain, $\tilde{C}_t$, candidate memory representing potential energy pattern update, C_{t-1}, previous memory, and $\odot$, element-wise multiplication to ensure selective gating. This function ensures that Stockholm's energy system retains relevant demand features—such as winter heating peaks or weekend consumption dips—while dynamically removing obsolete information, thus maintaining an efficient memory representation for forecasting purposes.

Having predicted the expected energy demand, the system must now adaptively regulate supply. This is governed by a Deep Reinforcement Learning model using Proximal Policy Optimization (PPO), where the agent learns to optimize energy dispatching decisions in response to real-time grid conditions and predicted demand spikes. The DRL-based energy regulation policy is defined as:

$$\pi_\theta(a_t \mid s_t) = \arg\max_\theta \mathbb{E}_t \left[\frac{\pi_\theta(a_t \mid s_t)}{\pi_{\theta_{\text{old}}}(a_t \mid s_t)} \cdot A_t \right]$$

where $\pi_\theta(a_t \mid s_t)$, updated energy distribution policy given current grid state, a_t, action representing distribution decisions across zones, s_t, system state comprising forecasted demand and real-time supply levels, $\pi_{\theta_{\text{old}}}(a_t \mid s_t)$, previous policy to compare updates, A_t, advantage function evaluating benefit of action over average, and $\frac{\pi_\theta}{\pi_{\theta_{\text{old}}}(a_t \mid s_t)}$, PPO policy ratio ensuring stable updates. This PPO-based function enables the smart grid to distribute energy more efficiently by evaluating how well each action (e.g., supply increase to a district) improves the balance between predicted demand and current supply. It avoids drastic policy changes while favoring beneficial updates.

To precisely quantify the learning process, we define the advantage function A_t, expressing the relative value of taking a particular action:

$$A_t = Q(s_t, a_t) - V(s_t)$$

where $Q(s_t, a_t)$, expected cumulative reward for taking action a_t in state s_t, and $V(s_t)$, value function representing average expected reward in state s_t. This advantage function determines whether the chosen energy distribution strategy outperforms the expected baseline, guiding the PPO policy to prefer only those actions that demonstrably improve energy balance.

The final reward function evaluates the effectiveness of the energy allocation at each step, integrating cost and supply-demand balance:

$$R(s_t, a_t) = \gamma \cdot E_{\text{eff}}(s_t, a_t) - \lambda \cdot C(a_t)$$

where $R(s_t, a_t)$, immediate reward, $E_{\text{eff}}(s_t, a_t)$, energy efficiency from matching predicted demand with supply, $C(a_t)$, energy cost for current allocation, γ, reward gain factor, and λ, penalty coefficient for cost. This function penalizes excessive supply during low demand and rewards optimal alignment, ensuring that Stockholm's energy system minimizes waste and operational costs while guaranteeing reliability.

This LSTM–DRL hybrid model demonstrates how Stockholm can dynamically predict demand and optimize energy distribution policies in real time. The LSTM component ensures that the AI accurately anticipates usage spikes by learning from historical patterns, while the PPO-based DRL component adaptively regulates energy allocation to avoid shortages or overproduction. The seamless mathematical linkage between sequential forecasting, memory retention, policy refinement, and reward optimization ensures that energy flow across the city remains stable, efficient, and responsive to fluctuating consumption patterns—solidifying Stockholm's position as a pioneer in sustainable AI-driven energy management.

Framework: Predictive Energy Demand and Dispatch Optimization Framework (PEDDOF)

Layered AI Architecture for Smart Grid Energy Management with Demand Forecasting

Layer	Main functionality	AI module(s) and components	Operational scope (Input → Processing → Output)
Layer 1	Energy data aggregation and external factors integration	Energy data integration and contextual feature engineering module	Historical energy consumption, weather data, occupancy levels → Input preparation for demand forecasting → Energy demand and external factors representation
Layer 2	Energy demand forecasting with temporal dependencies	Long Short-Term Memory (LSTM)	Time-series consumption data, temperature, and occupancy → LSTM-based prediction model → Forecasted energy demand at future time steps
Layer 3	Adaptive energy dispatch with supply-demand optimization	Deep Reinforcement Learning (DRL)	Predicted energy demand and real-time grid status → Policy learning to optimize energy dispatching → Dispatch actions for energy allocation across zones
Layer 4	Policy refinement and stability for energy distribution	Proximal Policy Optimization (PPO)	Policy performance ratios, advantage function → Clipped objective optimization → Updated policy for stable and efficient grid control
Layer 5	Grid response and continuous learning loop	Grid telemetry analytics and continuous policy retraining module	Optimized energy dispatch policy → Grid reallocation and monitoring → Feedback for retraining and continuous policy learning

1.2.2.2 Improved Resource Allocation

With the insights gained from predictive analytics, Stockholm can allocate resources more effectively across its energy infrastructure. The AI system identifies which areas may experience high demand and directs energy resources to those regions accordingly. This intelligent allocation not only improves the reliability of energy supply but also reduces operational costs by minimizing energy losses during transmission and distribution (Badidi, 2022). To translate this allocation mechanism into a mathematical formulation, we introduce an AI-driven energy resource allocation function. This function governs the dynamic distribution of energy across various zones by integrating the demand forecasts generated by Long Short-Term Memory (LSTM) networks with the decision-making process powered by Deep Reinforcement Learning (DRL). The approach seamlessly aligns proactive demand forecasting with real-time policy optimization, ensuring that energy resources are allocated efficiently across the network based on both predictive insights and learned actions. This integration of forecasting and decision-making creates a robust, data-driven system for intelligent energy distribution, capable of responding to fluctuating demand patterns while minimizing inefficiencies and transmission losses.

To begin the process of energy management, the system first forecasts the demand for energy using a Long Short-Term Memory (LSTM) network. This approach allows the system to learn from historical data and effectively predict future energy needs. By maintaining a memory of past data, the model can predict demand more accurately, taking into account various time-dependent factors. This forecasted demand provides the input necessary for the next decision-making process in energy allocation.

$$h_t = \sigma(W_h h_{t-1} + W_x x_t + b)$$

where h_t, current hidden state, h_{t-1}, previous hidden state, x_t, current input, W_h, weight matrix for the hidden state, W_x, weight matrix for the input, b, bias term, and σ, nonlinear activation function. This function enables the LSTM to model temporal dependencies in the energy demand data.

Next, once the demand is forecasted, the system must decide how to optimally distribute the energy to various zones. This decision-making process is powered by Proximal Policy Optimization (PPO), a reinforcement learning algorithm. PPO helps to fine-tune the actions that allocate energy in a way that maximizes long-term performance while maintaining stability in policy updates. The algorithm ensures that the learned policy doesn't deviate too drastically from previous decisions, ensuring more stable and reliable decisions in energy allocation.

$$L^{\text{CLIP}}(\theta) = \mathbb{E}_t\left[\min\left(r_t(\theta)\hat{A}_t, \text{clip}(r_t(\theta), 1-\epsilon, 1+\epsilon)\hat{A}_t\right)\right]$$

where $L^{\text{CLIP}}(\theta)$, clipped objective function, $r_t(\theta)$, probability ratio, $\hat{A}_t$, advantage estimate, and ϵ, clipping parameter. This function computes the clipped surrogate objective, which optimizes the policy π_θ by minimizing the clipped advantage $\hat{A}_t$, ensuring that the new policy does not deviate significantly from the previous policy. This function helps ensure stability while maximizing the cumulative reward for energy distribution decisions.

Once the system is capable of forecasting demand and making optimized decisions, the next step is to allocate the energy efficiently across various zones. The allocation depends on the predicted demand and the policy actions derived from PPO. This stage ensures that the energy distributed to each zone is proportional to its demand while minimizing losses during transmission. This layer integrates the learned policy and the forecasted demand to ensure that the energy distribution aligns with the actual needs of each zone.

$$A_t^{(z)} = \phi \cdot D_t^{(z)} + \psi \cdot \pi_\theta\left(a_t^{(z)} \mid s_t^{(z)}\right) - \eta \cdot L^{(z)}$$

where $A_t^{(z)}$, total allocated energy at time t for zone z, $D_t^{(z)}$, LSTM-forecasted energy demand for zone z, $\pi_\theta\left(a_t^{(z)} \mid s_t^{(z)}\right)$, PPO policy-derived distribution decision for zone z, $L^{(z)}$, estimated energy loss during transmission to zone z, ϕ, weight factor scaling predicted demand, ψ, importance weight on learned policy action, and η, penalty term for energy transmission loss. This function tightly integrates the demand forecasting output $D_t^{(z)}$ and the policy decision πθπθ, ensuring both predictive insight and real-time decision-making guide the allocation. The term $L^{(z)}$ reflects grid-specific loss factors such as resistance, distance, or infrastructure inefficiencies.

In real-world applications, transmission losses are a critical consideration when distributing energy. The system must account for these losses to ensure that energy reaches its destination efficiently. Transmission losses are influenced by infrastructure factors such as distance and resistance, and the function takes these into account by calculating losses based on the distances between energy sources and their destinations.

$$L^{(z)} = \sum_{i=1}^{N} \frac{R_i}{d_i^2}$$

where $L^{(z)}$, energy loss for zone z, R_i, resistance of the i-th segment, d_i, distance of the i-th segment, and N, total number of segments in the grid. This function ensures that the energy losses are accounted for during the allocation process to improve system efficiency.

Finally, the system incorporates a cost minimization layer to ensure that the total operational cost of energy distribution is as low as possible. This function evaluates both the cost of energy allocation and the penalties associated with transmission losses. The goal is to minimize the overall cost while still meeting the demand and maintaining efficiency in the energy grid.

$$C_t = \sum_{z=1}^{Z} \left[\omega \cdot C\left(A_t^{(z)}\right) + \lambda \cdot L^{(z)} \right]$$

where C_t, total operational cost at time t, $C\left(A_t^{(z)}\right)$, cost of energy allocation for zone z as a function of allocated energy $A_t^{(z)}$), Z, total number of city zones, ω, weight balancing allocation cost, and λ, coefficient representing sensitivity to losses. This cost function aggregates zone-wise operational costs and penalizes transmission inefficiencies. It draws a direct connection to the earlier immediate reward function $R(s_t, a_t)$ by serving as the system-level performance evaluator, guiding reinforcement learning reward updates.

This allocation model captures how Stockholm can intelligently redirect energy in response to predicted consumption patterns while minimizing losses. The function $A_t^{(z)}$ explicitly links the LSTM-predicted demand $D_t^{(z)}$ and the PPO-learned action πθπθ, preserving the mathematical narrative established earlier. Moreover, the citywide cost objective C_t acts as a cumulative measure of resource efficiency, integrating real-time optimization and spatial granularity into a coherent, city-scale decision-making system. The formulation reflects how modern AI architectures transform predictive analytics into operational excellence, ensuring that energy reaches exactly where it's needed—with minimal waste, maximum foresight, and optimal learning-guided decisions.

Framework: Energy Allocation & Optimization AI Framework (EAOAF)
Layered AI Architecture for Predictive Energy Demand and Resource Allocation

Layer	Main functionality	AI module(s) and components	Operational scope
Layer 1	Data collection: Gather historical and real-time energy demand data from different zones	Sensor and Meter Data Integration and Preprocessing Module	Input → Historical data, sensor readings → Output → Preprocessed input for AI models
Layer 2	Demand forecasting: Use LSTM to predict future energy demand for each zone	Long Short-Term Memory (LSTM)	Input → Historical demand data, weather data → Output → Predicted demand for each zone
Layer 3	Learning and decision-making: Use deep reinforcement learning (PPO) to optimize energy distribution	Deep Reinforcement Learning (PPO)	Input → Predicted demand, grid status → Output → Policy-derived energy allocation decisions
Layer 4	Energy resource allocation: Apply forecasted demand and policy actions to allocate energy	Energy allocation orchestration and dispatch engine	Input → Predicted demand, policy-derived actions → Output → Total allocated energy for each zone

(continued)

Layer	Main functionality	AI module(s) and components	Operational scope
Layer 5	Energy loss consideration: Account for transmission losses during allocation to improve efficiency	Network loss estimation and efficiency optimization module	Input → Transmission loss data → Output → Refined energy allocation minimizing losses
Layer 6	System cost minimization: Optimize total operational cost based on energy allocation and transmission losses	System cost modeling and multi-objective optimization module	Input → Energy allocation decisions, transmission loss data → Output → Total operational cost

1.2.2.3 Cost Savings and Environmental Benefits

The implementation of AI in Stockholm's energy management system has resulted in measurable cost savings for the city. By optimizing energy distribution and reducing reliance on carbon-intensive energy sources during peak periods, the system minimizes unnecessary expenditures associated with traditional generation methods. This strategic adaptation allows Stockholm to significantly lower its overall energy costs while maintaining supply reliability. Furthermore, the enhanced efficiency of the smart energy grid contributes meaningfully to environmental sustainability by reducing greenhouse gas emissions linked to peak-time fossil fuel production and inefficient distribution (Badidi, 2022). To support this dual objective—economic optimization and environmental responsibility—we introduce a predictive AI-driven model employing Extreme Gradient Boosting (XGBoost) to optimize real-time energy source scheduling under combined cost-emissions constraints.

The XGBoost model performs the prediction by fitting a set of decision trees in an ensemble framework, which can then be used to forecast future energy demand. The output of this model provides predicted demand for each zone, which is critical for efficient energy distribution and decision-making.

$$\hat{D}_t^{(z)} = \sum_{i=1}^{T} \alpha_i \cdot f_i(x_t)$$

where $\hat{D}_t^{(z)}$, predicted energy demand for zone z at time t, α_i, weight for the i-th decision tree, $f_i(x_t)$, output of the i-th decision tree based on input features x_t, and T, total number of trees. This function computes the predicted energy demand $\hat{D}_t^{(z)}$ by aggregating the outputs of all decision trees, weighted by their corresponding factors α_i, ensuring an accurate forecast based on past and present energy consumption data.

Following the demand forecasting, the next critical step is the energy source selection. This layer optimizes the energy source scheduling based on a cost-emissions trade-off, a function of the predicted demand, operational cost data, and emissions data from various energy sources. The selected energy source must not only fulfill the predicted energy demand but also minimize both costs and environmental impact, making this trade-off essential for the smart grid to meet both economic and sustainability goals. To accomplish this, the XGBoost model is applied to predict the most appropriate energy source for each zone, taking into account the multifaceted relationship between cost, emissions, and demand.

The XGBoost model ranks the available energy sources based on their cost-emissions performance, ensuring that the selected source balances both economic and ecological objectives.

$$S_t^{(z)} = \arg\min_{s \in S} \left[\alpha \cdot C_s^{(z)} + \beta \cdot E_s^{(z)} - \delta \cdot A_t^{(z)} \right]$$

where $S_t^{(z)}$, selected energy source at time t for zone z, S, the set of available energy sources (e.g., solar, wind, gas, hydro), $C_s^{(z)}$, operational cost per unit of source s in zone z, $E_s^{(z)}$, estimated emissions of source s, $A_t^{(z)}$, previously allocated energy, a, cost sensitivity factor, β, emissions penalty weight, and δ, allocation alignment weight. This decision function enables the AI system to rank all available sources by minimizing a cost-emissions trade-off, ensuring the selected source not only meets the required energy for zone z but does so with the least combined economic and ecological impact.

Once the energy sources have been selected, the cost-emissions trade-off optimization is applied. This layer further refines the source allocation by minimizing both operational costs and emissions, based on the real-time data of the energy system. The XGBoost model is again employed to optimize the allocation across all zones, ensuring that the energy selected for each zone provides the required amount while maintaining the lowest possible cost and emissions.

This step enables the system to perform an integrated optimization, where both cost and environmental impact are factored into the final allocation decision.

$$A_t^{(z)} = \arg\min_{s \in S} \left[\gamma \cdot C_s^{(z)} + \mu \cdot E_s^{(z)} \right]$$

where $A_t^{(z)}$, energy allocation for zone z at time t, S, set of available energy sources, $C_s^{(z)}$, operational cost of source s in zone z, $E_s^{(z)}$, emissions from source ss in zone z, γ, cost minimization weight, and μ, emissions reduction weight. This function performs a multi-objective optimization to minimize both cost and emissions for energy source selection across zones, ensuring the lowest economic and ecological impact.

The next step is the calculation of sustainability benefits, where the AI system evaluates the total environmental and economic gain. This function compares the optimized cost and emissions for each zone with the traditional baseline, highlighting the improvements achieved through AI-enhanced decision-making. The sustainability benefit score provides a measurable quantification of how much more efficient the system has become in terms of both cost and emissions.

This function serves as the basis for evaluating the effectiveness of the energy management system in meeting both economic and environmental objectives.

$$\mathcal{B}_t = \sum_{z=1}^{Z} \left[\kappa \cdot \left(C_{\text{baseline}}^{(z)} - C_{S_t^{(z)}} \right) + \mu \cdot \left(E_{\text{baseline}}^{(z)} - E_{S_t^{(z)}} \right) \right]$$

where $\mathcal{B}_t$, total benefit score for time t, $C_{\text{baseline}}^{(z)}$, traditional cost before AI optimization, $C_{S_t^{(z)}}$, cost of AI-selected source $S_t^{(z)}$, $E_{\text{baseline}}^{(z)}$, baseline emissions for zone z, $E_{S_t^{(z)}}$, emissions from selected source, κ, cost savings coefficient, and μ, emissions reduction coefficient. This function expresses in measurable units how the AI-enhanced decisions lower both costs and emissions compared to conventional practices. The benefit score $\mathcal{B}_t$ provides feedback to reinforcement learning agents in training, thus closing the optimization loop across forecasting, allocation, source selection, and sustainability gain.

The final layer involves reinforcement learning (RL), where the system continuously improves based on the sustainability gain feedback. By evaluating the sustainability benefits, the RL agents adjust their decision-making policies to further optimize energy allocation. This dynamic adjustment process ensures that the system adapts over time, consistently refining its approach to energy management based on both cost and emissions outcomes.

$$\pi_\theta(a_t | s_t) = \arg\max[\mathbb{E}_t[R_t]]$$

where $\pi_\theta(a_t| s_t)$, learned policy for action a_t at state s_t and R_t, expected reward at time t. This function optimizes the policy π_θ by selecting the action a_t that maximizes the expected reward R_t, ensuring that the system continually learns to make better energy allocation decisions in line with both economic and ecological goals.

The proposed AI-driven model demonstrates a comprehensive and integrated approach to optimizing energy source selection and sustainability within an urban energy management system. By *combining* advanced machine learning techniques, particularly Extreme Gradient Boosting (XGBoost), and Reinforcement Learning (RL), the model effectively addresses both economic and environmental objectives. The process begins with forecasting energy demand using XGBoost, which captures complex patterns in historical and external data. This demand forecast is then used to inform real-time energy source scheduling, optimizing for both cost and emissions reduction. The model further refines this decision-making by evaluating the trade-off between operational costs and environmental impact, ensuring that the chosen energy source minimizes both. The system continuously improves over time through feedback from sustainability gain calculations, allowing for dynamic adjustments in energy allocation policies. By employing cost savings and emissions reductions into a single framework, the model offers a systematic method for minimizing energy costs while promoting sustainability and aligning economic goals with ecological responsibility. This AI-based approach not only enhances operational efficiency but also positions energy management as an active participant in environmental stewardship.

Framework: AI-driven Energy Source Scheduling and Sustainability Optimization (AI-ESS-SO)

Layered AI Architecture for Real-Time Energy Source Selection and Sustainability Gain

Layer	Main functionality	AI module(s) and components	Operational scope
Layer 1	Data collection: Gather real-time energy consumption, cost, and emissions data for different zones	Energy consumption, cost, and emissions data integration and preprocessing service	Input → Real-time energy consumption data, historical cost and emissions data → Output → Preprocessed input for AI models
Layer 2	Demand forecasting: Use XGBoost for predicting future energy demand patterns across zones	Extreme Gradient Boosting (XGBoost)	Input → Historical energy demand data, weather data → Output → Predicted energy demand patterns for each zone
Layer 3	Source selection: Optimize energy source scheduling based on cost-emissions trade-off	Extreme Gradient Boosting (XGBoost)	Input → Predicted demand, operational cost data, emissions data → Output → Selected energy source for each zone
Layer 4	Cost-emissions trade-off optimization: Apply XGBoost model to minimize both operational costs and emissions	Extreme Gradient Boosting (XGBoost)	Input → Cost data, emissions data, forecasted energy demand → Output → Optimized energy source allocation
Layer 5	Sustainability benefit calculation: Evaluate total sustainability gain by comparing AI-scheduled costs and emissions to baseline	Sustainability benefit quantification and baseline comparison service	Input → Traditional cost and emissions data, AI-scheduled costs and emissions → Output → Sustainability gain score for each zone
Layer 6	Feedback and learning: Reinforcement learning agents adjust decision-making based on sustainability gain	Reinforcement learning	Input → Sustainability gain score, historical performance data → Output → Adjusted energy scheduling policy

Figure 1.4 shows the structure of Stockholm's AI-enhanced energy supply and demand balancing system. The central node, "AI-Enhanced Energy Supply and Demand Balancing in Stockholm," connects to three main components: "Machine Learning for Predictive Analysis," "Improved Resource Allocation," and "Cost

Fig. 1.4 AI-enhanced energy supply and demand balancing in Stockholm

Savings and Environmental Benefits." "Machine Learning for Predictive Analysis" includes analyzing historical energy usage data, predicting future demand and demand spikes, and making real-time adjustments in distribution. "Improved Resource Allocation" highlights identifying high-demand areas, targeting energy distribution, and reducing operational costs by minimizing energy losses. "Cost Savings and Environmental Benefits" emphasizes reducing energy expenditure, decreasing reliance on traditional energy sources during peak times, and lowering greenhouse gas emissions to support environmental sustainability.

1.3 Waste Management

The implementation of artificial intelligence (AI) in waste management systems is revolutionizing how cities handle waste, promote recycling, and minimize environmental impact. This analysis explores the innovative strategies adopted by

New York City and San Francisco in utilizing AI technologies to enhance their waste management processes, ultimately leading to improved sustainability outcomes.

1.3.1 Case Study 1: AI-Powered Waste Collection in New York City

New York City has implemented a cutting-edge AI-driven waste collection system designed to optimize the efficiency of its waste management operations. By integrating real-time data analytics and machine learning, the city aims to reduce operational costs while promoting sustainability.

1.3.1.1 Smart Bins with Sensors

One of the key components of New York City's AI waste management system is the deployment of smart bins equipped with sensors. These sensors monitor the fill levels of waste bins in real time and send data to a central system (Yigitcanlar & Cugurullo, 2020). This technology allows the city to track which bins are full and need collection, minimizing unnecessary pickups for bins that are not yet full. This not only saves fuel and reduces emissions from waste collection vehicles but also ensures that waste does not overflow, maintaining public cleanliness.

To optimize smart waste collection dynamically based on real-time bin-level data, we introduce a Deep Reinforcement Learning (DRL) framework using Proximal Policy Optimization (PPO). This model continuously learns the optimal waste collection policy by interacting with a dynamic urban environment, where the goal is to minimize collection costs, vehicle travel distance, and emissions, while preventing bin overflow. The DRL model's core structure begins with a state-action value function that determines the expected return for selecting a particular action given the current state of the waste bin network. The State-Action Value Function is defined as:

$$Q^*(s_t, a_t) = \mathbb{E}_\pi \left[\sum_{k=0}^{\infty} \gamma^k R(s_{t+k}, a_{t+k}) \right]$$

where s_t is the system's current state at time t (e.g., bin fill levels, vehicle locations), a_t is the action taken (e.g., dispatching a truck to a specific bin), $R(s_{t+k}, a_{t+k})$ is the immediate reward at future step $t + k$, $\gamma \in (0, 1)$ is the discount factor for future rewards, and π is the policy being followed. This value function evaluates the long-term reward of each waste collection decision across time, allowing the model to account for cascading effects, such as delayed collection resulting in bin overflow.

To learn the optimal policy π^* under uncertain and dynamic waste conditions, we use Proximal Policy Optimization (PPO), an actor-critic algorithm suitable for continuous decision-making environments. The policy optimization objective is expressed as:

$$L^{\text{PPO}}(\theta) = \mathbb{E}_t[\min(r_t(\theta)A_t, \text{clip}(r_t(\theta), 1 - \epsilon, 1 + \epsilon)A_t)]$$

where $r_t(\theta) = \frac{\pi_\theta(a_t \mid s_t)}{\pi_{\theta_{\text{old}}}(a_t \mid s_t)}$ is the probability ratio between the new and old policy, A_t is the advantage function that measures the relative value of action a_t at state s_t, and ϵ is a small parameter that limits policy deviation to ensure stable updates.

The reward function $R(s_t, a_t)$ is decomposed to model waste-specific operational goals:

$$R(s_t, a_t) = -\alpha D(s_t, a_t) - \beta E(s_t, a_t) - \delta O(s_t, a_t)$$

where $D(s_t, a_t)$ is the total driving distance required by the vehicle given the selected bins, $E(s_t, a_t)$ is the fuel-based energy cost, $O(s_t, a_t)$ is the overflow penalty if a bin is skipped past its threshold, and α, β, δ are weighting parameters calibrated according to city policy.

The advantage function A_t, required in PPO, is computed as:

$$A_t = R(s_t, a_t) + \gamma V(s_{t+1}) - V(s_t)$$

where $V(s_t)$ is the value function estimating the expected future reward from state s_t, helping the model identify whether the chosen action leads to better-than-average outcomes compared to the expected return. The policy π is optimized via PPO to maximize long-term waste management efficiency, using feedback from the advantage A_t, which is derived from the reward function $R(s_t, a_t)$, which in turn captures operational cost components like distance and emissions. The state-action value function $Q^*(s_t, a_t)$ acts as the fundamental learning signal for choosing and updating the collection policy.

This DRL-based model dynamically adjusts waste collection routes in real time by constantly evaluating sensor input from smart bins and optimizing the dispatch of vehicles in a way that minimizes overflow, fuel usage, and cost. The state-action framework ensures long-term reward maximization, while the PPO mechanism guarantees stable learning in uncertain environments. The reward decomposition allows city managers to directly influence the AI system by tuning trade-offs between driving distance, environmental impact, and cleanliness. In this way, the proposed AI framework enables an intelligent, self-improving waste management system aligned with urban sustainability goals and efficient service delivery.

Framework: AI-Driven Waste Management Optimization Framework (AI-WMOF)

Layered AI Architecture for Real-Time Smart Waste Collection Optimization

Layer	Main functionality	AI module(s) and components	Operational scope
Layer 1	Data collection: Gather real-time sensor data from waste bins	Device telemetry ingestion and edge gateway service	Input → Sensor data (fill levels) → Output → Preprocessed data for AI models
Layer 2	Preprocessing: Normalize and structure the data	Data normalization and feature engineering service	Input → Raw data (fill levels) → Output → Structured input for AI model
Layer 3	Learning: DRL with PPO for dynamic collection policy optimization	Deep Reinforcement Learning (DRL), Proximal Policy Optimization (PPO)	Input → Historical data, sensor streams → Output → Optimized collection policies
Layer 4	Optimization: Waste collection route optimization using PPO	Proximal Policy Optimization (PPO)	Input → Optimized collection policies → Output → Refined waste collection route
Layer 5	Real-time adjustment: Adjusting collection policies based on environmental data	Real-time policy adaptation and environmental context engine	Input → Dynamic environmental data (waste levels, vehicle locations) → Output → Adjusted collection schedule
Layer 6	Deployment: Implementing optimized collection policies and monitoring system performance	Route execution orchestrator and operations monitoring console	Input → Optimized collection route → Output → Real-time waste collection
Layer 7	Monitoring: Continuous performance assessment and system feedback	Performance analytics and continuous improvement module	Input → Performance data (collection times, emissions) → Output → Performance reports

1.3.1.2 Predictive Analytics for Route Optimization

By leveraging data collected from smart bins, the city employs predictive analytics to optimize waste collection routes. The AI algorithms analyze historical data and current bin statuses to generate the most efficient collection routes for waste collection trucks (Yigitcanlar & Cugurullo, 2020). This system reduces the number of miles driven by collection vehicles, leading to significant cost savings and a decrease in greenhouse gas emissions. To enable predictive optimization of waste collection routing based on smart bin data, we propose a Long Short-Term Memory (LSTM)-based predictive model embedded within a Deep Reinforcement Learning (DRL) framework. The LSTM network captures temporal patterns in waste generation from historical fill-level data, allowing the DRL agent to make preemptive routing decisions rather than reactive ones. This integration results in optimized routes that anticipate bin fill levels in advance, ensuring timely pickups and emission reduction.

The forecasting mechanism begins with the core function of the LSTM unit, which learns sequential dependencies across multiple time steps:

$$h_t = \sigma(W_h \cdot x_t + U_h \cdot h_{t-1} + b_h)$$

where h_t is the updated hidden state representing the predicted waste accumulation trend, x_t is the input at time t (e.g., bin fill level, time of day, location), h_{t-1} is the previous hidden state storing past temporal patterns, W_h and U_h are the input and recurrent weight matrices, b_h is the bias term, and σ is the activation function introducing nonlinearity for complex pattern learning. This temporal prediction h_t is passed to the DRL agent's decision-making module. The value of predictive routing lies in the agent's ability to integrate not just current fill levels but future projections into its route-planning policy. We define the next essential learning component as the updated state-action value function which now incorporates the LSTM prediction as a dynamic input:

$$Q^*_{\text{pred}}(s_t, a_t) = \mathbb{E}_\pi \left[\sum_{k=0}^{\infty} \gamma^k R(\hat{s}_{t+k}, a_{t+k}) \right]$$

where $\hat{s}_{t+k}$ is the predicted state at future time step $t + k$, derived from the LSTM output h_t, and all other terms remain consistent with the previously defined value function. This extension allows the agent to act on projected waste bin statuses rather than static values.

The predictive routing policy is optimized through the same PPO formulation previously defined, where the advantage estimation now incorporates predicted states. This directly impacts the reward modeling, which is restructured to include predictive distance optimization:

$$R(\hat{s}_t, a_t) = -\alpha D_{\text{pred}}(\hat{s}_t, a_t) - \beta E(\hat{s}_t, a_t) - \delta O(\hat{s}_t, a_t)$$

where $D_{\text{pred}}(\hat{s}_t, a_t)$ represents the forecasted total route length given the predicted fill status $\hat{s}_t$, $E(\hat{s}_t, a_t)$ estimates the future fuel cost, and $O(\hat{s}_t, a_t)$ is the penalty for future overflow, while α, β, δ remain tuning parameters. This formulation explicitly links LSTM's output with downstream optimization in the DRL pipeline.

This integrated LSTM-DRL model demonstrates the transformative capacity of predictive analytics in urban waste collection. The LSTM module continuously forecasts bin statuses, enabling the reinforcement learning agent to simulate future collection scenarios. By learning optimal actions based not only on current but also future conditions, the system preemptively adjusts vehicle routes, minimizing the distance traveled and fuel consumed. The predictive reward formulation allows municipal services to model what-if scenarios and avoid future inefficiencies.

Altogether, this framework embodies a high-performing, anticipatory AI system that aligns with sustainability, cost-efficiency, and real-time urban operational intelligence.

Framework: AI-Driven Waste Collection Routing Optimization Framework (AI-WCROF)

Layered AI Architecture for Predictive Waste Collection Routing Optimization

Layer	Main functionality	AI module(s) and components	Operational scope
Layer 1	Data collection: Gather real-time sensor data from waste bins	Sensor telemetry ingestion and edge gateway service	Input → Sensor data (fill levels) → Output → Preprocessed data for AI models
Layer 2	Preprocessing: Normalize and structure the data	Data normalization and feature engineering module	Input → Raw data (fill levels) → Output → Structured input for AI model
Layer 3	Learning: LSTM for temporal prediction and DRL for policy optimization	Long Short-Term Memory (LSTM), Deep Reinforcement Learning (DRL)	Input → Historical data, sensor streams → Output → Optimized routing policies based on predictive data
Layer 4	Optimization: Waste collection route optimization using PPO	Proximal Policy Optimization (PPO)	Input → Optimized routing policies → Output → Refined waste collection route
Layer 5	Deployment: Implementing optimized collection policies and monitoring system performance	Route execution orchestration and real-time operations monitoring module	Input → Optimized collection route → Output → Real-time waste collection
Layer 6	Monitoring: Continuous performance assessment and system feedback	Performance analytics and continuous improvement module	Input → Performance data (collection times, emissions) → Output → Performance reports

1.3.1.3 Enhancing Recycling Rates

The AI system also plays a crucial role in promoting recycling efforts. By analyzing waste composition data, the city can better understand the types of materials being discarded and develop targeted educational campaigns to encourage residents to recycle more effectively. For example, if data shows high levels of recyclable materials in general waste, the city can launch initiatives to educate the public on proper recycling practices, ultimately increasing recycling rates and reducing landfill contributions (Yigitcanlar & Cugurullo, 2020).

To support adaptive recycling strategies and inform targeted awareness campaigns, we introduce a high-dimensional classification model using Extreme

Gradient Boosting (XGBoost) for real-time waste composition analysis. This ensemble-based learning framework is ideal for processing mixed data from sensor arrays (e.g., smart bins, visual recognition systems) and categorizing discarded items into recyclable and non-recyclable classes with high precision.

The classification model processes multidimensional inputs x_i, such as material density, colorimetric patterns, and disposal timestamps, to predict the material class $\hat{y}_i \in \{0, 1\}$, where 1 indicates recyclable. The prediction function is modeled as an additive composition of weighted decision trees:

$$\hat{y}_i = \sum_{t=1}^{T} f_t(x_i), f_t \in \mathcal{F}+$$

where x_i is the feature vector for the i-th waste item, f_t represents the t-th decision tree in the ensemble, and $\mathcal{F}$ is the functional space of all possible regression trees. The model aggregates predictions from each f_t to estimate the final class $\hat{y}_i$, enabling probabilistic classification of recyclables.

To ensure the model remains generalizable and avoids overfitting across varied waste profiles, XGBoost minimizes the following regularized loss:

$$\mathcal{L} = \sum_{i=1}^{n} \ell(y_i, \hat{y}_i) + \sum_{t=1}^{T} \Omega(f_t)$$

where $\ell(y_i, \hat{y}_i)$ is the loss between the true class y_i and the predicted output $\hat{y}_i$, and $\Omega(f_t) = \gamma T + \frac{1}{2}\lambda\|w\|^2$ is the complexity penalty for each decision tree, with w representing the leaf weights. The terms γ and λ control tree complexity and regularization strength, respectively, ensuring model robustness across high-volume urban datasets.

Once categorized, the model output serves as a dynamic feedback layer for the AI system, enabling policy teams to generate targeted interventions. If the proportion of recyclable items classified as general waste exceeds a predefined threshold, the system triggers educational outreach.

To model this threshold-based policy activation mathematically, we define an alert function for misclassified recyclables:

$$\phi(t) = \begin{cases} 1 & \text{if } \frac{1}{N_t} \sum_{i=1}^{N_t} \mathbb{1}_{\{y_i = 1 \wedge \hat{y}_i = 0\}} > \theta \\ 0 & \text{otherwise} \end{cases}$$

where N_t is the number of items analyzed in time window t, $1\{\cdot\}$ is the indicator function capturing true recyclable items misclassified as non-recyclable, and θ is the threshold above which public education is deemed necessary. When $\phi(t) = 1$, targeted campaigns are automatically deployed based on location and waste stream metadata.

This XGBoost-based classification and decision support model empowers cities to extract actionable insights from waste composition data. It classifies individual waste items with high accuracy, incorporates model regularization to ensure resilience across waste variability, and links classification errors directly to policy triggers for public outreach. This real-time loop between classification and intervention enables cities to adaptively respond to recycling inefficiencies, reduce landfill contributions, and foster sustainable waste behavior through informed and responsive education campaigns.

Framework: AI-Driven Recycling Classification and Policy Framework (AI-RCPF)

Layered AI Architecture for Predictive Recycling Classification and Public Education

Layer	Main functionality	AI module(s) and components	Operational scope
Layer 1	Data collection: Gather real-time sensor data for waste composition analysis	Sensor data ingestion and compositional feature extraction module	Input → Waste sensor data (e.g., material density, visual patterns) → Output → Preprocessed data for classification model
Layer 2	Preprocessing: Normalize and structure the collected waste data	Data normalization and feature engineering pipeline	Input → Raw data (e.g., material features) → Output → Structured input for XGBoost classification model
Layer 3	Learning: XGBoost-based classification for recyclable vs non-recyclable materials	Extreme Gradient Boosting (XGBoost)	Input → Preprocessed waste data → Output → Classified waste items (recyclable vs non-recyclable)
Layer 4	Optimization: Policy generation based on misclassification thresholds	Policy generation and adaptive intervention engine	Input → Classified waste data → Output → Triggered interventions (e.g., public education campaigns)
Layer 5	Deployment: Implement targeted interventions and track effectiveness	Campaign deployment and outreach management system	Input → Intervention triggers → Output → Deployment of public education campaigns
Layer 6	Monitoring: Continuous assessment of waste recycling efficiency and campaign feedback	Recycling performance analytics and behavioral feedback module	Input → Feedback data (recycling behavior) → Output → Reports on campaign effectiveness

Figure 1.5 displays the structure of New York City's AI-driven waste collection system. The central node, "AI-Driven Waste Collection System in New York City," connects to three main components: "Smart Bins with Sensors," "Predictive Analytics for Route Optimization," and "Enhancing Recycling Rates." "Smart Bins with Sensors" includes real-time monitoring of bin fill levels, data transmission to a central system, and reduction of unnecessary pickups to save fuel and reduce emissions. "Predictive Analytics for Route Optimization" involves analyzing

Fig. 1.5 AI-driven waste collection system in New York City

historical data and current bin statuses, generating efficient collection routes, and achieving cost savings with decreased greenhouse gas emissions. "Enhancing Recycling Rates" covers analyzing waste composition data, developing targeted recycling education campaigns, and increasing recycling rates to reduce landfill contributions.

1.3.2 Case Study 2: AI Innovations in Waste Management in San Francisco

San Francisco is recognized for its ambitious zero waste goals, and AI is an integral part of the city's strategy to achieve these objectives. The city employs AI technologies in various aspects of its waste management system to enhance efficiency and sustainability.

1.3.2.1 Automated Sorting Systems

San Francisco has implemented AI-powered automated sorting systems at its recycling facilities. These systems utilize computer vision and machine learning algorithms to identify and sort recyclable materials more accurately than manual sorting processes (Badidi, 2022). By minimizing contamination in recycling streams, the city can improve the quality of recyclables and enhance the efficiency of the recycling process, ensuring that more materials are successfully diverted from landfills.

To enable high-accuracy classification of recyclable materials within sorting facilities, San Francisco's AI-based infrastructure incorporates Convolutional Neural Networks (CNNs), capable of interpreting visual waste features in real time. These models excel at extracting spatial hierarchies from image data, distinguishing subtle differences between materials such as plastic, metal, glass, and non-recyclables under varying lighting and motion conditions.

Each waste item is represented as an image input tensor $x_i \in \mathbb{R}^{h \times w \times c}$, where h, w, are the image height and width, and c is the number of channels. CNN performs a sequence of convolutions to transform these raw images into feature maps. The core function of a convolutional layer is defined as:

$$z_{i,j,k} = \sum_{m=1}^{M} \sum_{n=1}^{N} \sum_{c=1}^{C} w_{m,n,c,k} \cdot x_{i+m,j+n,c} + b_k$$

where $z_{i,\ j,\ k}$ is the output at position (i, j) in the kth feature map, $w_{m,\ n,\ c,\ k}$ is the filter weight for kernel size $M \times \text{N}$ and channel c, $x_{i\ +\ m,\ j\ +\ n,\ c}$ is the input pixel at the shifted position, and b_k is the bias term for feature map k. This function scans the input image to extract patterns such as edges, textures, or material-specific characteristics across multiple layers.

To quantify the model's performance and guide training, a cross-entropy loss function is used:

$$\mathcal{L} = -\sum_{i=1}^{N} \sum_{k=1}^{K} y_{i,k} \log\left(\hat{y}_{i,k}\right)$$

where $y_{i,\ k}$ is the true one-hot encoded label for item i, and $\hat{y}_{i,k}$ is the predicted probability for class k. This loss penalizes incorrect predictions more severely for high-confidence errors, enabling the CNN to refine classification accuracy over multiple iterations.

The resulting feature maps are passed through activation and pooling layers, then flattened and processed by a fully connected layer to generate the classification output $\hat{y}_i \in \{0, 1, \ldots, K\}$, where K is the total number of waste classes. The final classification output is defined by a softmax function:

$$\hat{y}_{i,k} = \frac{e^{s_{i,k}}}{\sum_{j=1}^{K} e^{s_{i,j}}}$$

where $s_{i,\ k}$ is the raw score for class k produced by the final dense layer before normalization. The softmax output $\hat{y}_{i,k}$ represents the probability that item i belongs to category k, allowing the sorting arm to make an accurate real-time selection decision.

This CNN-based vision model enables San Francisco's automated sorting systems to detect and classify waste items with millisecond-level latency and industrial-grade precision. The convolutional layers extract deep material features across spatial dimensions, while the softmax-based classification layer ensures probabilistic decision-making aligned with the system's operational constraints. As waste composition becomes more complex and variable, this AI framework ensures consistent material quality, reduces contamination in recycling streams, and drives operational efficiency in support of the city's zero waste targets.

Framework: AI-Driven Automated Recycling Sorting Framework (AI-ARSF)
Layered AI Architecture for High-Accuracy Recycling Classification and Sorting

Layer	Main functionality	AI module(s) and components	Operational scope
Layer 1	Data collection: Gather real-time sensor data (images) for waste sorting	Visual data acquisition and sensor fusion module	Input → Waste images (from camera or sensors) → Output → Preprocessed data for CNN classification model
Layer 2	Preprocessing: Normalize and structure the image data for classification	Image preprocessing and feature extraction pipeline	Input → Raw waste image data → Output → Structured input for CNN model
Layer 3	Learning: CNN-based classification for recyclable vs non-recyclable materials	Convolutional Neural Network (CNN)	Input → Preprocessed image data → Output → Classified waste items (recyclable vs non-recyclable)
Layer 4	Optimization: Softmax-based classification for material prediction	Probabilistic classification and material identification module	Input → Classified waste features → Output → Predicted material category (with probabilities)
Layer 5	Deployment: Implement sorting actions and track system performance	Robotic sorting control and execution management system	Input → Predicted material categories → Output → Sorting actions (robotic arms or bins) and real-time monitoring

(continued)

Layer	Main functionality	AI module(s) and components	Operational scope
Layer 6	Monitoring: Continuous assessment of system performance and accuracy	Performance analytics and adaptive calibration module	Input → Performance data (sorting accuracy, processing time) → Output → System performance reports

1.3.2.2 Real-Time Monitoring and Feedback

Similar to New York City, San Francisco uses smart bins equipped with sensors to monitor waste levels. The AI system collects data on bin fill levels and provides real-time feedback to waste management teams, allowing them to adjust collection schedules and routes dynamically (Badidi, 2022). This responsive approach ensures that the city can efficiently manage waste collection while minimizing disruptions to the community and environmental impacts.

To operationalize real-time responsiveness in urban waste logistics, we introduce a Deep Reinforcement Learning (DRL) model using Multi-Agent Reinforcement Learning (MARL), where each waste collection vehicle functions as an autonomous agent receiving live sensor data from the smart bin network. This approach enables continuous coordination among agents to dynamically optimize routing based on system-wide waste status.

At each time step t, agent i receives an observation otioti derived from bin fill levels and traffic status within its operating region. The policy π_i maps observations to actions such as rerouting, delaying, or initiating collection. Each agent maximizes its expected reward by updating a decentralized value function, while also incorporating shared environment signals. The global reward is defined as:

$$R_t^{\text{global}} = \frac{1}{N} \sum_{i=1}^{N} \left[-\alpha D_t^i - \beta E_t^i - \delta O_t^i \right]$$

where N is the number of collection agents, D_t^i is the distance traveled by agent i, E_t^i is the energy/fuel consumed, O_t^i is the overflow penalty for bins not serviced in time, and α, β, δ are weight coefficients reflecting municipal priorities. This function aggregates the cost and performance of all agents to guide coordinated optimization.

Each agent maintains a localized action-value function $Q^i\left(o_t^i, a_t^i\right)$ that learns to associate environmental conditions with optimal waste collection decisions. To prevent instability in decentralized training, we introduce a centralized critic QcentralQcentral for policy evaluation across all agents:

$$Q^{\text{central}}(\mathrm{o}_t, \mathrm{a}_t) = \mathbb{E}\left[\sum_{k=0}^{\infty} \gamma^k R_{t+k}^{\text{global}} \mid \mathrm{o}_t, \mathrm{a}_t \right]$$

where $o_t = (o_t^1, \ldots, o_t^N)$ and $a_t = (a_t^1, \ldots, a_t^N)$ are the full observation and action vectors from all agents at time t, and $\gamma \in (0, 1)$ is the reward discount factor. This centralized function allows for policy gradients to be computed with respect to the overall system outcome, while preserving decentralized execution at inference time.

The real-time feedback loop operates by continuously streaming bin fill levels into each agent's policy network. If a spike in fill level is detected in a specific subregion, the responsible agent autonomously adapts its route using its learned policy, while updates in Q^{central} propagate improved strategies across the entire agent network.

This MARL-based real-time monitoring model captures the dynamic nature of urban waste generation and optimally distributes collection effort across multiple agents. Each vehicle acts autonomously but cooperatively, learning from both local sensor input and shared environmental feedback. The centralized critic ensures stability and convergence in a multi-agent setting, allowing for real-time adaptation without central command. This architecture supports efficient, flexible, and eco-conscious waste collection, ensuring that urban responsiveness is embedded directly into the operational logic of smart city services.

Framework: AI-Driven Multi-Agent Waste Collection Framework (AI-MAWCF)

Layered AI Architecture for Dynamic and Responsive Urban Waste Collection

Layer	Main functionality	AI module(s) and components	Operational scope
Layer 1	Data collection: Gather real-time sensor data from smart bins	Sensor telemetry integration and edge gateway service	Input → Waste sensor data (bin fill levels, traffic status) → Output → Preprocessed data for agent decision-making
Layer 2	Learning: Multi-agent reinforcement learning for waste collection route optimization	Deep Reinforcement Learning (DRL), Multi-Agent Reinforcement Learning (MARL)	Input → Sensor data, observations from multiple agents → Output → Optimized routes and actions for each agent
Layer 3	Optimization: Coordinated reward function across agents for global optimization	Multi-agent coordination and global reward shaping engine	Input → Local agent actions and rewards → Output → Global reward for system-wide coordination
Layer 4	Deployment: Real-time decision-making and autonomous execution of collection actions	Real-time dispatch orchestration and autonomous execution controller	Input → Optimized routes and actions → Output → Real-time waste collection and system monitoring
Layer 5	Monitoring: Continuous assessment of system performance and feedback	Performance analytics, monitoring, and policy feedback module	Input → System performance data (waste collection efficiency, energy use) → Output → Performance reports and policy adjustments

1.3.2.3 Public Engagement and Education

AI technologies in San Francisco also facilitate public engagement initiatives. The city uses data analytics to understand community behavior regarding waste disposal and recycling (Badidi, 2022). By analyzing trends in waste generation and disposal practices, the city can tailor educational programs and outreach efforts to address specific challenges and promote better waste management practices among residents. This targeted approach enhances community participation in sustainability efforts and fosters a culture of environmental stewardship.

To support adaptive education campaigns based on changing waste behaviors, we introduce a clustering model using K-Means, which helps the city group communities by their disposal patterns. This makes it easier to deliver tailored messages and programs to each group based on actual behavior, not assumptions.

Let each neighborhood or zone be described by a feature vector x_i, which includes average waste generated, recycling rate, contamination level, and seasonal trends. The K-Means algorithm groups all areas into K clusters by minimizing the total distance between each point and its assigned cluster center:

$$\min_{\{C_k\}_{k=1}^{K}} \sum_{k=1}^{K} \sum_{x_i \in C_k} \|x_i - \mu_k\|^2$$

where x_i is the data point for area i, C_k is the set of areas assigned to cluster k, and μ_k is the center of that cluster. The model tries to make each group as similar as possible inside and different from others.

After clustering, the city can assign specific outreach campaigns to each group. For example, areas with low recycling but high contamination can receive focused education on sorting waste. Meanwhile, high-performing areas can be used as examples for others.

To keep the system responsive, the clustering is updated regularly. When new data shows that a neighborhood's behavior has changed, it is reassigned to a more fitting cluster. The assignment rule is:

$$x_i \rightarrow \arg \min_{k} \|x_i - \mu_k\|^2$$

This rule simply says: assign area i to the nearest cluster center. This helps the city adjust its programs quickly based on real data.

This K-Means model gives city planners a simple, effective way to group communities based on how they handle waste. Instead of applying one-size-fits-all messages, the city can focus its educational efforts where they are needed most. The model helps increase recycling, reduce contamination, and build trust by showing residents that the city understands their specific needs. This makes education smarter, more efficient, and more impactful.

Framework: AI-Driven Community Waste Management Framework (AI-CWMF)

Layered AI Architecture for Adaptive Waste Management Education and Outreach

Layer	Main functionality	AI module(s) and components	Operational scope
Layer 1	Data collection: Gather community-specific waste data (recycling rate, contamination level, etc.)	Community data integration and waste feature extraction module	Input → Community waste data (recycling rate, contamination level, waste volume, seasonal trends) → Output → Preprocessed data for clustering model
Layer 2	Preprocessing: Normalize and structure the data for clustering	Data normalization and feature engineering pipeline	Input → Raw community data → Output → Structured input for K-Means clustering model
Layer 3	Learning: K-Means clustering for grouping communities based on waste behaviors	K-Means Clustering	Input → Preprocessed data → Output → Clustered groups of communities based on waste management patterns
Layer 4	Optimization: Targeted educational campaigns based on clustering results	Personalized outreach planning and campaign orchestration engine	Input → Clustered community groups → Output → Tailored educational outreach campaigns for each group
Layer 5	Monitoring: Continuous assessment and re-clustering based on changing waste behaviors	Behavioral analytics dashboard and adaptive re-clustering service	Input → New waste data (changes in behaviors) → Output → Re-clustered groups and adjusted educational efforts

Figure 1.6 presents the structure of San Francisco's AI innovations in waste management. The central node, "AI Innovations in Waste Management in San Francisco," connects to three main components: "Automated Sorting Systems," "Real-Time Monitoring and Feedback," and "Public Engagement and Education." "Automated Sorting Systems" includes the use of computer vision and machine learning for material identification, accurate sorting to reduce contamination, and improved recycling quality. "Real-Time Monitoring and Feedback" features smart bins with sensors for monitoring waste levels, dynamic adjustments in collection schedules, and efficient waste collection with minimized community disruptions. "Public Engagement and Education" emphasizes data analytics to understand waste disposal behavior, customized educational programs, and increased community participation in sustainability efforts.

The case studies of New York City and San Francisco illustrate the transformative impact of AI technologies in urban waste management. By employing advanced data analytics, smart sensor technology, and automated sorting systems, both cities have enhanced their waste management processes, leading to improved efficiency and significant environmental benefits. The integration of AI not only streamlines waste collection and recycling operations but also fosters community engagement and

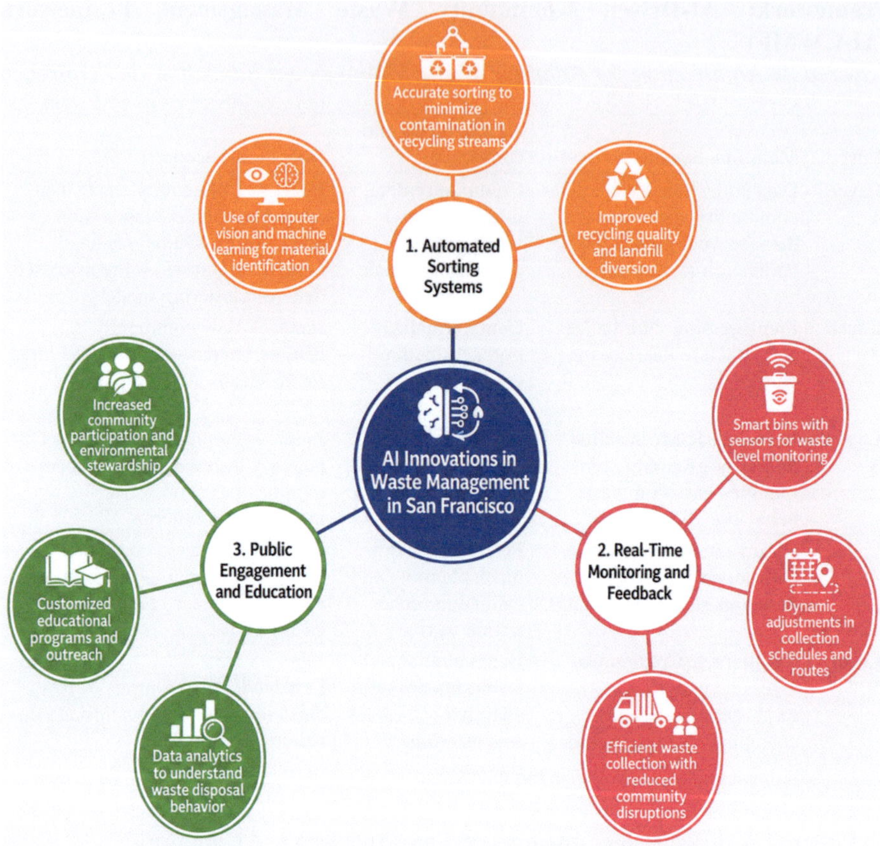

Fig. 1.6 AI innovations in waste management in San Francisco

promotes sustainable practices. As urban areas continue to grow and face increasing waste challenges, the innovative strategies employed by these cities serve as valuable models for other urban centers aiming to enhance their waste management systems through technological advancements.

1.4 Public Safety and Security

The integration of artificial intelligence (AI) into public safety and security systems is reshaping urban environments, enhancing the ability of cities to respond to emergencies, managing crime, and ensuring the safety of residents. This analysis examines the innovative strategies adopted by Dubai and Barcelona in leveraging AI technologies for public safety and security, highlighting their effectiveness and implications for urban governance.

1.4.1 *Case Study 1: AI-Enhanced Public Safety in Dubai*

Dubai has positioned itself as a global leader in the integration of technology into public safety and security operations. The city employs a comprehensive AI framework that enhances law enforcement capabilities and improves emergency response times.

1.4.1.1 Predictive Policing

The Dubai Police Force utilizes AI-driven predictive policing algorithms to analyze crime data and identify patterns. By examining historical crime records, social media activity, and other relevant datasets, the AI system can forecast potential crime hotspots (Badidi, 2022). This proactive approach allows law enforcement agencies to allocate resources more effectively, focusing patrols on areas predicted to experience higher crime rates, thereby preventing criminal activities before they occur.

To help police anticipate where crimes might happen, we use a type of AI model called Long Short-Term Memory (LSTM). This model is trained to understand how crime patterns evolve over time by studying past events. It looks at things like the location of incidents, time of day, and recent activities in an area. With this information, the system builds a memory of past crime behavior and uses it to predict future risks.

The model begins by updating its internal memory every time new crime data arrives. This update is calculated using the following equation:

$$h_t = \sigma(W_h \cdot x_t + U_h \cdot h_{t-1} + b_h)$$

This equation updates the hidden state h_t, which stores knowledge of crime trends up to time t. The model uses the new input data x_t (such as recent crimes or crowd activity), combines it with the previous memory h_{t-1}, and adjusts it using learned weights W_h and U_h. The result is passed through a function σ to help the model focus on the most important information. We use this equation so the model can remember long-term patterns, like which neighborhoods tend to see crime at certain times or during specific events.

Once the memory has been updated, the model uses it to make a prediction:

$$\hat{y}_t = \text{sigmoid}\left(W_y \cdot h_t + b_y\right)$$

This equation generates the predicted crime risk $\hat{y}_t$, which tells us how likely a crime is to happen at that moment and place. The number ranges between 0 (very low risk) and 1 (very high risk). We use this prediction to decide where to send patrols. The more accurate this prediction, the more effective the police response can be.

To help the model improve over time, we compare its predictions with what actually happens. If it predicted a high risk and no crime happened—or missed a real incident—it learns from that error using a loss function:

$$\mathcal{L} = -[y_t \log(\hat{y}_t) + (1 - y_t) \log(1 - \hat{y}_t)]$$

This equation calculates the difference between the true outcome y_t (1 if a crime happened, 0 if it didn't) and the prediction $\hat{y}_t$. If the prediction was far from reality, the model adjusts itself to do better next time. This is how the system becomes more accurate with more data.

This AI model helps Dubai's law enforcement move from reacting to crime toward preventing it. The LSTM structure allows the system to detect patterns in time—whether crime increases at night, in certain districts, or after certain events. The model predicts these risks in advance and helps officers prioritize where to go. By learning from past behavior and improving with every update, the system strengthens public safety while saving resources. This technology supports a smart, data-driven approach to urban security.

Framework: AI-Driven Predictive Policing Framework (AI-DPPF)

Layered AI Architecture for Proactive Crime Prevention and Resource Allocation

Layer	Main functionality	AI module(s) and components	Operational scope
Layer 1	Data collection: Gather real-time crime data from various sources	Secure data ingestion, deidentification, and consent management pipeline	Input → Historical crime records, social media data, sensor data → Output → Preprocessed data for LSTM model
Layer 2	Preprocessing: Normalize and structure the crime data for model input	Data normalization, spatio-temporal feature engineering, and outlier remediation module	Input → Raw crime data → Output → Structured input for LSTM model
Layer 3	Learning: LSTM-based prediction of crime risk at specific locations and times	Long Short-Term Memory (LSTM)	Input → Preprocessed data (crime data, location, time, events) → Output → Predicted crime risk for given area
Layer 4	Optimization: Adjust resource allocation based on predicted crime risk	Risk-aware resource allocation and patrol routing orchestration engine	Input → Predicted crime risk → Output → Patrol allocation and strategic planning
Layer 5	Monitoring: Continuous evaluation of the prediction accuracy and resource allocation effectiveness	Performance monitoring, fairness auditing, and human oversight governance module	Input → Performance data (actual crime outcomes, resource utilization) → Output → System feedback and updates

1.4.2 Smart Surveillance Systems

Dubai has implemented an extensive network of smart surveillance cameras equipped with AI capabilities. These cameras use facial recognition technology and advanced analytics to monitor public spaces in real time, identifying suspicious behaviors and alerting law enforcement (Yigitcanlar & Cugurullo, 2020). This system not only enhances public safety by facilitating rapid responses to incidents but also serves as a deterrent to potential criminals who are aware of the surveillance measures in place.

To further optimize this predictive policing system, we propose a Deep Learning-based Surveillance Framework that leverages Convolutional Neural Networks (CNNs) for real-time facial recognition and behavioral anomaly detection. Central to the system's functionality is its capacity to continuously learn from evolving patterns of human behavior, facial features, and environmental context. The framework incorporates Recurrent Neural Networks (RNNs) and Long Short-Term Memory (LSTM) networks to capture temporal dependencies in the observed activities.

Facial recognition and object detection are the foundational components of this model. In real time, the system identifies both faces and objects within the camera network's field of view. The YOLOv4 model is employed for object detection, enabling the system to detect and track entities such as people, vehicles, and other significant objects within each video frame.

The mathematical formulation that underpins the object detection process is as follows:

$$I'_{i,j} = \sum_{m=0}^{F_h - 1} \sum_{n=0}^{F_w - 1} \mathrm{F}_{m,n} \cdot I_{i+m,j+n}$$

where $I'_{i,j}$ is the output feature map, F is the filter applied to the input image, $I_{i+m,j+n}$ is the input image at position $(i + m, j + n)$, and F_h, F_w represent the height and width of the filter, respectively. This convolution operation helps in identifying key facial features by extracting them from the raw image data.

Once the facial features are recognized, the system will move to detect behavioral anomalies. This phase uses Long Short-Term Memory (LSTM) networks, which are particularly well-suited for sequential data, as they can capture long-term dependencies and patterns in human behavior over time. The LSTM's internal mechanisms, including the forget, input, and output gates, allow it to retain relevant information and discard irrelevant data over time, making it ideal for continuous surveillance.

To understand how the LSTM updates its internal state and detects patterns, we introduce the following set of equations. These equations represent the core functionality of the LSTM, where the cell state and hidden state evolve at each time step:

$$f_t = \sigma\left(W_f[h_{t-1}, x_t] + b_f\right)$$
$$i_t = \sigma(W_i[h_{t-1}, x_t] + b_i)$$
$$\tilde{C}_t = \tanh(W_C[h_{t-1}, x_t] + b_C)$$
$$C_t = f_t * C_{t-1} + i_t * \tilde{C}_t$$
$$o_t = \sigma(W_o[h_{t-1}, x_t] + b_o)$$
$$h_t = o_t * \tanh(C_t)$$

where x_t is the input at time step t, h_t is the hidden state at time t, C_t is the cell state, f_t, i_t, o_t are the forget, input, and output gates, respectively, W and b are the weight matrices and biases. This LSTM equation models the flow of information, allowing the system to detect anomalies in sequential data by learning from long-term dependencies.

The predictive policing component of the framework uses Recurrent Neural Networks (RNNs) to anticipate future behaviors based on historical patterns. By updating its hidden state over time, the RNN is able to predict the next most likely event in the observed sequence of behaviors, aiding law enforcement in anticipating and preventing crimes before they occur.

The RNN updates its hidden state using the following equation, which forms the foundation of the predictive model:

$$h_t = \tanh(W_h[h_{t-1}, x_t] + b_h)$$

where x_t is the input at time step t, h_t is the hidden state at time t, and W_h and b_h are the weight matrix and bias term, respectively. RNN's role in this system is to predict future actions based on previous inputs, allowing the framework to anticipate and respond to suspicious activities before they escalate.

This AI-driven framework integrates several techniques, such as CNNs for facial recognition, LSTMs for behavioral anomaly detection, and RNNs for predictive policing. These models work in tandem to create a robust surveillance system capable of detecting, predicting, and responding to suspicious behavior in real time. Each mathematical equation is carefully constructed to ensure the system's accuracy and adaptability in a dynamic urban environment. By using these advanced AI models, the framework provides a powerful tool for law enforcement agencies, enhancing their ability to proactively intervene in potential incidents. The rigorous mathematical foundation of the model ensures that it is both reliable and reproducible, setting a new standard for AI applications in urban safety and surveillance.

Framework: AI-Driven Predictive Policing Framework (AI-DPPF)
Layered AI Architecture for Proactive Crime Prevention and Resource Allocation

Layer	Main functionality	AI module(s) and components	Operational scope
Layer 1	Data collection: Gather real-time video data from surveillance cameras	Video ingestion deidentification and spatio-temporal feature extraction pipeline	Input → Raw video data → Output → Preprocessed input for AI models
Layer 2	Facial recognition: Applying CNN for face detection in video feed	Convolutional Neural Networks (CNN)	Input → Video frames → Output → Detected faces for further processing
Layer 3	Behavioral anomaly detection: Using LSTM for temporal behavior pattern recognition	Long Short-Term Memory (LSTM)	Input → Facial recognition data → Output → Identified patterns of suspicious behavior
Layer 4	Predictive policing: Using RNN for forecasting future suspicious behaviors	Recurrent Neural Networks (RNN)	Input → Behavioral data → Output → Predicted future suspicious activities
Layer 5	Optimization: Using AI-driven model outputs for law enforcement resource allocation	Risk-aware resource allocation optimization and human oversight governance engine	Input → Detected anomalies, resource data → Output → Optimized resource deployment plan

1.4.2.1 Emergency Response Optimization

Dubai's emergency response infrastructure employs Artificial Intelligence (AI) to dynamically route emergency service vehicles based on real-time urban data. By integrating traffic congestion levels, road closures, GPS trajectories, and incident urgency, the system continuously recalibrates routing strategies to minimize response delays during high-stakes events (Badidi, 2022). This optimization process is governed by a reinforcement learning (RL) framework—specifically, a Proximal Policy Optimization (PPO) algorithm—that adaptively selects routing actions to ensure timely intervention while accounting for systemic constraints.

At the core of this model lies the state-action value function, which estimates the long-term effectiveness of route-selection actions within the emergency response domain:

$$Q^*(s_t, a_t) = \mathbb{E}_\pi \left[\sum_{k=0}^{\infty} \gamma^k R(s_{t+k}, a_{t+k}) \right]$$

Here, the model looks at the current situation s_t, like traffic or roadblocks, and decides an action a_t, such as choosing a new route. It estimates how good that decision will be in the long run by considering not only the immediate benefit but also the future impact. The reward R reflects how fast and safe the route is, and γ is a factor that gives slightly less importance to future outcomes, helping the model stay focused on quick responses.

To ensure that decisions stay practical and don't change too drastically from one step to the next, we apply Proximal Policy Optimization (PPO), which updates the AI's routing policy with care. The PPO objective helps the model improve steadily:

$$L^{\text{PPO}}(\theta) = \mathbb{E}_t[\min(r_t(\theta)A_t, \text{clip}(r_t(\theta), 1-\epsilon, 1+\epsilon)A_t)]$$

The term $r_t(\theta)$ compares how the new routing policy performs compared to the previous one, while A_t tells us how much better or worse the action was than expected. The clipping keeps the learning steps from being too large, so the system improves steadily and safely.

The advantage function guiding PPO is computed as:

$$A_t = R(s_t, a_t) + \gamma V(s_{t+1}) - V(\text{s}_t)$$

where $V(s_t)$ is the value function estimating expected cumulative rewards from state s_t. This helps the model learn whether routing changes genuinely lead to faster or more effective emergency responses.

The reward function used to optimize routing reflects the real-world priorities of emergency operations:

$$R(s_t, a_t) = -\alpha T(s_t, a_t) - \beta C(s_t, a_t)$$

The time function $T(s_t, a_t)$ measures how long it takes to reach the target, while $C(s_t, a_t)$reflects the cost of delays, such as health risks or traffic disruptions. The weights α and β allow the city to balance speed and impact depending on the emergency level.

By constantly learning from real-time data and past performance, this model helps emergency vehicles find faster, smarter routes. It reacts quickly to changes in road conditions and adjusts decisions based on what works best. This means ambulances, fire units, and police teams can get to the scene faster, save lives more efficiently, and avoids unnecessary delays, all while using city resources wisely.

Framework: AI-Driven Emergency Vehicle Routing Framework (AI-DERVF)
Layered AI Architecture for Proactive Emergency Response and Routing Optimization

Layer	Main functionality	AI module(s) and components	Operational scope
Layer 1	Data collection: Gather real-time urban data including traffic congestion, road closures, GPS trajectories, and incident urgency	Urban data integration and telemetry ingestion service	Input → Raw urban data (traffic, road closures, GPS) → Output → Preprocessed input for the AI model

(continued)

Layer	Main functionality	AI module(s) and components	Operational scope
Layer 2	Preprocessing: Transform raw urban data into actionable input for reinforcement learning model	Spatio-temporal feature engineering and data quality validation module	Input → Raw urban data → Output → Actionable input for reinforcement learning model
Layer 3	Learning: Proximal policy optimization (PPO) for dynamically selecting the optimal route	Proximal Policy Optimization (PPO)	Input → Preprocessed data → Output → Optimal route-selection decisions
Layer 4	Optimization: Steady improvements in routing decisions with PPO objective to avoid drastic changes	Proximal Policy Optimization (PPO)	Input → Routing decisions → Output → Improved routing strategy with stable learning
Layer 5	Monitoring and evaluation: Continuous assessment of routing decisions and emergency vehicle performance	Real-time performance analytics and post-incident evaluation dashboard	Input → Routing performance data → Output → Performance reports and refined strategies

Figure 1.7 illustrates the structure of Dubai's AI-enhanced public safety framework. The central node, "AI-Enhanced Public Safety in Dubai," branches into three main components: "Predictive Policing," "Smart Surveillance Systems," and "Emergency Response Optimization." "Predictive Policing" includes analyzing historical crime data and social media, forecasting potential crime hotspots, and allocating resources effectively to prevent crimes. "Smart Surveillance Systems" encompasses a network of AI-equipped cameras, facial recognition and behavior analytics, and real-time alerts for law enforcement to ensure rapid response. "Emergency Response Optimization" focuses on analyzing data from traffic and location information, determining the fastest routes for emergency responders, and reducing response times during critical incidents.

1.4.3 Case Study 2: AI Innovations in Public Safety in Barcelona

Barcelona is also at the forefront of integrating AI into its public safety and security initiatives, focusing on enhancing the efficiency and effectiveness of law enforcement operations.

1.4.3.1 Smart Policing Initiatives

Barcelona has increasingly adopted AI technologies to enhance the effectiveness of its law enforcement strategies. By leveraging crime analytics, the city is able to

Fig. 1.7 AI-enhanced public safety in Dubai

systematically identify spatial-temporal patterns in criminal activity—especially in domains such as theft—and direct police patrols toward areas of elevated risk (Yigitcanlar & Cugurullo, 2020). This data-driven approach to policing not only increases operational efficiency but also contributes to proactive crime prevention by anticipating incidents before they occur.

To formalize this intelligent targeting mechanism, we deploy a Support Vector Machine (SVM) model for binary risk classification across urban zones. The objective is to distinguish between high-risk and low-risk areas based on historical and real-time crime indicators such as location, time of day, incident density, and crime typology. The SVM model defines a hyperplane that separates these two risk classes with maximum margin, ensuring robust generalization and reduced misclassification under urban uncertainty.

The decision function governing classification is expressed as:

$$f(x) = \text{sign}(w \cdot x + b)$$

Here, x is the input data point (e.g., time of theft, location, frequency), w is the weight vector that defines the direction of the boundary, and b is the bias term that shifts the boundary. If the result is positive, the area is considered high risk; if negative, low risk. This helps officers know where to concentrate efforts.

The model is trained to find the best boundary by solving an optimization problem that balances classification accuracy with simplicity. The goal is to keep the model from being too sensitive to noise or outliers:

$$\min_{w,b,\xi} \frac{1}{2}\|w\|^2 + C\sum_{i=1}^{n} \xi_i \quad \text{subject to :} \quad y_i(w \cdot x_i + b) \geq 1 - \xi_i, \xi_i \geq 0$$

where $\|w\|^2$ represents the geometric margin regularization promoting generalization, ξ_i are slack variables allowing soft violations for non-separable cases, $C \in \mathbb{R}^+$ is a regularization constant balancing the trade-off between margin maximization and classification error, and $y_i \in \{-1, +1\}$ are the true class labels for training samples x_i.

To maintain operational relevance, the model is retrained periodically using updated crime data ingested from law enforcement systems and citizen reporting platforms. The reclassification of spatial zones based on recent input xx allows for real-time adaptive patrol deployment, enabling Barcelona's police force to dynamically shift personnel and resources based on evolving crime risk distributions.

This SVM-based AI framework transforms urban crime data into actionable intelligence, empowering law enforcement agencies to prioritize high-impact zones. By minimizing false positives through margin optimization and accounting for incomplete data via slack variables, the model ensures precision-targeted patrolling. Its integration into citywide decision systems fosters a transition from reactive to preventive policing—establishing Barcelona as a vanguard in AI-enhanced public safety governance.

Framework: AI-Driven Crime Risk Classification Framework (AI-DCRCF)
Layered AI Architecture for Proactive Crime Prevention and Patrol Optimization

Layer	Main functionality	AI module(s) and components	Operational scope
Layer 1	Data collection: Gather historical and real-time crime data, including location, time, and incident density	Secure data ingestion, deidentification, and geospatial feature extraction module	Input → Raw crime data (location, time, frequency) → Output → Preprocessed input for SVM

(continued)

Layer	Main functionality	AI module(s) and components	Operational scope
Layer 2	Preprocessing: Transform raw crime data into structured inputs for the SVM model	Spatio-temporal data normalization and feature engineering pipeline	Input → Structured crime data → Output → Actionable input for SVM model
Layer 3	Learning: Support vector machine (SVM) for binary classification of urban zones into high-risk and low-risk areas	Support vector machine (SVM)	Input → Preprocessed crime data → Output → High-risk vs low-risk zone classification
Layer 4	Optimization: Regularization and slack variables used to balance classification accuracy and margin maximization	Support vector machine (SVM)	Input → Classification performance → Output → Regularized model with balanced margin and reduced noise
Layer 5	Monitoring and evaluation: Continuous reclassification of zones based on updated crime data for adaptive patrol deployment	Performance monitoring, fairness auditing, and human oversight governance module	Input → Updated crime data → Output → Refined patrol deployment strategy

1.4.3.2 Integrated Security Platforms

Barcelona has embraced the concept of integrated security platforms to unify its emergency response ecosystem. By leveraging AI technologies that facilitate real-time communication and inter-agency collaboration, the city enhances both the speed and precision of emergency interventions (Badidi, 2022). These platforms eliminate organizational silos by enabling dynamic data sharing across law enforcement, emergency medical services, and fire response units, forming a cohesive and intelligent urban safety network.

To mathematically operationalize this multi-agency coordination, we propose an AI-driven platform built upon Graph Neural Networks (GNNs). Each public safety unit is modeled as a node v_i in a graph $\mathcal{G} = (\mathcal{V}, \mathcal{E})$, where edges $(i, j) \in \mathcal{E}$ represent communication or task dependencies. The node features encode the operational status of each unit, including current capacity, location, response latency, and incident type.

The real-time inter-agent information exchange is governed by the node state update rule:

$$h_i^{(t+1)} = \sigma\left(\sum_{j \in \mathcal{N}(i)} W \cdot h_j^{(t)} + b\right)$$

This function updates the internal state $h_i^{(t+1)}$ of unit i by aggregating real-time information from all its connected units $j \in N(i)$. The system adjusts behavior dynamically based on what other units are doing—improving communication and cooperation during critical incidents.

To ensure collaboration, the model uses a score that evaluates the overall quality of information flow across the system:

$$S = \sum_{(i,j) \in E} \omega_{ij} \cdot \left\| h_i - h_j \right\|^2$$

This expression helps identify where communication is lagging. If two units are supposed to be aligned but their states h_i and h_j are too different, the system sees a problem. The model then adjusts the coordination process to reduce confusion or delay.

As incidents unfold, the platform dynamically adjusts the graph structure $\mathcal{G}$ to reflect emergent operational demands—e.g., new edge formations during multi-agency incidents or weight recalibration in response to changing priorities. The GNN continually re-aggregates message-passing pathways to ensure globally optimal situational awareness and task distribution.

This GNN-based formulation transforms Barcelona's emergency infrastructure into an adaptive, graph-structured intelligence system. It tightly integrates decentralized unit information into a collective decision-making process, allowing seamless cooperation during high-pressure scenarios. The system's ability to minimize coordination errors while adapting in real time exemplifies how AI can enhance resilience, response precision, and systemic situational control in urban public safety.

Framework: AI-Driven Multi-Agency Coordination Framework (AI-DMACF)
Layered AI Architecture for Adaptive Coordination and Task Distribution in Urban Public Safety

Layer	Main functionality	AI module(s) and components	Operational scope
Layer 1	Data collection: Gather real-time data on unit capacity, location, response latency, and incident type	Operational telemetry ingestion, unit capacity registry, and incident taxonomy service	Input → Raw unit data (capacity, location, latency) → Output → Structured node features for GNN
Layer 2	Preprocessing: Transform raw data into structured node features for the GNN model	Graph-structured feature engineering and dynamic topology construction module	Input → Raw data → Output → Preprocessed input for the GNN model
Layer 3	Learning: Graph Neural Network (GNN) for real-time inter-agent communication and coordination	Graph Neural Network (GNN)	Input → Preprocessed node features → Output → Real-time updates on coordination and communication

(continued)

Layer	Main functionality	AI module(s) and components	Operational scope
Layer 4	Optimization: Score function to evaluate and optimize information flow across the system	Graph Neural Network (GNN)	Input → Inter-agent communication data → Output → Optimized information flow and coordination
Layer 5	Monitoring and evaluation: Dynamic graph updates to adjust task distribution and ensure situational awareness	Real-time coordination dashboard and adaptive task reallocation engine	Input → Operational updates → Output → Recalibrated task distribution and adaptive unit coordination

1.4.3.3 Community Engagement Through Technology

Barcelona also prioritizes community engagement in its public safety efforts. The city employs AI-driven mobile applications that allow residents to report safety concerns and suspicious activities directly to the police (Yigitcanlar & Cugurullo, 2020). This participatory approach not only empowers citizens to take an active role in their community's safety but also provides law enforcement with valuable insights into local issues, facilitating more targeted responses.

For instance, to strengthen public safety through real-time communication with residents, we propose an AI-driven community engagement model based on Natural Language Processing (NLP). The model would be integrated into mobile applications to help law enforcement quickly understand and classify the content of citizen reports. These reports may describe anything from noise complaints to suspicious behavior. Since residents write freely in their own words, the system needs to extract meaning from unstructured text.

The AI begins by converting each user message into a numeric format that the model can work with. This is done through word embeddings, which turn language into mathematical vectors while preserving meaning. Then, a lightweight transformer-based model analyzes the sequence of words to identify the type and urgency of the report. We define the classification function as:

$$\hat{y}_i = \text{softmax}(W \cdot f_{\text{NLP}}(x_i) + b)$$

In this equation, x_i is the input message written by a resident, and $f_{\text{NLP}}(x_i)$ is the output of the transformer model applied to that message. The matrix W and bias term b are adjusted during training to improve accuracy. The softmax function converts the model's output into a list of probabilities that indicate how likely the message belongs to different predefined categories (such as "public safety," "disturbance," "vandalism," or "non-urgent").

Once categorized, the model estimates the urgency of the message based on key words or phrases that commonly signal threats or emergencies. The urgency scoring process can be expressed as:

$$u_i = \sum_{j=1}^{m} \alpha_j \cdot \mathbb{1}_{\{w_j \in \text{alert terms}\}}$$

Here, each word w_j in the message is checked against a list of predefined alert terms—such as "weapon," "injured," or "fire." The indicator function 1 returns 1 when there is a match, and the term α_j assigns more weight to words considered more urgent. This produces a total urgency score u_i for the report, allowing the system to flag and prioritize critical messages for faster review.

The system can then link reports to city maps using location data and automatically highlight zones where incidents are rising. This helps city authorities adapt resource deployment in real time.

By integrating Natural Language Processing into mobile platforms, this AI-driven model would give citizens a faster and more effective voice in shaping local safety. At the same time, it offers law enforcement structured, high-quality information that can lead to quicker and more accurate responses. This kind of innovation reflects how digital tools will strengthen civic participation and make safety systems in future cities more inclusive, responsive, and intelligent.

Framework: AI-Driven Community Engagement Framework (AI-DCEF)

Layered AI Architecture for Real-time Community Safety Reporting and Task Prioritization

Layer	Main functionality	AI module (s) and components	Operational scope
Layer 1	Data collection: Gather unstructured text data from residents via mobile applications	Citizen data ingestion API	Input → Raw text data from citizens → Output → Preprocessed input for the NLP model
Layer 2	Preprocessing: Convert text data into structured numeric vectors using word embeddings	Text vectorization engine	Input → Text data → Output → Structured word embeddings for NLP model
Layer 3	Learning: Transformer-based model for classifying and categorizing citizen reports	Transformer-based Model (NLP)	Input → Word embeddings → Output → Categorized citizen report (e.g., 'public safety', 'disturbance')
Layer 4	Optimization: Softmax function to assign probabilities to different report categories	Softmax function (for classification)	Input → Classifier output → Output → Probability distribution for report categories
Layer 5	Monitoring and evaluation: Real-time urgency scoring and dynamic resource allocation	Urgency scoring and dispatch module	Input → Urgency indicators → Output → Urgency scores and task prioritization for response teams

Fig. 1.8 AI innovations in public safety in Barcelona

Figure 1.8 illustrates the structure of Barcelona's AI innovations in public safety. The central node, "AI Innovations in Public Safety in Barcelona," connects to three main components: "Smart Policing Initiatives," "Integrated Security Platforms," and "Community Engagement through Technology." "Smart Policing Initiatives" includes using data analytics to assess crime trends, developing targeted interventions, and adjusting patrols to prevent crimes like thefts. "Integrated Security Platforms" focuses on an AI-enabled platform for real-time information sharing, enhancing coordination among law enforcement agencies, and improving response capabilities during emergencies. "Community Engagement through Technology" features AI-driven mobile applications for resident reports, empowering citizens to report safety concerns, and providing insights into local issues to facilitate targeted law enforcement responses.

The case studies of Dubai and Barcelona highlight the transformative potential of AI technologies in enhancing public safety and security. By implementing predictive policing, smart surveillance systems, and integrated communication platforms, both cities have significantly improved their capacity to respond to crime and emergencies. The proactive and data-driven approaches adopted by these urban centers not only enhance public safety but also foster a collaborative relationship between law enforcement and the community. As urban areas continue to face evolving security challenges, the strategies employed by Dubai and Barcelona serve as valuable models for other cities aiming to leverage technology for improved public safety outcomes.

1.5 Community Engagement and Governance

The integration of artificial intelligence (AI) in community engagement and governance is revolutionizing how urban centers interact with their citizens. In particular, the city of Tokyo exemplifies the effective use of AI-driven platforms to foster participatory governance, enhance public service delivery, and empower residents to take an active role in their communities. This analysis explores the innovative strategies implemented by Tokyo to improve community engagement through AI technologies.

1.5.1 Case Study 1: AI-Driven Platforms for Community Engagement in Tokyo

Tokyo has leveraged the power of artificial intelligence (AI) to create integrated platforms that facilitate real-time citizen feedback on local issues and urban services. These platforms empower residents to report various urban challenges, from infrastructure issues to service inefficiencies, and enable city officials to respond promptly to community needs. By employing AI technologies, these platforms not only streamline communication between the public and government authorities but also enhance the responsiveness of city governance, ensuring that citizens' concerns are addressed swiftly and accurately. The ultimate goal is to improve urban living conditions and foster a participatory form of governance.

To operationalize this AI-driven feedback platform, we propose an Artificial Intelligence model that utilizes Natural Language Processing (NLP) for real-time analysis of unstructured citizen feedback. This approach requires the conversion of free-text reports into structured data that can be categorized, prioritized, and assigned to relevant city departments for action.

At the core of this system lies an NLP-based model that uses word embeddings to convert textual feedback into numerical vectors, followed by a transformer-based architecture for contextual understanding. This model will classify citizen reports into predefined categories such as "infrastructure," "traffic," or "public services" and further assign urgency scores based on key terms or phrases.

Initially, each citizen's feedback message is converted into numerical vectors through word embeddings. This process involves using an embedding matrix E, where each word w_i in the message is mapped to a vector e_i. These word embeddings preserve the semantic meaning of the text while making it compatible with AI algorithms.

$$x_{\mathrm{i}} = Ew_i$$

where x_i is the numeric vector representing word w_i, E is the embedding matrix, and w_i is the word in the feedback message. This function transforms the raw textual feedback into a numerical format suitable for further processing in AI models.

The next step involves passing these word embeddings through a transformer-based model to capture the contextual relationships between words. The transformer model uses self-attention mechanisms to learn dependencies between words across different parts of the message. The model output, h_i, represents the transformed feature vector of the input message.

$$h_i = \mathrm{Transformer}(x_i)$$

where h_i is the output of the transformer for the input x_i, which is the word embedding of the citizen's feedback. The function transforms the embedding into a more contextually aware representation of the text, preserving the meaning and relationships between words.

Once the contextualized feature vectors are obtained from the transformer, we apply a softmax function to classify the feedback into predefined categories. The output of the softmax function provides a set of probabilities p_j that represent the likelihood of the feedback belonging to each category, such as "infrastructure," "traffic," or "public service."

$$y_i = \mathrm{softmax}(Wh_i + b)$$

where y_i represents the vector of probabilities for each class, W is the weight matrix, h_i is the transformed feature vector, and b is the bias term. This function converts the model's output into probabilities, which are then used to categorize the report into one of the predefined classes.

To prioritize citizen reports based on urgency, the model calculates an urgency score u_i by identifying specific "alert terms" in the feedback message. Each word wjwj is checked against a predefined list of alert terms, such as "fire," "injury," or "accident," and the model assigns weights α_j to these words based on their perceived urgency. The urgency score is computed by summing the weighted scores for all identified terms in the message.

$$u_i = \sum_{j=1}^{m} \alpha_j \cdot 1_{\{w_j \in \text{alert terms}\}}$$

where u_i is the urgency score for report i, w_j are the words in the feedback message, α_j is the urgency weight for word w_j, and $1_{\{w_j \in \text{alert terms}\}}$ is the indicator function that returns 1 if w_j is in the list of alert terms and 0 otherwise. This function assigns urgency scores based on the presence of critical terms in the feedback, helping prioritize high-risk reports.

Finally, the feedback, once categorized and scored for urgency, is passed to the task assignment module, where the appropriate city department (e.g., infrastructure, traffic, medical services) is assigned to handle the report. The system dynamically adjusts resource allocation based on the urgency score and classification, ensuring that high-priority issues are addressed first.

$$\text{Task Assignment} = \text{Argmax}(y_i) \ \text{ with urgency score } u_i$$

where y_i is the classification output from the softmax function, and u_i is the urgency score. This function assigns the report to the department with the highest priority, ensuring that the most urgent issues are dealt with promptly.

This AI-driven community engagement model transforms unstructured citizen feedback into actionable intelligence for city officials. Through the use of Natural Language Processing (NLP), word embeddings, transformer-based models, and urgency scoring, the system efficiently categorizes reports, prioritizes them based on urgency, and ensures that they are addressed by the appropriate city department. This model not only improves responsiveness but also fosters a more participatory form of governance, where citizens actively contribute to shaping the safety and development of their communities.

Framework: AI-Driven Community Engagement Framework (AI-DCEF)
Layered AI Architecture for Real-time Community Safety Reporting and Task Prioritization

Layer	Main functionality	AI module (s) and components	Operational scope
Layer 1	Data collection: Gather unstructured text data from citizens via mobile applications	Citizen data ingestion and preprocessing module	Input → Raw text data from citizens → Output → Preprocessed input for the NLP model
Layer 2	Preprocessing: Convert text data into structured numeric vectors using word embeddings	Text embedding and feature extraction engine	Input → Text data → Output → Structured word embeddings for NLP model
Layer 3	Learning: Transformer-based model for classifying and categorizing citizen reports	Transformer-based Model (NLP)	Input → Word embeddings → Output → Categorized citizen report (e.g., 'public safety', 'disturbance')

(continued)

Layer	Main functionality	AI module (s) and components	Operational scope
Layer 4	Optimization: Softmax function to assign probabilities to different report categories	Softmax function (for classification)	Input → Classifier output → Output → Probability distribution for report categories
Layer 5	Monitoring and evaluation: Real-time urgency scoring and dynamic resource allocation	Real-time urgency scoring and task allocation module	Input → Urgency indicators → Output → Urgency scores and task prioritization for response teams

1.5.1.1 Citizen Reporting Applications

Tokyo has developed mobile applications powered by AI that allow residents to report local issues, such as potholes, broken streetlights, and other infrastructure concerns, directly to city authorities (Yigitcanlar & Cugurullo, 2020). The AI system analyzes incoming reports to categorize and prioritize them based on urgency and potential impact. For instance, reports of safety hazards can be flagged for immediate attention, ensuring that critical issues are addressed promptly.

For instance, to support this smart, responsive system of citizen-government interaction, we propose an AI-driven governance model using a multilayered classification and prioritization framework based on deep learning. The core of the system is a neural model that reads incoming reports, classifies them by issue type, and ranks them based on severity, urgency, and spatial proximity to sensitive zones (e.g., schools, hospitals, or major roads).

The classification process begins with a deep neural network that receives each report as an input vector x_i, which contains both textual features (e.g., keywords) and metadata (e.g., time, location, and media attachments). The issue type is predicted using:

$$\hat{y}_i = \text{softmax}(W_1 \cdot \text{ReLU}(W_0 \cdot x_i + b_0) + b_1)$$

In this equation, the input x_i passes through two dense layers with ReLU activation to capture nonlinear patterns. The softmax function then produces probabilities for different classes such as “road issue,” “lighting,” “sanitation,” or “safety.” The system selects the most likely class, helping route the report to the correct department.

Next, the softmax function converts these scores into a probability distribution, where the value for each class represents the likelihood that the report belongs to that class.

$$y_i = \text{softmax}(Wh_i + b)$$

where y_i is the vector of probabilities for each class, W is the weight matrix, h_i is the feature vector generated by the neural network, b is the bias term, and the softmax function ensures that the sum of the probabilities across all classes is 1. This function transforms the raw output of the DNN into a probability distribution, allowing for clear decision-making on which class the citizen report belongs to.

Once categorized, the system assigns a priority score that guides how quickly each report should be addressed. This priority is based on impact factors such as the location's sensitivity and the estimated risk level described in the message. We define the score as:

$$p_i = \lambda_1 \cdot r_i + \lambda_2 \cdot d_i + \lambda_3 \cdot z_i$$

Here, r_i a risk rating derived from keywords (e.g., "collapse," "fire," "danger"), d_i is the density of similar issues nearby (indicating an ongoing problem), and z_i measures how close the issue is to sensitive infrastructure. The weights λ_1, λ_2, λ_3 allow city administrators to adjust the model's priorities based on local policy.

Once the priority score is calculated, the next step is to determine the urgency by identifying if any alert terms (such as "fire," "danger," etc.) are present in the citizen's report. This can be seen as an additional layer to further refine the priority of issues based on key terms that signal an emergency. This is where the formula for urgency based on alert terms comes into play.

$$u_i = \sum_{j=1}^{m} \alpha_j \cdot 1_{\{w_j \in \text{alert terms}\}}$$

where u_i is the urgency score for report i, w_j are the words in the feedback message, α_j is the urgency weight for word w_j, and $1_{\{w_j \in \text{alert terms}\}}$ is the indicator function that returns 1 if w_j is in the list of alert terms and 0 otherwise. (e.g., "fire," "weapon," "danger"), and 0 otherwise. This function aggregates the weighted presence of urgent words across the report, producing a total urgency score that reflects the criticality of the report based on the alert terms.

The platform regularly updates as new reports arrive. Similar reports are grouped into clusters using geospatial data, so large-scale issues like blackout zones or flooding can be detected automatically and escalated without delay.

This proposed deep learning-based model would help—any future-oriented city—manage citizen feedback with more speed, accuracy, and fairness. AI takes on the burden of sorting, evaluating, and prioritizing public reports, freeing up human decision-makers to focus on resolution. In doing so, the city becomes not just smarter, but more inclusive and accountable—building a truly participatory governance system that reflects the needs of its people in real time.

Framework: AI-Driven Citizen Feedback Governance Model (AI-CFGM)
Layered AI Architecture for Efficient Citizen Feedback and Task Prioritization

Layer	Main functionality	AI module(s) and components	Operational scope
Layer 1	Data collection: Gather unstructured text data from citizens via mobile applications	Citizen data intake, consent management, and preprocessing module	Input → Raw text data from citizens → Output → Preprocessed input for the DNN model
Layer 2	Preprocessing: Convert text data into structured numeric vectors using word embeddings	Text embedding, feature engineering, and vectorization engine	Input → Text data → Output → Structured word embeddings for DNN model
Layer 3	Learning: Deep Neural Network (DNN) for classifying and categorizing citizen reports	Deep Neural Network (DNN)	Input → Word embeddings → Output → Categorized citizen report (e.g., 'road issue', 'lighting')
Layer 4	Optimization: Softmax function to assign probabilities to different report categories	Softmax function (for classification)	Input → Classifier output → Output → Probability distribution for report categories
Layer 5	Monitoring and evaluation: Real-time urgency scoring and dynamic resource allocation	Real-time urgency scoring, priority ranking, and task allocation module	Input → Urgency indicators → Output → Urgency scores and task prioritization for response teams

1.5.1.2 Real-Time Data Collection and Analysis

The AI-driven platform not only facilitates the reporting process but also aggregates data from multiple sources, including social media, sensor networks, and community feedback (Badidi, 2022). By analyzing this data, Tokyo's governance systems can identify patterns in community concerns and prioritize resource allocation accordingly. This real-time data analysis enables city officials to make informed decisions and allocate resources more effectively to address pressing issues.

For instance, to support this type of intelligent decision-making based on continuous input from multiple data streams, we propose an AI model based on a Multi-Source Attention Network (MSAN). This model combines structured and unstructured data—from sensors, social media, and citizen reports—into a unified system that detects trends and triggers rapid responses in urban governance.

Each data source provides a distinct type of information: sensor feeds are numeric and high-frequency, social media posts are textual and event-driven, while citizen reports are semi-structured. The model learns how to weigh and combine these signals through an attention mechanism that highlights the most important inputs depending on the current situation. The fused representation is computed using the following formulation:

$$v_t = \sum_{s=1}^{S} \alpha_t^{(s)} \cdot h_t^{(s)}$$

Here, $h_t^{(s)}$ is the hidden representation from source s at time t, and $\alpha_t^{(s)}$ is the attention weight that determines how important this source is in the current context. The final vector v_t captures the combined view of the urban situation at time t, allowing the system to respond based on the full spectrum of inputs.

Once this combined view is formed, the system classifies the nature of the issue and suggests an appropriate response level:

$$\hat{y}_t = \text{softmax}(W \cdot v_t + b)$$

The output $\hat{y}_t$ reflects a set of possible response actions—such as "immediate dispatch," "monitor," or "inform department." The model chooses the one that best fits the real-time situation. The weights W and bias b are tuned so that the classification reflects local governance policies and past decision outcomes.

The system continuously adapts as new data flows in. Attention scores shift dynamically depending on what matters most: if sensors detect unusual air quality while social media remains quiet, the model will favor the sensor data. If a protest trend begins online, the balance shifts in the other direction. This makes the platform flexible and responsive.

This proposed multi-source attention network would give cities a powerful tool to synthesize vast streams of live urban data into clear, actionable insights. Rather than treating all inputs equally or relying solely on predefined thresholds, the system learns over time which sources offer the best signals in different situations. This enables city leaders to anticipate problems, allocate resources efficiently, and make decisions based on a real-time, multidimensional understanding of urban dynamics.

Framework: Multi-Source Attention Network for Urban Governance (MSAN)
Layered AI Architecture for Real-Time Urban Data Analysis and Decision-Making

Layer	Main functionality	AI module(s) and components	Operational scope (Input → Processing → Output)
Layer 1	Data collection and preprocessing	Multi-source data ingestion, cleansing, and harmonization pipeline	Input → Sensor data, social media posts, citizen reports → Output → Preprocessed data for AI model
Layer 2	Learning: Multi-source data fusion and attention mechanism	Multi-Source Attention Network (MSAN)	Input → Preprocessed data from multiple sources → Output → Combined urban data representation

(continued)

Layer	Main functionality	AI module(s) and components	Operational scope (Input → Processing → Output)
Layer 3	Classification and response decision-making	Softmax classifier	Input → Combined data representation → Output → Response action classification (e.g., dispatch, monitor, inform department)
Layer 4	Adaptation and continuous model improvement	Online feedback learning and continuous model retraining engine	Input → Feedback data from real-time events → Output → Model retraining and adaptation for future predictions

1.5.1.3 Enhanced Communication and Transparency

The use of AI in community engagement also promotes transparency in governance. By providing a platform for residents to report issues and track the status of their complaints, Tokyo fosters a sense of accountability among city officials (Yigitcanlar & Cugurullo, 2020). Citizens can see how their reports are being handled, which enhances trust in local government and encourages further participation. This transparency is essential in creating a collaborative relationship between residents and government authorities.

For instance, to support this transparent interaction between residents and city authorities, we propose an AI-driven feedback tracking model based on a Time-Aware Recurrent Neural Network (T-RNN). This model is designed to monitor the lifecycle of each citizen-submitted report, respond to inactivity or delays, and improve communication flow by automatically evaluating when to send updates or reminders.

Each report is denoted as r_i and consists of a time-ordered sequence of status updates, metadata, and timestamps. We define it as:

$$r_i = \left\{x_1^{(i)}, x_2^{(i)}, \ldots, x_T^{(i)}\right\}$$

Here, $x_1^{(i)}$ represents the features of report r_i at time t, including the current status ("submitted," "acknowledged," "in progress," etc.), delay time since the last update, the communication channel used, and optional citizen feedback or media attachments.

This sequence is processed through a Gated Recurrent Unit (GRU), which maintains a contextual memory of the report's evolution over time:

$$h_t^{(i)} = \text{GRU}\left(x_t^{(i)}, h_{t-1}^{(i)}\right)$$

the output $h_t^{(i)}$ captures the entire history and current communication status of report r_i up to time t. This memory allows the model to recognize delays, unusual silences, or unresolved steps in the feedback loop.

From the final state $h_T^{(i)}$, the model computes a transparency score T_i, which estimates how effectively the platform is keeping the citizen informed about their report:

$$T_i = \sigma\left(W \cdot h_T^{(i)} + b\right)$$

This score ranges from 0 to 1. A low score signals insufficient communication or stalled resolution for report r_i, prompting the system to automatically generate a progress update or escalate the issue. A high score indicates that updates have been timely and the citizen is well-informed, requiring no immediate action.

These transparency scores can also be aggregated across city districts or service categories, allowing the municipality to publish real-time dashboards that reflect response times, report closure rates, and overall communication effectiveness—without compromising privacy.

This proposed feedback tracking model demonstrates how AI can turn individual reports into transparent communication channels. By learning when updates are expected and proactively responding to communication gaps, the system strengthens trust between residents and local government. In future cities, this kind of intelligent transparency will be key to ensuring that civic participation is not only encouraged but respected with real, visible outcomes.

Framework: AI-Driven Feedback Tracking for Community Engagement (T-RNN). *Layered AI Architecture for Transparent Citizen Communication and Feedback Loop.*

Layer	Main functionality	AI module(s) and components	Operational scope (Input → Processing → Output)
Layer 1	Data collection and preprocessing	Citizen data intake, consent management, and preprocessing service	Input → Citizen reports, status updates, communication data → Output → Preprocessed data for AI model
Layer 2	Learning: Time-Aware Recurrent Neural Network (T-RNN)	Time-Aware Recurrent Neural Network (T-RNN)	Input → Time-ordered report data with status updates → Output → Processed report history and communication status
Layer 3	Transparency score computation and feedback generation	Sigmoid function, weighted sum	Input → Final state of report data → Output → Transparency score (0–1) indicating communication effectiveness
Layer 4	Dashboard generation and reporting for city authorities	Real-time transparency analytics dashboard and reporting service	Input → Aggregated transparency scores across districts → Output → Real-time dashboards with communication performance metrics

1.5.1.4 Empowerment Through Participation

By utilizing AI technologies, Tokyo empowers its citizens to play an active role in local governance. The ability to report issues and provide feedback fosters a sense of community ownership and responsibility (Badidi, 2022). When residents feel that their voices are heard and their concerns are addressed, they are more likely to engage in other civic activities, thereby strengthening community ties and promoting a culture of collaboration.

For instance, to support this participatory process and turn civic input into meaningful engagement, we propose an AI-driven model based on Graph-Based Sentiment Aggregation (GBSA). This model analyzes community feedback submitted through digital platforms—such as surveys, open comments, and real-time reports—and aggregates sentiment across citizens who are connected by location, topic, or interaction history.

We define the citizen feedback network as a graph:

$$G = (V, E)$$

where $V = \{v1, v2, \ldots, vn\}$ is the set of nodes, with each node $v_i \in V$ representing a citizen who has submitted feedback, $E \subseteq V$ is the set of edges, where an edge $e_{ij} \in E$ connects citizens v_i and v_j based on contextual similarity (e.g., geographic proximity, similar feedback content, or shared engagement channels).

Each node v_i is associated with a feedback message x_i, which is processed by a sentiment extraction function to produce a numerical sentiment score:

$$s_i = f_{\text{sentiment}}(x_i)$$

Here, x_i is the feedback text submitted by a citizen, and s_i is the resulting sentiment score (e.g., -1 for negative, 0 for neutral, +1 for positive). This step helps the system understand how people feel—not just what they report.

All submissions are then organized into a graph structure $G = (V, E)$, where each node $v_i \in V$ represents a citizen, and edges $e_{ij} \in E$ represent similarity in location, topic, or community group. Each node carries its sentiment score. This graph allows the AI model to propagate sentiment signals across communities by leveraging structural relationships between citizens. The connections help the model understand not just isolated opinions but how emotions and concerns are distributed and reinforced across the civic network.

The aggregated sentiment around a topic or area is then computed by smoothing the individual scores across the graph:

$$\bar{s}_i = \frac{1}{|\mathcal{N}(i)|} \sum_{j \in \mathcal{N}(i)} s_j$$

Here, $\mathcal{N}(i)$ represents the neighborhood (i.e., connected citizens) of node v_i, and $\bar{s}_i$ is the average sentiment from the community around that citizen or topic. This helps the city understand shared emotions—not isolated opinions—so they can respond at the community level.

As more feedback is collected, the model identifies engagement clusters with strong positive or negative sentiments, guiding where to invest in participatory projects, neighborhood outreach, or policy adjustments.

This proposed graph-based sentiment model transforms digital participation into measurable civic energy. It helps city leaders hear not just individual concerns but collective voices, revealing where residents feel empowered and where trust needs rebuilding. By turning engagement into a visible, structured process, AI strengthens inclusion and ensures that future urban governance responds to both words and emotions expressed by the community.

Framework: Graph-Based Sentiment Aggregation for Civic Engagement (GBSA)

Layered AI Architecture for Community Feedback Analysis and Sentiment Aggregation

Layer	Main functionality	AI module(s) and components	Operational scope (Input → Processing → Output)
Layer 1	Data collection and preprocessing	Citizen feedback ingestion, cleansing, and sentiment feature preprocessing pipeline	Input → Citizen feedback (e.g., surveys, comments, reports) → Output → Preprocessed data for sentiment analysis
Layer 2	Learning: Graph-Based Sentiment Aggregation (GBSA)	Graph-Based Sentiment Aggregation (GBSA)	Input → Preprocessed citizen feedback → Output → Sentiment scores across the citizen network
Layer 3	Sentiment aggregation and propagation across the community	Graph-based sentiment diffusion and community aggregation engine	Input → Sentiment scores of connected citizens in the graph → Output → Aggregated community sentiment for each node
Layer 4	Policy and engagement decision-making	Policy decision support and targeted engagement recommendation engine	Input → Aggregated sentiment scores → Output → Recommendations for participatory projects, neighborhood outreach, and policy adjustments

1.5.1.5 Integration with Smart City Initiatives

The AI-driven community engagement platform in Tokyo is part of a broader smart city initiative that aims to leverage technology to improve urban living conditions. By integrating community feedback into the decision-making process, the city can develop and implement policies that are more aligned with the needs and preferences of its residents (Yigitcanlar & Cugurullo, 2020). This alignment is critical for creating sustainable urban environments that reflect the values and aspirations of the community.

For instance, to support this integration between citizen participation and multi-sector urban planning, we propose a Policy Recommendation Engine based on Federated Multi-Modal Learning (FMML). In this approach, each city department d_j—such as transport, sanitation, energy, or housing—trains its own local AI model f_j using internal datasets and domain-specific citizen input. These local models remain decentralized to protect data privacy, but their learned parameters are shared with a central aggregator to build a comprehensive global understanding of urban needs.

Each department trains its model on local data $\mathcal{D}_j$ and periodically sends its model weights (not raw data) to a central server. The central policy representation model fglobalfglobal is computed by aggregating all departmental contributions:

$$f_{\text{global}} = \frac{1}{M} \sum_{j=1}^{M} f_j(\mathcal{D}_j)$$

Here, $f_j(\mathcal{D}_j)$ represents the model trained by department d_j on its private dataset $\mathcal{D}_j$, and M is the total number of participating departments. This aggregation creates a shared understanding that supports interdepartmental coordination, enabling the city to design policies that reflect the full scope of citizen needs and service contexts.

Meanwhile, each resident feedback instance is multi-modal, containing a combination of inputs: textual descriptions, geolocation, and optional voice input. Each input is processed through a multi-modal encoder:

$$z_i = f_{\text{enc}}\left(x_i^{\text{text}}, x_i^{\text{geo}}, x_i^{\text{voice}}\right)$$

The output z_i is a unified vector representation of resident report i, capturing what was said, where it happened, and how it was communicated. These vectors are matched against departmental decision policies derived from $\mathrm{f}_{\text{global}}$ to produce ranked policy recommendations or targeted service actions.

To maintain accountability, the system also records how often each type of resident feedback leads to an action:

$$A_k = \frac{1}{N_k} \sum_{i=1}^{N_k} \mathbb{1}_{\{z_i \Rightarrow \text{policy}_k\}}$$

This expression computes the action rate A_k for policy k, showing how many resident inputs z_i (from total N_k) were actually translated into that policy. This enables the city to transparently show which community concerns are turning into measurable outcomes.

This proposed federated learning and multi-modal engagement model connects citizen voices directly to smart city planning while preserving data privacy and departmental autonomy. It allows governments to shape responsive and inclusive policies based on real, distributed insights—helping cities evolve as living systems that grow with the communities they serve.

Framework: Federated Multi-Modal Learning for Smart City Policy Recommendations (FMML)

Layered AI Architecture for Community Engagement and Urban Policy Development

Layer	Main functionality	AI module(s) and components	Operational scope (Input → Processing → Output)
Layer 1	Data collection and preprocessing	Multi-modal data ingestion, consent management, and feature normalization module	Input → citizen feedback (text, geolocation, voice) → output → preprocessed feedback data for multi-modal encoding
Layer 2	Learning: Federated Multi-Modal Learning (FMML)	Federated Multi-Modal Learning (FMML)	Input → Departmental feedback data → Output → Local models for each department
Layer 3	Model aggregation and policy coordination	Federated learning aggregation	Input → Local departmental models → Output → Global policy representation for citywide coordination
Layer 4	Policy recommendation and engagement action	Policy recommendation ranking and action orchestration engine	Input → Global policy representation → Output → Ranked policy recommendations and actionable service actions

Figure 1.9 represents the structure of Tokyo's AI-driven platforms for community engagement. The central node, "AI-Driven Platforms for Community Engagement in Tokyo," branches into five main components: "Citizen Reporting Applications," "Real-Time Data Collection and Analysis," "Enhanced Communication and Transparency," "Empowerment through Participation," and "Integration with Smart City Initiatives." "Citizen Reporting Applications" involve mobile applications for reporting local issues, AI-based categorization and prioritization of reports, and flagging safety hazards for immediate response. "Real-Time Data Collection and Analysis" includes aggregating data from social media and sensors, identifying community concern patterns, and allocating resources accordingly. "Enhanced Communication and Transparency" features tracking the status of reports, promoting transparent issue handling, and encouraging citizen participation. "Empowerment through Participation" highlights fostering community ownership, enabling active citizen roles in governance, and strengthening community ties. "Integration with Smart City Initiatives" aligns community feedback with smart city goals, incorporates it into decision-making, and develops policies aligned with residents' values.

Fig. 1.9 AI-driven platforms for community engagement in Tokyo

1.6 Conclusion

This chapter demonstrates that artificial intelligence has evolved into the operational and cognitive backbone of sustainable urbanism, reshaping how cities perceive, manage, and regenerate their systemic flows of mobility, energy, and waste. Through rigorous case studies of Barcelona, Singapore, Amsterdam, Stockholm, New York City, and San Francisco, the chapter establishes that the convergence of predictive modeling, reinforcement learning, and adaptive optimization constitutes a reproducible digital framework for environmental governance. Each city becomes a living laboratory where data-driven intelligence is not merely deployed but embedded within the metabolism of urban systems—governing intersections, energy grids, and waste streams as dynamic, self-learning ecosystems.

In synthesis, the comparative analyses reveal that Deep Reinforcement Learning (DRL), Long Short-Term Memory (LSTM) networks, Extreme Gradient Boosting (XGBoost), and Proximal Policy Optimization (PPO) algorithms form the algorithmic infrastructure of sustainable urban transformation. These models collectively operationalize foresight, adaptability, and accountability—qualities once absent from traditional infrastructure planning. By encoding environmental costs, such as emissions, energy losses, and material inefficiencies, directly into reward structures, the presented frameworks transform ethical and ecological imperatives into quantifiable optimization objectives.

The architectures proposed—ranging from the DRL-Based Emission-Conscious Urban Signal Optimization (DRL-EUSO) to the AI-Driven Waste Management Optimization Framework (AI-WMOF)—constitute a unified methodological doctrine for intelligent urban governance. They demonstrate how cities can internalize sustainability into algorithmic decision-making, transitioning from static control regimes to adaptive, learning infrastructures. This paradigm shift marks the emergence of "cognitive urban ecosystems" wherein AI does not merely manage complexity—it learns from it, anticipates its evolution, and continually aligns efficiency with equity and ecological balance.

This transformation reflects a deeper epistemological realignment: intelligence becomes a civic resource, and data becomes a moral medium through which cities reconcile progress with preservation. The mathematical precision of these models is thus not an abstraction but an ethical framework—quantifying the interplay between human behavior, machine learning, and planetary responsibility. In final analysis, this chapter illustrates that the sustainable city of the future is neither a fixed blueprint nor a reactive organism—it is an adaptive intelligence, perpetually recalibrating the rhythms of urban life toward balance, resilience, and shared prosperity.

References

Badidi, E. (2022). Edge AI and Blockchain for Smart Sustainable Cities: Promise and Potential. Sustainability, 14, 7609. https://doi.org/10.3390/su14137609

Yigitcanlar, T., & Cugurullo, F. (2020). The Sustainability of Artificial Intelligence: An Urbanistic Viewpoint from the Lens of Smart and Sustainable Cities. Sustainability, 12, 8548. https://doi.org/10.3390/su12208548

Chapter 2
AI-Driven Urban Intelligence: Integrated Frameworks for Sustainable and Human-Centric Smart City Governance

This chapter explores the transformative integration of Artificial Intelligence (AI) into the design, governance, and operational logic of modern urban ecosystems. It focuses on two emblematic regions in Poland—the Silesian and Lesser Poland Voivodships—which stand as living laboratories of how algorithmic intelligence, when properly governed, can bridge sustainability, inclusivity, and ethical digital transformation. By drawing on these distinct socio-economic and industrial landscapes, the chapter demonstrates how AI-powered urban frameworks transcend technological experimentation to become instruments of regional resilience and human advancement.

Through a structured exploration of layered AI architectures, the chapter introduces a series of specialized computational frameworks that redefine how cities learn, adapt, and act. The first model, the Extreme Gradient Boosting-based Sustainability Prediction framework (XGBoost-SP), applies machine learning to forecast environmental and industrial sustainability outcomes using high-dimensional urban data. The Deep Reinforcement Learning with Proximal Policy Optimization (DRL-PPO) framework then extends this adaptive logic into the domain of real-time traffic management and smart mobility, enabling self-optimizing control systems that reduce congestion, emissions, and travel time. In the field of urban health and security, the Long Short-Term Memory and Convolutional Neural Network framework (LSTM-CNN) integrates time-sequenced predictive analytics with intelligent surveillance, generating early warnings for health crises and safety incidents.

Energy efficiency and environmental management are addressed through a hybrid Long Short-Term Memory and Extreme Gradient Boosting (LSTM-XGBoost) model, which fuses temporal demand forecasting with anomaly detection to ensure stability in smart grid operations. The environmental monitoring component advances further with the Graph Neural Network and Gated Recurrent Unit framework (GNN-GRU), which uses interconnected sensor networks to track and forecast urban air and water quality across space and time. Finally, the chapter culminates with the Fairness-Constrained Learning for Smart City Inclusivity framework (FCL-SCI), an ethically grounded AI model that embeds fairness constraints directly

D. Sargiotis, *Blueprints for Future Living*,
https://doi.org/10.1007/978-3-032-24178-8_2

into predictive and decision-making processes, ensuring equitable access to public resources and preventing algorithmic bias.

Each of these architectures operates as an autonomous yet interconnected layer of what the chapter defines as "AI-Driven Urban Intelligence": a comprehensive ecosystem where data collection, federated learning, optimization, and ethical governance function in harmony. By employing techniques such as Federated Learning for decentralized policy aggregation, Bayesian optimization for hyperparameter tuning, and fairness-aware reward structuring, the frameworks collectively construct a scientific foundation for sustainable, transparent, and adaptive city management.

What distinguishes this chapter is its rigorous mathematical formalization coupled with its normative vision. It moves beyond descriptive smart city case studies, proposing reproducible, mathematically explicit AI systems capable of being deployed across different geopolitical and socio-technical contexts. Every equation, from the XGBoost regularization function to the fairness-constrained objective formulation, is presented as a step toward operationalizing ethical intelligence in urban governance.

For international scholars, engineers, and policymakers, this chapter offers a unique synthesis of deep technical precision and ethical foresight. It provides a clear roadmap for designing AI ecosystems that not only optimize performance but also institutionalize fairness, transparency, and human-centricity. The Silesian and Lesser Poland Voivodships thus emerge not as isolated case studies but as prototypes for globally transferable AI governance models—demonstrating that the future of intelligent cities will not be defined solely by data abundance or computational power, but by the wisdom with which intelligence is aligned to human values and planetary sustainability.

2.1 Silesian and Lesser Poland Voivodships

The case study of Silesian and Lesser Poland Voivodships in Poland serves as an exemplar of integrating Artificial Intelligence (AI) within the management of smart cities, particularly in a regional context. This analysis delves into how these regions apply AI technologies across various domains to enhance urban efficiency, sustainability, and livability, while addressing the critical socio-technical dynamics that arise in the development of smart city infrastructures (Skubis et al. 2024).

2.1.1 Context and Background

Poland's Silesian and Lesser Poland Voivodships are two distinct regions with unique economic and industrial characteristics, making their approach to AI adoption in smart city development both complex and region-specific. The Silesian Voivodship, historically known for its focus on traditional industries such as manufacturing and energy production, contrasts with the more diversified approach of Lesser Poland, which includes significant applications in life sciences and Information and Communication Technologies (ICT). This distinction provides a broad spectrum for analyzing the impact and potential of AI in different regional economies and urban settings (Skubis et al. 2024).

For instance, to enable real-time adaptation of smart city policies tailored to region-specific industrial footprints, we introduce a state-of-the-art forecasting framework using Extreme Gradient Boosting (XGBoost). This model was selected due to its outstanding performance in high-dimensional, heterogeneous datasets typical of urban environments, and its proven success in sustainability forecasting and optimal resource allocation, particularly relevant to the industrialized Silesian and diversified Lesser Poland regions. The algorithm dynamically processes urban sensor data, regional industry emissions, and energy usage patterns, allowing for predictive modeling of environmental sustainability outcomes. We define the XGBoost-based sustainability impact prediction as:

$$\hat{y}_i = \sum_{t=1}^{T} f_t(x_i), f_t \in \mathcal{F}$$

where, $\hat{y}_i$, predicted sustainability metric for region i, x_i, vector of regional features including emissions, energy use, industrial activity, public transportation load, $\mathcal{F}$, set of regression trees used by XGBoost, f_t, individual regression tree learning function at boosting step t, and T, total number of trees in the ensemble model. This formulation allows the AI to learn nonlinear, multi-factorial dependencies between industrial patterns and sustainability indicators, dynamically adjusting regional forecasts through a sequence of optimized decision trees. Each tree f_t incrementally refines predictions by minimizing the residuals from previous iterations, enabling precise estimation of real-world sustainability metrics across heterogeneous urban areas.

To prevent model overfitting and ensure policy robustness under variable socio-technical conditions, a regularized objective function is introduced:

$$\mathcal{L}(\phi) = \sum_{i=1}^{n} l(y_i, \hat{y}_i) + \sum_{t=1}^{T} \Omega(f_t)$$

where $\mathcal{L}(\phi)$, overall objective loss of the XGBoost model, $l(y_i, \hat{y}_i)$, convex loss function measuring the error between true and predicted sustainability scores, and $\Omega(f_t)$, regularization function penalizing complexity of the tree f_t to improve generalizability across different city environments.

The $\Omega(f_t)$ is further defined as:

$$\Omega(f_t) = \gamma T + \frac{1}{2}\lambda \sum_{j=1}^{T} w_j^2$$

here, γ, penalty term for the number of leaves in each tree, λ, regularization parameter controlling leaf weight shrinkage, w_j, weight assigned to the j-th leaf in tree f_t, T, number of terminal nodes (leaves).

This regularization term balances the model's flexibility and predictive accuracy, ensuring reliable forecasting even in the face of differing policy environments between regions. By adjusting for both tree complexity and weight magnitude, the framework enhances interpretability for urban planners while maintaining resilience to overfitting from sparse or skewed regional data.

To optimize hyperparameters and ensure sustainable policy learning remains adaptive, Bayesian optimization is employed to search the parameter space:

$$\theta^* = \arg \max_{\theta \in \Theta} \mathbb{E}[R(\theta)]$$

where θ^*, optimal parameter set for XGBoost configuration, Θ, parameter search space including learning rate, tree depth, and regularization terms, and $\mathbb{E}[R(\theta)]$, expected reward representing forecasting accuracy under validation data for sustainability metrics. This hyperparameter tuning ensures that region-specific models are customized for local emission profiles, urban density, and socio-economic behavior, allowing governments to implement predictive policies that are precise, efficient, and grounded in high-resolution data patterns.

Next, we synthesize individual departmental models into a cohesive, global policy representation that reflects the comprehensive understanding of urban sustainability. By aggregating the models trained on department-specific data, we can create a unified model that encapsulates insights from various sectors, ensuring that the resulting policy recommendations are informed by the collective data of the entire city. This process is essential for promoting interdepartmental coordination and making data-driven decisions at the municipal level. The aggregation of these individual models can be mathematically represented by the following equation:

$$f_{\text{global}} = \frac{1}{M} \sum_{j=1}^{M} f_j(D_j)$$

where f_{global}, global policy model, M, total number of participating departments, $f_j(D_j)$, departmental model trained by department d_j on dataset D_j, and D_j, department-specific dataset. This function aggregates the predictions of all local models to generate a global policy model that ensures coordinated decision-making across departments.

This AI-driven model demonstrates a sophisticated approach to urban governance by integrating the expertise of various departments within a city to create a comprehensive, data-informed policy framework. By leveraging federated learning, each department trains its own model on region-specific data—whether related to transport, energy, or environmental factors—ensuring that each sector's unique needs are addressed without compromising data privacy. The key innovation of this framework lies in its ability to aggregate the local models into a unified global model, thereby enabling cohesive policy recommendations that are responsive to the dynamic needs of the entire urban ecosystem. The model's multilayered design facilitates the real-time adaptation of sustainability strategies, optimizing resource allocation and policy effectiveness across different urban settings. Through the integration of XGBoost, Bayesian Optimization, and Federated Learning, this framework allows for the continuous refinement of urban policies based on real-time data, ensuring that the resulting strategies are adaptable to evolving conditions and that they reflect the diverse socio-economic and industrial characteristics of each city sector. Ultimately, this AI-driven model exemplifies the potential of advanced machine learning techniques to transform urban governance, enhancing decision-making, sustainability, and interdepartmental coordination while maintaining robust privacy standards.

Framework: XGBoost-Based Sustainability Prediction for Smart Cities (XGBoost-SP)

Layered AI Architecture for Real-Time Sustainability Forecasting and Policy Development

Layer	Main functionality	AI module(s) and components	Operational scope (Input → Processing → Output)
Layer 1	Data collection and preprocessing	Data integration and quality assurance system	Input → Urban sensor data, regional industry emissions, energy usage → Output → Preprocessed data for XGBoost model
Layer 2	Learning: XGBoost sustainability prediction model	Extreme Gradient Boosting (XGBoost)	Input → Preprocessed regional data (emissions, energy, industrial activity) → Output → Sustainability predictions for regions
Layer 3	Model aggregation for interdepartmental coordination	Federated learning aggregation	Input → Local departmental models → Output → Global sustainability policy model
Layer 4	Policy decision support and adaptation	Decision intelligence and policy optimization engine	Input → Global sustainability model → Output → Policy recommendations, sustainability forecasts, and strategic actions

2.1.2 Traffic Management and Smart Mobility

One of the critical areas where AI has shown significant impact in these regions is in traffic management. Using AI-driven predictive analytics, the cities within the Silesian and Lesser Poland regions can manage traffic flow more effectively, reduce congestion, and optimize transportation systems. AI-powered real-time traffic monitoring systems analyze data from smart cameras, enabling the identification of congestion points, anomalies, and accidents. This allows for quicker responses and the optimization of traffic lights and routes, significantly improving urban mobility and reducing the environmental footprint of transportation (Skubis et al. 2024).

For instance, to support smart traffic systems that adapt in real time to changing urban conditions, we propose an advanced Deep Reinforcement Learning (DRL) model. This approach is best suited for dynamic and uncertain environments like traffic networks. Specifically, Proximal Policy Optimization (PPO) is used to train an agent that makes continuous decisions about route optimization and signal timing based on traffic flow data. PPO ensures stable and safe learning, which is essential when applying AI to real-world city traffic infrastructure. The model is governed by the following value function for expected future rewards:

$$V^{\pi}(s_t) = \mathbb{E}_{\pi}\left[\sum_{k=0}^{\infty} \gamma^k R(s_{t+k}, a_{t+k})\right]$$

where $V^{\pi}(s_t)$, value of state s_t under policy π, $\mathbb{E}_{\pi}$, expected value given the policy, γ, discount factor between 0 and 1, and (s_{t+k}, a_{t+k}), reward received after taking action a_{t+k} in state s_{t+k}. This function estimates how good it is to be in a given traffic state, taking into account not just immediate benefits (e.g., reduced congestion) but also the long-term effects on the transportation system. It helps the AI agent decide whether it's worth modifying a traffic signal or recommending a reroute.

The agent's goal is to select actions that improve traffic conditions by maximizing an objective function. PPO improves learning stability by restricting how much the new policy can deviate from the old one. The PPO objective is:

$$\mathcal{L}^{\mathrm{PPO}}(\theta) = \mathbb{E}_t[min(r_t(\theta)A_t, \mathrm{clip}(r_t(\theta), 1-\epsilon, 1+\epsilon)A_t)]$$

where $\mathcal{L}^{\mathrm{PPO}}(\theta)$, PPO objective function to maximize, $r_t(\theta)$, policy ratio between new and old policy probabilities for action a_t, A_t, advantage function showing how much better an action is compared to average, and ϵ, clipping threshold that controls how much the policy can change.

This function ensures that the AI learns to improve traffic strategies without making abrupt or unsafe changes, which is crucial in a live urban traffic environment.

To compute the immediate benefit of each traffic action, we introduce the reward function:

$$R(s_t, a_t) = \eta_1 \cdot f_{\text{speed}}(s_t, a_t) - \eta_2 \cdot f_{\text{delay}}(s_t, a_t) - \eta_3 \cdot f_{\text{pollution}}(s_t, a_t)$$

where $R(s_t, a_t)$, total reward from action a_t in state s_t, $f_{\text{speed}}(s_t, a_t)$, estimated average vehicle speed as a benefit term, $f_{\text{delay}}(s_t, a_t)$, traffic delay cost function, $f_{\text{pollution}}(s_t, a_t)$, function measuring emissions based on idle time and congestion, and η_1, η_2, η_3 are weighting factors that balance priorities between mobility, time, and environmental impact.

Each sub-function here provides specific feedback to the AI: maximizing average speed encourages fluid traffic, penalizing delay keeps the system responsive, and minimizing pollution supports sustainability. These terms are dynamically computed using sensor inputs and traffic camera data, updated in real time.

This AI-driven framework empowers smart cities to adapt their traffic systems continuously, learning from past traffic patterns while responding instantly to new events. The Deep Reinforcement Learning model with PPO creates a closed-loop control system that balances traffic flow, reduces emissions, and improves public service responsiveness. By optimizing decisions in both the short and long term, the model helps cities in the Silesian and Lesser Poland Voivodships implement smarter, more sustainable mobility systems—showing how AI will actively shape urban transportation for tomorrow's cities.

Framework: Deep Reinforcement Learning with PPO for Smart Traffic Management (DRL-PPO)

Layered AI Architecture for Real-Time Traffic Optimization and Urban Mobility

Layer	Main functionality	AI module(s) and components	Operational scope (Input → Processing → Output)
Layer 1	Data collection and preprocessing	Data fusion and signal processing system	Input → Traffic data (sensor readings, camera feeds) → Output → Preprocessed data for traffic flow modeling
Layer 2	Learning: Deep Reinforcement Learning (DRL) with Proximal Policy Optimization (PPO)	Deep Reinforcement Learning (DRL) with Proximal Policy Optimization (PPO)	Input → Preprocessed traffic flow data → Output → Optimized traffic control decisions (signal timing, routing)
Layer 3	Traffic strategy optimization and stability	Adaptive traffic control and stability engine	Input → Traffic control decisions → Output → Balanced traffic flow, optimized signal timing, reduced congestion
Layer 4	Policy and environmental impact assessment	Environmental impact analytics and policy evaluation module	Input → Traffic and environmental data → Output → Recommendations for sustainable, eco-friendly traffic policies

2.1.3 Healthcare and Public Safety

AI also plays a pivotal role in healthcare management, particularly in predictive health analytics and emergency response systems. AI algorithms analyze large datasets of patient information to predict disease outbreaks, optimize resource allocation, and personalize healthcare services. In the realm of public safety, AI-driven surveillance systems with facial recognition and object detection capabilities enhance security in public spaces by identifying potential threats and enabling faster law enforcement responses. This integration of AI in healthcare and public safety directly contributes to the well-being and safety of the regions' citizens, aligning with human-centric smart city goals (Skubis et al. 2024).

To support predictive health analytics and emergency response within smart cities, we propose an AI-driven model based on Recurrent Neural Networks (RNN), specifically using Long Short-Term Memory (LSTM). This approach is selected because it can learn patterns from time-sequenced health data, such as emergency admissions, reported symptoms, or sensor-based vital signs, and use them to make timely health predictions. LSTM is ideal for capturing both short-term signals (e.g., sudden spikes in symptoms) and long-term dependencies (e.g., seasonal disease trends). The prediction of emergency demand at each time step is modeled as:

$$h_t = \sigma(W_h \cdot x_t + U_h \cdot h_{t-1} + b_h)$$

where h_t, hidden state representing health risk features at time t, x_t, input vector including patient symptoms, demographic and spatial data at time t, h_{t-1}, hidden state from the previous time step, W_h, input weight matrix, U_h, recurrent weight matrix, b_h, bias term, and σ, activation function that adds nonlinearity. This function updates the AI model's memory about patient behavior and population risk factors, continuously learning from new medical data. It helps predict how risk levels change over time based on both current conditions and past trends.

To retain relevant long-term medical information and discard irrelevant data, the LSTM updates its memory using the following cell state update:

$$c_t = f_t \cdot c_{t-1} + i_t \cdot \tilde{c}_t$$

where c_t, memory cell state at time t, f_t, forget gate that determines what past health data to discard, c_{t-1}, memory from previous time, i_t, input gate that selects which new data to keep, and $\tilde{c}_t$, candidate memory content from new health input. This cell memory allows the AI model to learn which health signals should be remembered long term, such as chronic illness trends or regional outbreak risks. It filters out noise while keeping essential health information that may signal future emergencies.

The final prediction, such as the number of emergency incidents or outbreak probability, is generated from the output gate:

$$\hat{y}_t = \tanh(c_t) \cdot o_t$$

where $\hat{y}_t$, predicted health alert or emergency intensity, $\tanh(c_t)$, compressed memory output ensuring stability, and o_t, output gate that controls how much of the memory influences the prediction. This output allows local healthcare systems to anticipate medical needs in specific areas, adjusting ambulance deployment, medical supplies, or hospital readiness before crises escalate.

To support safety and security in public spaces, AI-enhanced surveillance integrates with real-time detection and response using Convolutional Neural Networks (CNNs). These models identify objects (e.g., weapons, unattended bags) or match facial features against public watchlists. The classification probability is given by:

$$\hat{y}_i = \text{softmax}(W_c \cdot f_i + b_c)$$

where $\hat{y}_i$, probability that image i matches a known object or person, f_i, extracted image features, W_c, classification weights, b_c, bias term, and softmax function that converts outputs to probabilities over all categories.

This model enables rapid alerting when visual threats or persons of interest are detected in public zones, allowing authorities to act immediately.

By combining LSTM for health prediction and CNN for safety monitoring, this AI framework equips smart cities with tools to act early and protect citizens efficiently. In healthcare, LSTM continuously tracks medical risks, guiding hospital resource use and preventing crisis overload. In public safety, CNN ensures that surveillance is not only passive but intelligent and actionable. Together, these models help transform healthcare and safety into in real-time, human-centric services—making smart cities safer and more responsive for everyone.

Framework: LSTM for Predictive Health Analytics and CNN for Public Safety (LSTM-CNN)

Layered AI Architecture for Healthcare and Public Safety in Smart Cities

Layer	Main functionality	AI module(s) and components	Operational scope (Input → Processing → Output)
Layer 1	Data collection and preprocessing	Health data integration and signal normalization system	Input → Health data, emergency reports, sensor data → Output → Preprocessed data for predictive health analytics
Layer 2	Learning: LSTM for predictive health analytics and CNN for public safety	Long Short-Term Memory (LSTM), Convolutional Neural Networks (CNN)	Input → Time-sequenced health data, surveillance camera images → Output → Health predictions, safety alerts, threat detection
Layer 3	Risk assessment and emergency response	Emergency intelligence and resource coordination engine	Input → Predicted health risks, detected threats → Output → Emergency response actions, resource allocation
Layer 4	Public safety monitoring and decision support	Public safety analytics and decision-support platform	Input → Surveillance alerts, traffic data → Output → Policy recommendations, safety protocols

2.1.4 Energy Efficiency and Environmental Management

In the context of energy efficiency, AI technologies in these voivodships are used to optimize energy consumption through smart grids and predictive energy management systems. By utilizing AI algorithms, the regions can forecast energy demand, manage the distribution of renewable energy, and detect anomalies in energy infrastructure that could lead to inefficiencies or failures. This not only reduces energy wastage but also supports the sustainable development goals (SDGs) by promoting cleaner, more efficient energy use (Skubis et al. 2024). For instance, to support energy forecasting and optimize smart grid management in these regions, we propose an AI model built on Long Short-Term Memory (LSTM) neural networks integrated with XGBoost-based anomaly detection. This hybrid approach is selected for its ability to handle both temporal energy demand patterns and unexpected deviations in real-time consumption, ensuring smooth grid operations and sustainable energy use. We begin with the LSTM-based model, which predicts future energy demand by learning from time series consumption data. The hidden state of the LSTM network at each time step is updated as follows:

$$h_t = \sigma(W_x x_t + W_h h_{t-1} + b)$$

where h_t, hidden state at time$t$ representing learned features of energy consumption trends, x_t, current input such as temperature, energy price, and grid load, h_{t-1}, hidden state from the previous step storing recent memory, W_x, W_h, weight matrices that transform input and recurrent memory, b, bias term to improve stability, and σ, nonlinear activation that introduces flexibility into learning. This function captures both short-term and historical consumption behaviors. The learned stateh_t then directly contributes to the memory cell, which refines what the model should retain or forget. The memory cell is updated as:

$$c_t = f_t \cdot c_{t-1} + i_t \cdot \tilde{c}_t$$

where c_t, updated cell memory that accumulates energy usage knowledge, c_{t-1}, previous memory cell, f_t, forget gate controlling which outdated consumption data to remove, i_t, input gate deciding which new consumption signals to keep, and $\tilde{c}_t$, candidate content generated from h_t.

This process allows the LSTM to maintain a selective memory of past energy usage, which directly informs the predicted future energy load. The final forecast is given by:

$$\hat{y}_t = \tanh(c_t) \cdot o_t$$

where $\hat{y}_t$, predicted energy demand at time t, $\tanh(c_t)$, scaled memory signal, and o_t, output gate filtering what information influences the prediction.

The next component of the system focuses on real-time anomaly detection in energy infrastructure. For this, we apply Extreme Gradient Boosting (XGBoost) trained on residual patterns between actual energy usage and LSTM predictions. This allows the system to flag deviations from expected behavior.

The predicted residual error from XGBoost is defined as:

$$\epsilon_t = \sum_{k=1}^{K} f_k(r_t), f_k \in \mathcal{F}$$

where ϵ_t, predicted residual at time t indicating potential anomaly, $r_t = y_t - \hat{y}_t$, the raw residual between real and predicted energy values, f_k, the k-th regression tree in the ensemble, K, total number of trees used in XGBoost, and $\mathcal{F}$, the space of regression functions. This function evaluates whether the actual usage y_t significantly deviates from the forecast $\hat{y}_t$. If the predicted residual ϵ_t crosses a dynamic threshold, the model triggers an alert for inspection or automatic reallocation of energy.

To avoid overfitting and preserve model generalization in a live urban environment, the XGBoost loss function includes a regularization term:

$$\mathcal{L}(\phi) = \sum_{t=1}^{T} l(y_t, \hat{y}_t) + \sum_{k=1}^{K} \Omega(f_k)$$

where $\mathcal{L}(\phi)$, total training loss for the hybrid model, $l(y_t, \hat{y}_t)$, prediction error loss at time t, and $\Omega(f_k)$, penalty term on the complexity of each tree f_k to reduce overfitting.

The logic connecting these functions is essential. The LSTM forecasts the energy demand based on time-dependent patterns. The residuals between predictions and actual values, passed to XGBoost, allow the system to detect irregularities such as energy leaks, grid stress, or equipment failure. Regularization ensures this detection system remains robust in diverse conditions found in different voivodships.

This AI framework—using LSTM for energy forecasting and XGBoost for anomaly detection—enables city managers to optimize renewable energy distribution, reduce losses, and act fast when unexpected conditions appear. In the Silesian and Lesser Poland voivodships, where energy needs vary between industrial and residential zones, this model ensures that smart grids operate efficiently and sustainably. More importantly, the integration of predictive analytics with real-time anomaly detection directly supports energy-related SDGs and strengthens the infrastructure for future-ready cities.

Framework: LSTM for Energy Forecasting and XGBoost for Anomaly Detection (LSTM-XGBoost)

Layered AI Architecture for Energy Management and Smart Grid Optimization

Layer	Main functionality	AI module(s) and components	Operational scope (Input → Processing → Output)
Layer 1	Data collection and preprocessing	Energy data integration and signal normalization system	Input → Energy consumption data, temperature, energy price → Output → Preprocessed data for LSTM model
Layer 2	Learning: LSTM for energy demand forecasting and XGBoost for anomaly detection	Long Short-Term Memory (LSTM), Extreme Gradient Boosting (XGBoost)	Input → Time-sequenced energy data, residual error from LSTM predictions → Output → Energy demand forecast, anomaly detection
Layer 3	Energy forecasting and smart grid optimization	Grid dispatch optimization and demand response orchestrator	Input → Energy demand forecasts, anomaly detection results → Output → Optimized energy distribution, grid operation adjustments
Layer 4	Sustainability policy and energy efficiency	Sustainability analytics and efficiency policy engine	Input → Optimized energy distribution, grid adjustments → Output → Sustainability policy recommendations, energy-saving strategies

Moreover, environmental management in the Silesian and Lesser Poland Voivodships benefits from AI's ability to monitor air and water quality in real time, allowing for proactive interventions to prevent environmental degradation. AI systems collect and analyze data from IoT-enabled sensors distributed throughout the cities, offering insights into pollution levels, water quality, and overall environmental health. This data-driven approach to environmental management ensures that these regions are better equipped to address challenges related to climate change and resource management, making their urban environments more resilient and sustainable (Skubis et al. 2024).

For instance, to manage environmental quality in real time and across complex urban terrains, we propose the use of Graph Neural Networks (GNNs), an advanced AI architecture that is particularly suited for processing data from spatially connected sensor networks. In these voivodships, IoT-enabled environmental sensors form a network of interconnected nodes, each capturing data such as air pollutants (PM2.5, NO_2), water pH, temperature, or turbidity. These nodes are naturally modeled as a graph, where connections between nearby sensors help detect spatial patterns and regional anomalies more accurately than isolated analysis.

The first step of the model is to update the hidden state of each environmental sensor node by aggregating data from its neighbors, capturing local dependencies. The GNN layer is defined as:

$$h_v^{(l+1)} = \sigma\left(\sum_{u \in \mathcal{N}(v)} W^{(l)} h_u^{(l)} + b^{(l)}\right)$$

where $h_v^{(l+1)}$, updated feature representation for sensor node v at layer $l + 1$, $\mathcal{N}(v)$, set of neighboring sensor nodes around node v, $W^{(l)}$, weight matrix that learns the contribution of neighboring sensor data at layer l, $h_u^{(l)}$, feature vector of neighbor node u, $b^{(l)}$, bias term for stability, and σ, nonlinear activation function. This function allows each sensor node (air or water) to enhance its understanding by learning from surrounding data points, effectively modeling how pollution or water quality varies across spatial locations in the city.

Once node features are aggregated, the model computes a regional environmental health score by pooling and projecting the updated node representations:

$$\hat{y} = \text{MLP}\left(\text{READOUT}\left(\left\{h_v^{(L)} | v \in V\right\}\right)\right)$$

where $\hat{y}$, predicted environmental quality index or pollution alert level for the entire region, $h_v^{(L)}$, final layer representation of each node v, V, set of all sensor nodes in the city, READOUT(·), permutation-invariant aggregation function (e.g., sum, mean, or max), and MLP, multilayer perceptron that outputs the final environmental prediction. This function takes into account the full structure of the sensor network and transforms the learned information into a city wide indicator of air or water quality. When this value crosses defined safety thresholds, the system can automatically trigger warnings or corrective actions.

To continuously adapt and detect environmental anomalies that might signal sudden pollution events or sensor failures, we include a temporal update mechanism over the graph using a forecasting function:

$$\hat{h}_v^{(t+1)} = f_{\text{GRU}}\left(h_v^{(t)}, x_v^{(t)}\right)$$

where $\hat{h}_v^{(t+1)}$, predicted future state of node v, $h_v^{(t)}$, current node representation, $x_v^{(t)}$, observed sensor values at time t, and f_{GRU}, gated recurrent unit (GRU) function that models time evolution of pollution or water quality patterns.

This temporal layer allows the system not only to detect current environmental conditions but also to forecast near-future values based on learned trends. If forecasts show sudden increases in pollutants or unexpected drops in water quality, it gives the city time to intervene before critical thresholds are breached.

This integrated AI model—built on Graph Neural Networks with temporal forecasting using GRUs—enables smart environmental management for the cities. It learns both the spatial layout of environmental sensor networks and the time-dependent patterns of pollution and water quality. The model provides authorities with a dynamic and localized understanding of environmental conditions, enabling proactive, not reactive, actions. In doing so, it supports resilience to climate threats and aligns directly with sustainability goals, ensuring that cities not only monitor their environmental health but also learn how to protect it intelligently.

Framework: GNN for Environmental Quality Monitoring and GRU for Forecasting (GNN-GRU)
Layered AI Architecture for Smart Environmental Management

Layer	Main functionality	AI module(s) and components	Operational scope (Input → Processing → Output)
Layer 1	Data collection and preprocessing	Environmental data integration and sensor quality assurance system	Input → Environmental sensor data (air quality, water pH, temperature) → Output → Preprocessed data for GNN and GRU
Layer 2	Learning: GNN for environmental quality and GRU for temporal forecasting	Graph Neural Networks (GNN), Gated Recurrent Units (GRU)	Input → Preprocessed environmental data, time-sequenced sensor data → Output → Environmental health score, future predictions
Layer 3	Anomaly detection and real-time alerts	Real-time environmental anomaly detection and alerting engine	Input → Environmental health score, detected anomalies → Output → Alerts for action, anomaly-driven intervention
Layer 4	Sustainability and policy recommendations	Sustainability analytics and policy recommendation platform	Input → Anomalies and forecasted trends → Output → Policy recommendations, proactive environmental measures

2.1.5 *Human-Centric Approach*

A notable aspect of the AI deployment in these regions is the emphasis on a human-centric approach to smart city development. This approach prioritizes inclusivity, ensuring that the benefits of AI technologies are equitably distributed across different societal groups, and that technological advancements do not exacerbate existing inequalities. The case study highlights the importance of ethical AI deployment, where transparency, citizen engagement, and inclusivity are central to urban management. This is particularly relevant in the context of ensuring that marginalized communities also benefit from the advancements in smart city technologies and are not left behind in the digital transformation (Skubis et al. 2024). To ensure that smart city technologies reflect fairness and inclusivity, we introduce an AI model based on Fairness-Constrained Learning, which is designed to balance accuracy in prediction with equity across sensitive social attributes such as income level, age, or geographic location. This approach is especially critical when smart city algorithms are used in resource allocation, public service access, or decision-making systems that affect diverse populations. We define the general form of the prediction function as:

$$\hat{y}_i = f_{\mathrm{w}}(x_i)$$

where $\hat{y}_i$, predicted outcome for individual or region i, x_i, feature vector including environmental data, infrastructure quality, and socio-demographic variables, and f_{w}, AI prediction function parameterized by weights ww. This model makes predictions relevant to urban services (e.g., energy access or transit availability), but to ensure ethical outcomes, it must go beyond accuracy. Therefore, the training process includes an additional fairness constraint applied during optimization:

$$\min_{\mathrm{w}} \sum_{i=1}^{n} \ell(f_{\mathrm{w}}(x_i), y_i) \text{ subject to } \left| \mathbb{E}_{S=0}[\hat{y}] - \mathbb{E}_{S=1}[\hat{y}] \right| \leq \delta$$

where $\ell(f_{\mathrm{w}}(x_i), y_i)$, loss function measuring the prediction error compared to ground truth y_i, $S \in \{0, 1\}$, binary indicator of sensitive group membership (e.g., marginalized vs. non-marginalized), $\mathbb{E}_{S=k}[\hat{y}]$, average prediction for group k, and δ, small positive threshold that controls allowable disparity between groups. This constraint ensures that the model treats different social groups fairly by limiting how much predicted outcomes can differ between them. It prevents the AI from making biased decisions that favor one group over another, especially in resource-limited settings like low-income districts or underrepresented communities.

To maintain this fairness dynamically across the city's evolving needs, a fairness-adjusted reward function is applied when the model operates in reinforcement or resource allocation environments:

$$R_i = r_i - \lambda \cdot \left| \mathbb{E}_{S=0}[\hat{y}] - \mathbb{E}_{S=1}[\hat{y}] \right|$$

where R_i, adjusted reward for prediction or decision at instance i, r_i, original utility of the decision (e.g., efficiency or service improvement), and λ, fairness penalty coefficient that determines how strongly fairness is prioritized relative to performance. This function promotes policies that remain effective while actively reducing disparities. It reflects the trade-off between maximizing service quality and ensuring all communities benefit equally. Importantly, it allows planners to control the level of fairness enforcement through the value of λ, adapting the model to their ethical priorities.

By incorporating fairness directly into the learning process, this AI-driven model supports a human-centric smart city vision, ensuring that no societal group is systematically disadvantaged. In the Silesian and Lesser Poland Voivodships, where socio-economic disparities exist between districts, this approach allows technology to enhance equity, not reinforce inequality. Through predictive learning constrained by ethical fairness, the model helps urban decision-makers align AI deployments with principles of inclusivity, transparency, and justice—ensuring that digital transformation truly benefits all residents.

Framework: Fairness-Constrained Learning for Smart City Inclusivity (FCL-SCI)

Layered AI Architecture for Ethical AI in Urban Resource Allocation

Layer	Main functionality	AI module(s) and components	Operational scope (Input → Processing → Output)
Layer 1	Data collection and preprocessing	Socio-demographic data integration and bias-aware preprocessing system	Input → Socio-demographic data, environmental data, infrastructure quality → Output → Preprocessed data for fairness-based learning
Layer 2	Learning: fairness-constrained prediction and optimization	Fairness-constrained learning	Input → Preprocessed data (e.g., income, age, geographic data) → Output → Fair predictions, equitable resource allocation
Layer 3	Fairness-adjusted decision-making	Equity-aware decision orchestration engine	Input → Fair predictions → Output → Resource allocation decisions, public service access
Layer 4	Sustainability policy and equity assessment	Sustainability and equity impact analytics platform	Input → Allocation decisions, equity assessments → Output → Policy recommendations, fairness-driven adjustments

Figure 2.1 depicts the structure of AI integration in smart city management across the Silesian and Lesser Poland Voivodships. At the center, the main node represents the AI-driven management framework in these regions, branching into six key areas: "Context and Background," "AI in Urban Services," "Traffic Management," "Healthcare and Public Safety," "Energy and Environment," and "Human-Centric Approach." "Context and Background" highlights the distinct economic focus of the Silesian Voivodship on manufacturing and energy, and of the Lesser Poland Voivodship on life sciences and ICT. "AI in Urban Services" emphasizes sustainability, inclusivity, and transparency in governance. "Traffic Management" includes AI analytics for traffic flow, congestion management, and real-time incident response. "Healthcare and Public Safety" involves predictive health analytics, emergency response, and AI-driven surveillance. "Energy and Environment" focuses on smart grids for efficient energy use and real-time environmental monitoring. "Human-Centric Approach" underscores inclusivity, ethical AI deployment, and citizen engagement.

The Silesian and Lesser Poland Voivodships provide a comprehensive case study on the integration of AI in smart city management, showcasing the transformative potential of AI across various domains such as traffic management, healthcare, energy efficiency, and environmental management. By adopting a human-centric approach and emphasizing inclusivity, transparency, and sustainability, these regions serve as models for other urban areas looking to leverage AI technologies for smart city development. The comparative analysis of these voivodships underscores the importance of regional context in the deployment of AI, highlighting how localized strategies can align with broader global trends in smart city innovation.

Fig. 2.1 AI integration in smart city management in Silesian and lesser Poland Voivodships

2.2 Conclusion

This chapter articulates a transformative paradigm in urban governance—one in which Artificial Intelligence (AI) evolves from a computational instrument into a systemic architecture for sustainable, transparent, and human-centric city management. Through the dual case study of the Silesian and Lesser Poland Voivodships, the analysis demonstrates that algorithmic intelligence, when grounded in fairness and contextual adaptation, becomes a new civic infrastructure—capable of reconciling industrial complexity, social equity, and environmental stewardship within a unified digital ecosystem.

The chapter's succession of frameworks establishes a methodologically coherent and ethically rigorous foundation for the operationalization of AI across diverse urban sectors. The Extreme Gradient Boosting-based Sustainability Prediction (XGBoost-SP) framework revealed how data-driven learning can dynamically forecast environmental and industrial trends, equipping regional policymakers with predictive sustainability intelligence. The Deep Reinforcement Learning with Proximal Policy Optimization (DRL-PPO) framework extended this predictive capacity into the temporal and behavioral dimensions of mobility, introducing self-adaptive traffic optimization that aligns with both efficiency and ecological restraint. In parallel, the Long Short-Term Memory and Convolutional Neural Network (LSTM-CNN) model illuminated the power of temporal and spatial deep learning for anticipatory healthcare analytics and intelligent public safety systems—fusing civic protection with data ethics.

Equally pivotal, the Long Short-Term Memory and Extreme Gradient Boosting (LSTM-XGBoost) framework advanced energy resilience through hybrid anomaly-aware forecasting, forging a scientific path toward adaptive smart grids and carbon-neutral infrastructure. Complementarily, the Graph Neural Network and Gated Recurrent Unit (GNN-GRU) model expanded the epistemic frontier of environmental management by modeling the interdependencies of urban sensor networks in space and time, turning environmental monitoring into a continuously learning organism of ecological awareness. The analytical trajectory culminated in the Fairness-Constrained Learning for Smart City Inclusivity (FCL-SCI) framework—an ethically binding structure that operationalizes fairness as a mathematical constraint within machine learning optimization, ensuring that algorithmic intelligence remains aligned with distributive justice, inclusion, and citizen dignity.

The ensemble of these AI architectures constructs a coherent meta-framework for AI-Driven Urban Intelligence. Each system—through its respective mathematical formulation, optimization dynamics, and federated learning mechanisms—demonstrates that sustainable governance emerges not from isolated technological solutions, but from their orchestration into a symbiotic, adaptive, and ethically supervised intelligence fabric. The chapter proves that federated aggregation, Bayesian hyperparameter tuning, and fairness-aware optimization are not mere computational enhancements; they represent a structural transformation of how cities learn, deliberate, and act across institutional silos.

From an epistemological perspective, the chapter contributes to the global discourse on intelligent urbanism by fusing formal AI modeling with normative urban ethics. It establishes reproducible mathematical protocols—from convex loss regularization in Extreme Gradient Boosting to fairness-penalized optimization in ethical learning—that are not only technically robust but philosophically defensible. This synthesis between algorithmic precision and human-centric governance positions the presented frameworks as transferable models for European and international smart city agendas, aligning with Sustainable Development Goals (SDGs), the European Union's Green Deal, and the ethical imperatives of the AI Act.

In final analysis, the Silesian and Lesser Poland Voivodships emerge as archetypal exemplars of digital transition: territories where industrial heritage and computational foresight converge into a living laboratory for post-industrial urban intelligence. The layered architectures presented herein constitute more than technical blueprints—they embody a new social contract between data, governance, and humanity. They show that the true measure of a smart city lies not in the quantity of its data nor the sophistication of its code, but in the moral architecture that governs its algorithms. By embedding fairness, transparency, and adaptive learning into the urban fabric, this chapter redefines the intelligent city as an ethical system—where technology serves civilization, not the reverse, and where the evolution of AI becomes inseparable from the evolution of human responsibility.

Reference

Skubis, I., Wolniak, R., & Grebski, W. W. (2024). AI and human-centric approach in smart cities management: Case studies from Silesian and Lesser Poland Voivodships. *Sustainability, 16*(8279). https://doi.org/10.3390/su16188279

Chapter 3
AI-Empowered Digital Twins for Smart City Governance: Comparative Case Studies of Dubai, Herrenberg, West Cambridge, Málaga, and Ålesund

This chapter opens as a panoramic exploration of how Artificial Intelligence (AI) and Digital Twin (DT) technologies are converging to redefine the logic, ethics, and mechanics of twenty-first century urban governance. It examines five emblematic cities, Dubai, Herrenberg, West Cambridge, Málaga, and Ålesund, each of which has evolved into a living laboratory where data, algorithms, and civic values intertwine to form intelligent, adaptive, and ethically grounded systems of decision-making. The discussion moves beyond the rhetoric of "smart cities" to articulate a paradigm of cognitive urbanism, in which urban infrastructure, ecology, economy, and social life are mirrored, learned, and optimized through a continuously evolving digital counterpart.

At its conceptual core, the chapter defines the Digital Twin (DT) not merely as a static replica of physical infrastructure, but as a self-learning, reasoning, and evolving digital organism that synchronizes in real time with its physical counterpart through multi-source data streams. These streams are analyzed and acted upon by advanced AI architectures, including Reinforcement Learning (RL), Graph Neural Networks (GNN), Gated Recurrent Units (GRU), Spatial-Temporal Graph Attention Networks (ST-GAT), Deep Q-Learning (DQL), Long Short-Term Memory (LSTM), Multi-Objective Reinforcement Learning (MORL), Gaussian Process Regression (GPR), Extreme Gradient Boosting (XGBoost), Bidirectional Encoder Representations from Transformers (BERT), and Federated Learning (FL). Through these computational engines, the DT becomes the city's cognitive architecture, continuously learning from real-world behavior, testing policy alternatives, and aligning technological performance with human purpose.

Yet the institutional uptake of AI-empowered urban digital twins is not determined solely by predictive accuracy or computational sophistication. Adoption hinges on whether the twin's inferences are regarded as legitimate, contestable, and accountable within the civic order, and this legitimacy is produced through governance rather than through technology alone. Accordingly, governance is treated here as a design layer of the digital twin itself, spanning data stewardship, model assurance, transparency and contestability of decisions, and operational

D. Sargiotis, *Blueprints for Future Living*,
https://doi.org/10.1007/978-3-032-24178-8_3

security. The chapter therefore introduces a compact governance blueprint and then uses the comparative cases to show how specific governance controls can be instantiated in practice.

Because the digital twin increasingly functions as a policy instrument rather than a neutral simulation, governance becomes the condition under which its outputs may be taken up as legitimate inputs to public decision-making. In practice, legitimacy is assembled through a small set of operational controls, notably data stewardship, model assurance, decision traceability with oversight, and participation with contestability that translate ethical commitments into verifiable procedures and system properties. Recent work on urban digital twins has repeatedly emphasized that privacy, security, accountability, and public trust are not peripheral constraints but determining variables of long-term viability.

The comparative cases that follow illustrate these controls as concrete instantiations rather than as abstract desiderata. Dubai foregrounds governance as an explicit optimization target, coupling human-centric objectives with privacy and fairness constraints at the model level and sustaining participation through NLP-mediated feedback loops. Herrenberg, by contrast, emphasizes participatory intelligence by embedding citizen discourse into planning simulations, thereby turning deliberation into a computationally encoded governance channel. West Cambridge provides a technically rigorous pathway for privacy-preserving analytics in distributed infrastructures, demonstrating how formal privacy guarantees can be coupled to operational safeguards aligned with regulatory expectations. Málaga operationalizes "trust as infrastructure" by making privacy preservation and encrypted aggregation architectural primitives, thereby reframing compliance as a design principle rather than as an external requirement. Ålesund highlights the democratic accountability dimension by treating inclusion, long-term engagement, and civic learning as core success conditions of the smart city transition.

Viewed collectively, this blueprint clarifies the book's central claim in operational terms: AI governance in smart cities is not a separate chapter of compliance, but the mechanism through which digital twins acquire the social authority to act as civic epistemic institutions.

Dubai is introduced as the pioneer of value-driven artificial intelligence, operationalizing happiness, fairness, and inclusion as measurable governance metrics within its celebrated "Happiness Agenda." The chapter presents how Dubai's Utility-Regularized Policy Model for Human-Centered Smart Cities (URPM-HCSC) and Real-Time Urban Adaptation Model for Smart City Well-being (RUAM-SCW) embed emotional satisfaction and social equity directly into the optimization functions of public policy. By deploying Federated Differential Privacy Governance Models (FDPGM) to safeguard personal data and Fairness-Constrained Adaptive Policy Optimization Models (FCAPOM) to guarantee equitable service access, Dubai redefines digital governance as an ethical computation of collective well-being. The city's Dynamic Civic Engagement Reinforcement Model (DCERM) demonstrates how Natural Language Processing (NLP) can sustain citizen participation by transforming public feedback into actionable intelligence for real-time decision systems.

Herrenberg, a small German town, embodies the European model of participatory intelligence and localized sustainability. Its AI-Enhanced Urban System Modeling (AIEUSM) framework combines Graph Neural Networks (GNN) with Gated Recurrent Units (GRU) to simulate dynamic relationships among mobility, energy, and environment, transforming the town into a graph of interdependent systems that can learn, predict, and adapt. Through Spatial-Temporal Graph Attention Networks (ST-GAT), Deep Q-Learning (DQL), and Gaussian Process Regression (GPR), the town anticipates traffic congestion, energy demand, and environmental risks before they occur. Citizens are not passive data points but active co-designers. Bidirectional Encoder Representations from Transformers (BERT) processes open feedback and discussions, integrating public sentiment into planning simulations. Herrenberg thus becomes a replicable model for smaller cities seeking to balance technological sophistication with civic intimacy and democratic inclusiveness.

West Cambridge, positioned at the crossroads of academia and municipal innovation, demonstrates how research ecosystems can serve as engines of explainable and privacy-preserving governance. Its Spatio-Temporal Attention Models (STAM) and Spatio-Temporal Urban Planning Simulation (STUPS) frameworks enable the anticipation of urban dynamics such as mobility patterns, energy use, and environmental fluctuations. The Federated Edge-Based Predictive Maintenance Framework (FEPMF) decentralizes analytics across local nodes, ensuring minimal latency and maximal data sovereignty. Here, Proximal Policy Optimization (PPO) algorithms fine-tune service delivery, while embedded explainability modules make every model decision traceable and interpretable. West Cambridge exemplifies how an academic-industrial alliance can transform research outputs into live urban applications that are transparent, ethical, and sustainable.

Málaga, representing Southern Europe's Mediterranean innovation frontier, introduces a model of data-driven sustainability and cultural intelligence. The city's AI-Integrated Climate-Adaptive Urban Twin (AI-CAUT) links Multi-Objective Reinforcement Learning (MORL) with Internet of Things (IoT) sensor networks to balance environmental protection, tourism management, and mobility optimization. Using Extreme Gradient Boosting (XGBoost) and Gaussian Process Regression (GPR), Málaga predicts fluctuations in air quality, energy consumption, and coastal micro-climates, allowing adaptive policy interventions that protect the environment while enhancing urban comfort. Its Fairness-Aware Service Allocation Model (FASAM) ensures equitable access to public services across different socio-economic neighborhoods, while Cultural Engagement Sentiment Analysis (CESA) employs Natural Language Processing (NLP) to assess citizen perceptions of cultural heritage and inclusivity. Málaga thus embodies a Mediterranean synthesis of smart governance and social empathy; it is an urban model where sustainability, creativity, and fairness operate within one computational loop.

Ålesund, situated along Norway's coastal fjords, closes the comparative narrative with a vision of resilient intelligence under uncertainty. Its Multi-Agent Reinforcement Learning (MARL) systems coordinate maritime operations, transportation networks, and resource distribution across complex topographies. Bayesian Long Short-Term Memory (B-LSTM) networks and Spatio-Temporal Environmental

Inference (STEI) frameworks provide uncertainty-aware climate forecasting, while Soft-Actor-Critic (SAC) policies implement entropy-regularized decision-making to balance efficiency, safety, and ecological preservation. Ålesund's Federated Resilience Data Platform (FRDP) ensures that local authorities, research institutions, and citizens co-manage the same knowledge base without compromising privacy or sovereignty. Through bilingual engagement portals and real-time visualization dashboards, public voice becomes an integral parameter of environmental policy. Ålesund thereby demonstrates how even compact cities can deploy AI-empowered Digital Twins (DTs) to maintain equilibrium between environmental volatility, social trust, and institutional foresight.

For the international reader, researchers, technologists, civic leaders, and policy architects, this chapter offers more than comparative analysis; it presents a strategic grammar for the intelligent city. It demonstrates that the true measure of urban intelligence lies not in the volume of data processed but in the moral coherence with which algorithms align public good, environmental care, and democratic inclusion. Whether through Dubai's happiness optimization, Herrenberg's participatory learning, West Cambridge's interpretability, Málaga's climate-cultural synthesis, or Ålesund's resilient autonomy, the chapter shows that AI-empowered Digital Twins (DTs) can transform cities into self-aware, self-correcting organisms that think with their citizens and evolve for their descendants.

In essence, this chapter functions both as an analytical synthesis and as a normative orientation. It reflects how five diverse cities have already encoded empathy, equity, and foresight into their computational architectures, while also inviting future cities to treat these commitments as design constraints rather than rhetorical aspirations.

3.1 Governance Architecture for AI-Empowered Urban Digital Twins

3.1.1 Operational Definition: Governance as Roles, Processes, Artifacts, and Controls

Operationally, governance for the AI components embedded in urban (local) digital twins can be treated as a socio-technical arrangement through which responsible AI principles are made actionable across designing, executing, monitoring, and evaluating AI applications, rather than remaining at the level of high-level guidance. In this sense, Papagiannidis et al. conceptualize responsible AI governance as a configuration of structural, procedural, and relational practices that collectively operationalize responsible AI beyond aspirational statements (Papagiannidis et al. 2025; Sargiotis 2024). In the LDT setting discussed by Mureddu et al., the urgency of such operationalization is sharpened by the integration of AI and ML into public-sector decision processes, which may yield less predictable and less explainable outcomes and thereby intensify demands for transparency, privacy protection, and accountable human involvement (Mureddu et al. 2025).

For the purposes of a governance architecture, these practice families can be instantiated as four mutually reinforcing elements: (1) roles, as the concrete expression of structural practices; (2) processes, as the routinization of procedural practices across lifecycle activities; (3) artifacts, as the documentary evidence required to render accountability inspectable; and (4) controls, as the monitoring, enforcement, and corrective mechanisms that keep systems within demonstrably accountable bounds. This mapping does not replace Papagiannidis et al.'s tripartite framework; rather, it specifies how structural, procedural, and relational practices are commonly materialized into implementable components in organizational and public-sector contexts where auditability and evidentiary scrutiny are salient (Papagiannidis et al. 2025).

First, governance must be understood in terms of roles, namely the explicit allocation of decision rights, responsibilities, and oversight mandates to identifiable actors. Papagiannidis et al. describe structural practices as specifying key decision-makers and articulating the rights and responsibilities of stakeholders engaged in AI development and deployment, including the clarification of roles and decision-making arrangements to support accountability and transparency (Papagiannidis et al. 2025). In LDT deployments, this structural question is tightly coupled to platform and data control. Mureddu et al. stress the need to balance clarity and accountability in areas such as data ownership, access rights, and usage restrictions, while remaining adaptable to innovation pressures (Mureddu et al. 2025). They further show that governance varies across LDT governance scenarios, differentiating models by who controls data resources and how openness and participation are organized, from government-controlled to citizen-engaging and ecosystem-driven arrangements (Mureddu et al. 2025).

Second, governance is procedural: it consists of repeatable processes that routinize oversight and evaluation across the AI lifecycle. Papagiannidis et al. situate procedural practices in processes for data and model management, pipeline evaluation, and human–AI interaction, and they explicitly connect responsible AI governance to robust compliance monitoring and enforcement as well as incident-response and crisis-management procedures that define detection, assessment, escalation pathways, coordination, and stakeholder communication (Papagiannidis et al. 2025). In the LDT context, Mureddu et al. similarly argue that practitioners require LDT-specific rules and processes for data collection, storage, and access, alongside technical standards and interoperability guidance for formats, protocols, and interfaces (Mureddu et al. 2025).

Third, governance requires artifacts, because accountability in audit-facing and high-stakes settings depends on evidence that can be inspected. Königstorfer and Thalmann argue that traditional software documentation is insufficient to provide the evidence auditors require for AI systems and that the lack of AI-specific documentation guidance constitutes an adoption barrier in regulated contexts (Konigstorfer and Thalmann 2022). They define documentation, in software-engineering terms, as an artifact that records architectural decisions in an accessible and non-ambiguous form for relevant stakeholders, including auditors (Konigstorfer and Thalmann 2022). This evidentiary function also appears within Papagiannidis et al.'s structural practices, where documentation is presented as necessary for drafting, reviewing, and approving rights and responsibilities, with mechanisms for regular review and updating as needs evolve (Papagiannidis et al. 2025). In LDT governance, Mureddu et al. likewise emphasize transparency measures that operationalize disclosure about data usage, storage, sources, automated processing, and data quality, thereby supplying precisely the kinds of inspectable artifacts on which accountability claims can credibly rest (Mureddu et al. 2025).

Fourth, governance must include control-oriented practices that enable monitoring, enforcement, and remediation under operational pressure, keeping AI-enabled decision support within demonstrably accountable bounds. In Papagiannidis et al., such control capacity is expressed within procedural practices through compliance monitoring and enforcement, together with incident-response and crisis-management processes that organize detection, assessment, escalation, coordination, and communication when harms or deviations arise (Papagiannidis et al. 2025). A comparable logic appears in Mureddu et al.'s discussion of local digital twins, where privacy and security requirements are operationalized through measures such as access control, encryption, and audit routines, alongside transparency and regulatory compliance expectations for public-sector deployment (Mureddu et al. 2025). From an auditability standpoint, Königstorfer and Thalmann further indicate that AI-specific documentation functions as an evidentiary basis that enables external scrutiny and verification of accountability-relevant properties, thereby supporting the practical enforceability of governance in regulated settings (Konigstorfer and Thalmann 2022).

3.1.2 Lifecycle-Oriented Governance Blueprint

A lifecycle-oriented governance blueprint for Local Digital Twins (LDTs) is most plausibly construed as an ordering of institutional responsibilities, technical commitments, and socio-legal safeguards that must remain intelligible as an LDT moves from early design choices to sustained operation and cross-domain reuse. Insofar as the LDT is framed as a "city service aggregator" that must integrate heterogeneous urban services and data sources, governance is implicated from the outset, because reusability and interoperability are treated as enabling conditions for practical integration rather than optional refinements appended after deployment (Croket et al. 2025). Under this framing, lifecycle orientation follows not from an abstract preference for "end-to-end" management, but from the recognition that the city-scale integrative ambition of an LDT continually reopens questions of coordination, control, and legitimacy as new domains, actors, and capabilities are incorporated (Croket et al. 2025).

Within the same conceptual landscape, Minimum Interoperability Mechanisms (MIMs) can be read as a governance-relevant scaffold that renders interoperability governable over time by specifying what must minimally hold for the ecosystem to remain composable. Croket et al. (2025) summarize a set of MIM subjects and functions spanning, inter alia, contextual linkage of datasets and streams, consistency of data models, contractual arrangements for appropriate use, trust, transparency, security, geospatial consistency, indicators, analytics, and resource information sharing. Interpreted as lifecycle devices, these mechanisms can be mobilized to discipline the progressive broadening of an LDT from a bounded solution into a reusable urban capability, because they articulate the conditions under which new components, datasets, and services may be integrated without collapsing into ad hoc, non-transferable local arrangements (Croket et al. 2025). The governance implication is that "minimal" should be treated as an evolving baseline that supports expansion while constraining fragmentation, rather than as a one-time technical checklist that is exhausted at go-live (Croket et al. 2025).

A lifecycle blueprint also requires an explicit approach to standards and capabilities as an ecology, not as a single architectural decision. Raes et al. (2025) emphasize the breadth of relevant standards across encoding, built-environment exchange, visualization, and user interaction, while also foregrounding the practical difficulty created by an abundance of standards and technologies in smart city and IoT contexts. Their discussion of "pivotal points of interoperability" suggests that governance can be strengthened by deliberately identifying the few concepts and component standards that repeatedly anchor integration, thereby reducing the integration burden when new components are introduced into an existing deployment (Raes et al. 2025). In this optic, lifecycle governance is partly a capability to curate and update a standards profile as the LDT's scope expands, while maintaining continuity at those points where interoperability is most strategically leveraged (Raes et al. 2025).

This curation imperative becomes more concrete where metadata and cataloging infrastructures are treated as necessary conditions for reuse. Raes et al. (2025) argue that the point at which multiple datasets from different sources become available demands a metadata registry and an asset catalog capable of meeting both human-readable and machine-to-machine requirements, including the standardization of metadata for datasets and computational assets such as algorithms and simulation models. A lifecycle-oriented governance blueprint may therefore position such cataloging as a persistent governance object, because discoverability, traceability, and controlled reuse depend on maintaining these registries as the ecosystem evolves, rather than treating them as static documentation produced during initial development (Raes et al. 2025). What is governed here is not merely data access, but the conditions under which assets remain interpretable, comparable, and safely reusable across time and institutional boundaries (Raes et al. 2025).

A further lifecycle layer is introduced when interoperability commitments are tied to concrete building blocks and maturity expectations. Raes et al. (2025) present an LDT building-block view in which specific components are associated with relevant MIMs and differentiated maturity characterizations, including cases where existing standards are "0, and where no existing standard is available. Read as governance guidance, this mapping supports staged investment and risk management: it helps distinguish domains where conformance can be demanded early from domains where governance must tolerate experimentation, prototype variability, and negotiated interfaces until standards or stable conventions emerge (Raes et al. 2025). Lifecycle orientation here is not rhetorical; it is a method for sequencing interoperability ambitions in line with what the standardization landscape can credibly support at different moments (Raes et al. 2025).

Security and controlled data use constitute an additional governance strand that is explicitly articulated as design guidance rather than post hoc compliance. Croket et al. (2025) develop "security by design" principles for LDTs that emphasize prudent data sharing, including the injunction to share only what is necessary, to bring algorithms and models to where data resides and where feasible, and to share insights over raw data when appropriate. They further connect secure operation to identity and access management, framing open identity management and robust access controls as part of the architectural environment required for trustworthy data use in an LDT context (Croket et al. 2025). In lifecycle terms, these prescriptions imply that governance must continuously align security practices with expanding data portfolios and stakeholder participation, because each new integration path potentially reconfigures exposure, risk, and permissible use (Croket et al. 2025).

At the scale-up horizon, governance is also conditioned by organizational strategy and stakeholder politics, not only by technical architecture. McAleer and Glidden (2025) argue that scaling an LDT to represent broader urban systems demands a comprehensive approach grounded in collaboration across stakeholders and is impeded by recurring challenges associated with data, privacy, security, and organizational capacity. They specify governance as operating through four interconnecting strategies: understanding stakeholder needs and requirements, ensuring collaboration and coordination among stakeholders, developing appropriate regulatory frameworks, and ensuring data privacy and security (McAleer and Glidden 2025). The lifecycle implication is that early-stage governance must secure alignment and acceptance through involvement and coordination, whereas later-stage governance must stabilize legitimacy through durable regulatory framing and operational privacy and security measures that can persist under expansion and increasing dependency (McAleer and Glidden 2025).

Legal and ethical constraints sharpen this picture by insisting that lifecycle governance must treat compliance and legitimacy as iterative, design-coupled processes. Mureddu et al. (2025) explicitly stage the regulatory debate as a tension between calls for new regulation and reliance on existing instruments such as GDPR, while also framing citizen privacy and cybersecurity as structurally implicated by the data gathering, integration, and analysis that LDTs presuppose. Their discussion of data commons as a more citizen-centric framework that goes beyond formal consent to consultation and involvement foregrounds governance choices about agency, participation, and boundary-setting over personal data use (Mureddu et al. 2025). Most consequentially for lifecycle orientation, their "gap analysis" framing advances compliance as a roadmapping and continuous-monitoring practice that includes allocation of roles and responsibilities, security and risk management, reviewing and updating processing, training and awareness, and periodic policy review integrated into design, thereby construing governance as an ongoing control loop rather than a one-off gatekeeping event (Mureddu et al. 2025).

From an integrative standpoint, a corrected lifecycle-oriented governance blueprint, grounded in the attached chapters, can be formulated as a staged alignment of interoperability mechanisms, standards curation, cataloging and asset governance, security-by-design practices, stakeholder coordination, and iterative legal-ethical oversight. The MIM framing offers a vocabulary for minimum ecosystem commitments (Croket et al. 2025), the standards and building-block analyses provide mechanisms for sequencing interoperability ambitions in line with what is technically and institutionally mature (Raes et al. 2025), the scale-up strategies clarify how governance must expand from technical stewardship to stakeholder and regulatory orchestration (McAleer and Glidden 2025), and the legal-ethical account anchors lifecycle governance in recurring assessment and policy revision under conditions of privacy, accountability, and risk (Mureddu et al. 2025).

3.1.3 Risk-to-Control Mapping for Adoption and Legitimacy

In AI-empowered urban digital twins, institutional uptake is materially conditioned by whether salient risks are translated into demonstrable controls and supporting evidence that can withstand legal–technical scrutiny over time (Jørgensen and Ma 2025). Jørgensen and Ma explicitly operationalize this translation through the Unified Digital Twin Compliance Framework and its companion instrument, the Digital Twin Compliance Evaluation Matrix (DTCEM), which converts compliance domains into qualitative indicators and assesses maturity on a four-level scale (Non-Compliant, Partially Compliant, Compliant, Fully Compliant) on the basis of evidentiary sufficiency rather than rhetorical commitment (Jørgensen and Ma 2025). This emphasis on evidence aligns with the adoption problem identified by Königstorfer and Thalmann, who show that, in regulated use cases, AI is often impeded because accountability cannot be demonstrated to auditors through conventional documentation, and AI-specific documentation guidance has historically been underdeveloped (Konigstorfer and Thalmann 2022).

A first locus of risk concerns data protection and privacy, because urban-scale twins tend to integrate extensive, heterogeneous, and often multi-party datasets, thereby expanding exposure pathways and amplifying compliance obligations (Jørgensen and Ma 2025). Within the compliance mapping logic advanced by Jørgensen and Ma, privacy risk is mitigated only insofar as privacy requirements are instantiated through concrete mechanisms and verifiable documentation, including purpose specification, data minimization, and privacy-by-design measures that are demonstrable rather than merely asserted (Jørgensen and Ma 2025). In this setting, accountability deficits intensify privacy risk, so controls that generate traceable evidence, such as audit trails and access logging, become integral to converting privacy claims into inspectable compliance signals (Jørgensen and Ma 2025).

A second locus of risk is cybersecurity and operational resilience, which becomes legitimacy-critical because compromised integrity or availability can undermine the public defensibility of twin-mediated decision support regardless of analytical performance (Jørgensen and Ma 2025). The risk-to-control mapping in Jørgensen and Ma treats security not as an optional technical enhancement but as a compliance domain whose satisfaction is evidenced through safeguards such as encryption, access control, continuous monitoring, incident handling, and business continuity planning, framed as operational obligations rather than aspirational ideals (Jørgensen and Ma 2025). The necessity of expressing such safeguards in assessable form is consistent with Königstorfer and Thalmann's analysis where systems are subject to certification, validation, or audit, and the core constraint is frequently the ability to provide auditors with accessible, unambiguous evidence that controls exist and are maintained in operation (Konigstorfer and Thalmann 2022).

A third locus concerns opacity, contested outcomes, and responsibility gaps, which arise when AI components generate inferences that meaningfully shape high-stakes public decisions yet remain difficult to explain, validate, or contest (Konigstorfer and Thalmann 2022). Under Jørgensen and Ma's compliance logic, transparency and traceability are not treated as abstract virtues but as requirements that must be supported by mechanisms such as logging, auditable records, and other traceability provisions that allow evaluation against compliance indicators (Jørgensen and Ma 2025). Complementarily, Königstorfer and Thalmann show why documentation itself functions as a primary accountability control; auditors and regulators rely on documentary artifacts to assess how a system was built, how it is governed, and how accountability is ensured, and the absence of AI-appropriate documentation becomes a structural barrier to adoption in regulated settings (Konigstorfer and Thalmann 2022).

A fourth locus concerns ethical and social harms, including unfairness and adverse downstream impacts, which are particularly destabilizing for public-sector legitimacy because they may be amplified across populations and politicized through scrutiny of distributive outcomes (Jørgensen and Ma 2025). In the compliance mapping advanced by Jørgensen and Ma, such risks are approached through obligations grouped under ethical responsibility and related domains, and they are evaluated through qualitative indicators that hinge on evidence of implemented mechanisms rather than on intention statements (Jørgensen and Ma 2025). Golpayegani and colleagues strengthen this evidentiary orientation by arguing, in the context of the EU AI Act, that documentation of high-risk systems and their risk management becomes central to demonstrating compliance; they therefore propose AI Cards as a framework that captures intended use, technical specifications, context of use, and risk management information in both human- and machine-readable forms (Golpayegani et al. 2024).

From an operational standpoint, the risk-to-control mapping becomes materially stronger when controls and their evidence are consolidated into structured documentation that supports conformity-oriented evaluation rather than remaining dispersed across informal narratives (Golpayegani et al. 2024). AI Cards are proposed precisely to make risk documentation interoperable and updatable, thereby supporting continuous compliance and facilitating the exchange of documentation across the AI value chain in ways aligned with regulatory expectations (Golpayegani et al. 2024). This documentation-centric approach also aligns with the empirical observation that, in regulated contexts, the risk level of an AI use case shapes both adoption feasibility and the scope of documentation demanded, since higher criticality magnifies the evidentiary burden required for accountability (Konigstorfer and Thalmann 2022). In this sense, the DTCEM's graded compliance levels and interpretive synthesis categories can be used to justify deployment readiness distinctions, for example, between configurations suitable only for research pilots and configurations robust enough for operational deployment, on the basis of demonstrated controls and documented evidence (Jørgensen and Ma 2025).

3.2 Dubai's "Happiness Agenda"

Dubai's "Happiness Agenda" stands out as a significant initiative aimed at enhancing the quality of life for its residents. This agenda aligns with the broader goals of the United Arab Emirates (UAE) government to position Dubai as a global leader in innovation, happiness, and well-being. This analysis explores the fundamental aspects of the Happiness Agenda, its implementation strategies, impacts, and implications for urban governance and citizen engagement.

3.2.1 Conceptual Framework of the Happiness Agenda

The Happiness Agenda represents a multidimensional framework for urban governance that positions citizen well-being as a fundamental goal of public administration and city planning. Rather than viewing happiness as a peripheral or subjective concept, this agenda integrates it as a measurable and actionable policy objective. It reflects a shift in governance paradigms—placing human experience and satisfaction at the center of institutional priorities—to ensure that urban development translates into genuine improvements in residents' quality of life (Dani et al. 2023). At the core of this framework are three interconnected pillars:

To translate well-being into a policy-relevant construct, the Happiness Agenda relies on a set of multidimensional metrics that span social, economic, psychological, and environmental domains. These metrics enable local authorities to identify what factors most significantly affect public satisfaction—ranging from access to green spaces and health services to economic opportunity and social inclusion. This structured measurement system facilitates a more accurate understanding of urban happiness and allows for targeted improvements across different population groups and neighborhoods.

The agenda leverages big data analytics to continuously monitor and respond to changing citizen needs and sentiments. Real-time data is collected from diverse sources, including household surveys, social media sentiment analysis, and citizen feedback platforms. These data streams are synthesized to form a dynamic portrait of public well-being, enabling urban decision-makers to adjust policies proactively. This evidence-based governance model allows cities to identify areas where happiness levels are declining and implement timely interventions (El-Agamy et al. 2024).

A defining characteristic of the Happiness Agenda is its emphasis on participatory governance. Residents are not merely the recipients of public services but active contributors to the design and evaluation of city initiatives. Through participatory platforms, co-creation workshops, and inclusive consultation mechanisms, citizens help shape the decisions that affect their daily lives. This participatory model fosters social trust, strengthens democratic legitimacy, and creates a shared sense of ownership over the urban future (Dani et al. 2023).

3.2.2 Implementation Strategies

The successful execution of the Happiness Agenda is supported by a set of integrated strategies that translate its conceptual vision into measurable improvements in everyday urban life. These strategies are rooted in the belief that technological innovation, service excellence, and cultural vitality are essential for enhancing the overall quality of life in smart cities. In the case of Dubai, the implementation of the agenda draws from both digital transformation and human-centered design to build a more responsive, inclusive, and livable urban environment.

To formally capture this dual alignment between technological optimization and resident well-being, we propose a utility-regularized policy model that integrates subjective well-being indicators directly into urban decision-making. The expected benefit of any smart city action or service intervention is modeled using a composite utility function:

$$U(a_t) = \alpha \cdot Q(a_t) + \beta \cdot H(a_t)$$

where $U(a_t)$ is the composite utility of action a_t, $Q(a_t)$ represents the objective service quality metrics (e.g., efficiency, response time, coverage), $H(a_t)$ is the predicted happiness or well-being impact of the action on the population, and α, β are weighting coefficients that balance technical performance with social value.

This utility function enables a human-centered AI policy framework, where urban actions are not evaluated solely by their technical efficiency but also by their contribution to the emotional and experiential dimensions of public life. By embedding predictive happiness models—trained on citizen sentiment, behavioral patterns, and satisfaction data—into the city's optimization engine, Dubai can ensure that smart infrastructure developments align with the broader civic goal of cultivating resident well-being. This model empowers the governance system to dynamically prioritize interventions that yield both operational and human-centric gains, operationalizing the Happiness Agenda as both a technological and ethical commitment.

Framework: Utility-Regularized Policy Model for Human-Centered Smart Cities (URPM-HCSC)

Layered AI Architecture for Smart City Well-being and Resource Allocation

Layer	Main functionality	AI module(s) and components	Operational scope (Input → Processing → Output)
Layer 1	Data collection and preprocessing	Data integration and preprocessing pipeline	Input → Socio-demographic data, environmental data, infrastructure quality → Output → Preprocessed data for fairness-based learning
Layer 2	Learning: fairness-constrained prediction and optimization	Fairness-constrained learning	Input → Preprocessed data (e.g., income, age, geographic data) → Output → Fair predictions, equitable resource allocation
Layer 3	Fairness-adjusted decision-making	Decision optimization engine for equitable resource allocation	Input → Fair predictions → Output → Resource allocation decisions, public service access
Layer 4	Sustainability policy and equity assessment	Policy evaluation and equity analytics platform	Input → Allocation decisions, equity assessments → Output → Policy recommendations, fairness-driven adjustments

Dubai's Happiness Agenda is closely aligned with its broader smart city transformation. The city adopts cutting-edge technologies such as the Internet of Things (IoT), Artificial Intelligence (AI), and digital twin platforms to monitor, simulate, and optimize urban systems in real time. These technologies enable predictive management of city functions—from energy distribution to waste collection—while also providing tools to simulate future scenarios and proactively resolve issues. Through this technological infrastructure, Dubai ensures that urban services remain adaptive and citizen-focused, reflecting a dynamic understanding of residents' needs and behaviors (Dani et al. 2023).

To ensure that this real-time, DT-enabled responsiveness remains institutionally defensible, we propose a city-level data-governance spine that operationalizes the "Layer 1" commitments of data collection and preprocessing in URPM-HCSC and the subsequent layered frameworks introduced below, including RUAM-SCW and AUSAM-SCW. Given that smart city digital twins typically rely on heterogeneous, continuously refreshed data streams whose reuse across agencies and use-cases can otherwise become opaque, the Dubai case would benefit from an explicit, auditable register that enumerates each stream and specifies its purpose of use, institutional or legal basis (including the consent stance where applicable), and a named stewardship responsibility for quality and provenance (Dani et al. 2023). This register should also define a disciplined access regime and treat baseline safeguards such as robust encryption and access controls as design-time requirements rather than retrofitted hardening (El-Agamy et al. 2024). In procedural terms, this configuration aligns with governance accounts that treat data governance as embedded in everyday data-management practices, where classification and encryption secure and organize data, and systematic checks mitigate bias and error propagation across the lifecycle (Papagiannidis et al. 2025). It also addresses the evidentiary problem that often constrains adoption in high-scrutiny settings because oversight and auditability depend on inspectable documentation rather than narrative assurances, the spine should mandate explicit data-preparation records (collection, cleaning, transformations, quality gates, and release decisions) that make the evidentiary chain underpinning downstream well-being prediction and allocation decisions reviewable (Konigstorfer and Thalmann 2022). From this standpoint, the spine functions as the administrative substrate that renders the later "challenge" frameworks implementable as controls, particularly FDPGM-SC for privacy-preserving analytics, FCAPOM-UG for fairness-constrained optimization, and DCERM for sustained civic participation, rather than leaving them as purely algorithmic desiderata.

To operationalize such responsiveness at scale, we propose a Real-Time Urban Adaptation Model (RUAM), which couples sensor-fed digital twins with an adaptive AI policy that learns continuously from environmental feedback. The system is modeled using a closed-loop reinforcement learning framework:

$$\pi^*(s_t) = \arg\max_{a_t} \mathbb{E}\left[\sum_{k=0}^{\infty} \gamma^k \left(R_{infra}(s_{t+k}, a_{t+k}) + \lambda \cdot R_{citizen}(s_{t+k}, a_{t+k})\right)\right]$$

where $\pi^*(s_t)$ is the optimal policy mapping state s_t to action a_t, $R_{infra}(s_{t+k}, a_{t+k})$ is the infrastructural performance reward (e.g., load balancing, energy efficiency, waste reduction), $R_{citizen}(s_{t+k}, a_{t+k})$ captures resident-centered metrics (e.g., response time satisfaction, livability impact), $\gamma \in (0, 1)$ is the discount factor, and λ is a balancing coefficient that amplifies the importance of citizen well-being in the optimization process.

This unified optimization mechanism ensures that every AI-driven intervention across urban systems considers not just operational efficiency but also citizen-centric outcomes. By continuously updating digital twins with real-time data and feeding insights into the reinforcement model, the system dynamically aligns infrastructure behavior with evolving citizen expectations. The inclusion of $R_{citizen}$ as a reward term reflects the broader ethos of the Happiness Agenda—where smart technologies are not ends in themselves but tools to enhance daily life, foster responsiveness, and sustain inclusive urban futures.

Framework: Real-Time Urban Adaptation Model for Smart City Well-being (RUAM-SCW)

Layered AI Architecture for Smart City Adaptation and Citizen Well-being

Layer	Main functionality	AI module(s) and components	Operational scope (Input → Processing → Output)
Layer 1	Data collection and preprocessing	Sensor data integration and processing pipeline	Input → Sensor data (e.g., energy, waste, traffic), resident feedback → Output → Preprocessed data for reinforcement learning
Layer 2	Learning: reinforcement learning for urban adaptation	Reinforcement Learning (RL)	Input → Preprocessed environmental data, sensor data → Output → Optimal urban adaptation policy, infrastructure adjustments
Layer 3	Policy implementation and monitoring	Adaptive policy execution and monitoring system	Input → Optimal urban adaptation policy → Output → Policy execution, real-time urban system adjustments
Layer 4	Citizen engagement and satisfaction	Citizen experience analytics and feedback platform	Input → Resident feedback, policy outcomes → Output → Citizen satisfaction metrics, future policy adjustments

A core implementation focus lies in optimizing public service delivery to directly enhance the day-to-day experiences of citizens. This includes improvements in transportation networks, emergency response systems, digital governance platforms, and healthcare access. The use of data analytics and automation helps streamline administrative processes, reduce waiting times, and deliver services more equitably across the city. By aligning service performance with user expectations, Dubai advances its goal of making public systems both efficient and deeply attuned to the well-being of its population (El-Agamy et al. 2024). To support this objective, we introduce an AI-Governed Urban Service Allocation Model (AUSAM) based on a constrained multi-objective optimization framework. This model allocates digital and physical services in real time, balancing efficiency with fairness:

$$\max_{\mathrm{a}_t} \left(\sum_{i=1}^{N} \eta_1 \cdot U_i(\mathrm{a}_t) - \eta_2 \cdot W_i(\mathrm{a}_t) \right) \quad \text{subject to} \quad \left| \mathbb{E}_{G=0}[U_i] - \mathbb{E}_{G=1}[U_i] \right| \leq \delta$$

where a_t is the vector of service allocation decisions at time t, $U_i(\mathrm{a}_t)$ denotes the utility derived by citizen i (e.g., accessibility, service speed), $W_i(\mathrm{a}_t)$ denotes the cost or waiting penalty, η_1, η_2 are tunable parameters reflecting priorities between utility maximization and delay minimization, $G \in \{0, 1\}$ represents group membership (e.g., underserved vs. general population), and δ is a small constant bounding fairness disparity between groups.

This model enables Dubai's service ecosystem to dynamically allocate resources where they are most impactful, while preserving procedural fairness across demographics. The constraint ensures that service gains do not disproportionately benefit one social group over another, embedding equity into the digital fabric of urban operations. As AI continuously learns from usage patterns, response metrics, and feedback, the allocation policy adapts to citizen behavior and expectations—ensuring that efficiency never comes at the expense of inclusivity.

Framework: AI-Governed Urban Service Allocation Model for Smart City Well-being (AUSAM-SCW)

Layered AI Architecture for Urban Service Allocation and Citizen Well-being

Layer	Main functionality	AI module(s) and components	Operational scope (Input → Processing → Output)
Layer 1	Data collection and preprocessing	Urban data integration and quality assurance pipeline	Input → Socio-demographic data, infrastructure data, service feedback → Output → Preprocessed data for service allocation model
Layer 2	Learning: fairness-constrained multi-objective optimization	Fairness-constrained learning	Input → Preprocessed data (e.g., accessibility, service speed, waiting times) → Output → Optimal service allocation decisions with fairness constraints
Layer 3	Service allocation and optimization	Real-time service orchestration and resource optimization engine	Input → Optimal service allocation policy → Output → Service delivery adjustments, real-time resource allocation
Layer 4	Fairness monitoring and policy adjustments	Fairness analytics and policy recommendation platform	Input → Resident feedback, allocation performance → Output → Fairness assessment, policy adjustment recommendations

Recognizing the psychological and social dimensions of happiness, the agenda prioritizes access to cultural, artistic, and recreational spaces. Dubai invests in public parks, community hubs, cultural festivals, and art installations that promote social cohesion, physical well-being, and emotional engagement. These spaces not only enhance the aesthetic and experiential quality of the urban environment but also foster inclusivity and belonging among residents of diverse backgrounds. In this way, the city supports a holistic vision of happiness that integrates mental, emotional, and communal well-being (Dani et al. 2023). To operationalize this vision, we propose an AI-based Cultural Infrastructure Impact Model (CIIM) that uses sentiment-aware spatial optimization to maximize urban emotional well-being. This model integrates data from social media, location-based surveys, and event attendance to learn and predict the affective response of communities to different cultural assets. The optimization objective is:

$$\max_{\mathrm{x}_t} \left(\sum_{j=1}^{M} \omega_j \cdot S_j(\mathrm{x}_t) - \lambda \cdot D_j(\mathrm{x}_t) \right)$$

where x_t is the set of AI-planned interventions at time t (e.g., opening a new cultural site, enhancing a festival), $S_j(x_t)$ represents the predicted average sentiment score (e.g., happiness, engagement) for community j, $D_j(x_t)$ is the disparity measure of cultural access across income or district boundaries, ω_j is the importance weight of each community based on density and vulnerability indices, and λ is a policy-controlled fairness regularization parameter.

Incorporating both positive emotional response and spatial equity into the optimization, this model ensures that cultural investments not only generate joy but do so inclusively. Real-time feedback from sensors, mobile app interactions, and NLP-based sentiment extraction allows the system to dynamically adjust plans—expanding programming in underserved areas or enhancing experiences that elicit the highest communal value. This functionally embeds emotional intelligence into city planning and reinforces the idea that happiness is not just a subjective experience, but a measurable and optimizable dimension of urban design.

Framework: Cultural Infrastructure Impact Model for Urban Well-being (CIIM-UW)

Layered AI Architecture for Smart City Cultural Optimization and Emotional Well-being

Layer	Main functionality	AI module(s) and components	Operational scope (Input → Processing → Output)
Layer 1	Data collection and preprocessing	Cultural data integration and sentiment processing pipeline	Input → Social media data, surveys, event attendance → Output → Preprocessed data for cultural infrastructure model
Layer 2	Learning: fairness-constrained optimization for sentiment-aware urban design	Fairness-constrained learning	Input → Preprocessed data (sentiment, accessibility, service quality) → Output → Optimized cultural investment decisions balancing sentiment and equity
Layer 3	Policy implementation and adjustment	Cultural investment execution and adaptive policy engine	Input → Optimized cultural investment policies → Output → Execution of interventions, dynamic adjustment based on citizen feedback
Layer 4	Emotional well-being and cultural equity monitoring	Emotional analytics and cultural equity assessment platform	Input → Citizen feedback, cultural program outcomes → Output → Emotional well-being metrics, policy equity assessments

Together, these strategies demonstrate how the Happiness Agenda is operationalized through a balanced integration of digital innovation, human-centered services, and cultural enrichment. The result is a governance model that not only responds to the functional needs of the city but also nurtures the emotional and social well-being of its people, reinforcing the idea that happiness is both a policy goal and a lived urban experience.

3.2.3 *Impacts of the Happiness Agenda*

The implementation of the Happiness Agenda has produced measurable and multidimensional outcomes that extend beyond emotional well-being, shaping Dubai's urban development, socio-economic vitality, and long-term sustainability strategy. These impacts underscore how the integration of happiness into governance is not merely symbolic but yields tangible benefits for individuals, communities, and the city as a whole.

To assess these diverse outcomes, we introduce a Causal Impact Estimation Model using a Structural Causal Model (SCM) framework. This model disentangles the direct and indirect effects of happiness-centered interventions on key urban indicators, such as public satisfaction, service utilization, and socio-environmental quality. The foundational structure of this approach is defined as:

$$Y_i = f(X_i, H_i, U_i)$$

where Y_i is the observed outcome variable for domain i (e.g., livability index, citizen health score, environmental satisfaction), X_i represents observable covariates such as baseline infrastructure, demographics, and economic indicators, H_i denotes the happiness-oriented intervention (e.g., access to wellness services, participatory programs, cultural amenities), and U_i captures unobserved confounders.

To formally identify the causal effect of the Happiness Agenda interventions H_i on urban outcomes Y_i, we compute the Average Treatment Effect (ATE) using do-calculus:

$$\text{ATE} = \mathbb{E}[Y_i \,|\, \text{do}(H_i = 1)] - \mathbb{E}[Y_i \,|\, \text{do}(H_i = 0)]$$

where $\text{do}(H_i = 1)$ and $\text{do}(H_i = 0)$ indicate counterfactual scenarios with and without the happiness intervention, respectively.

This casual modeling approach enables the municipality to simulate hypothetical changes in well-being policy and measure their differentiated impacts across sectors and populations. By controlling both observed and latent variables, the SCM ensures that policy evaluations are not merely correlational but reflect deep structural insights. This framework translates the abstract goal of "happiness" into quantifiable metrics, offering governance systems a scientifically robust mechanism to track how emotional, infrastructural, and economic well-being co-evolve in a smart city context.

Framework: Causal Impact Estimation Model for Urban Well-being (CIEM-UW)
Layered AI Architecture for Happiness-Centered Urban Governance

Layer	Main functionality	AI module(s) and components	Operational scope (Input → Processing → Output)
Layer 1	Data collection and preprocessing	Multi-source data integration and preprocessing pipeline	Input → Social media data, surveys, event attendance → Output → Preprocessed data for causal modeling
Layer 2	Learning: causal impact estimation using structural causal modeling (SCM)	Structural causal modeling (SCM)	Input → Preprocessed data (e.g., demographics, happiness interventions, infrastructure) → Output → Estimated causal effects of happiness interventions on urban outcomes
Layer 3	Policy simulation and impact prediction	Policy scenario simulation and outcome forecasting engine	Input → Causal impact estimations → Output → Simulation of policy changes, impact predictions on urban outcomes
Layer 4	Feedback and dynamic adjustment	Adaptive policy refinement and feedback optimization platform	Input → Resident feedback, policy outcomes → Output → Adjusted policies, reallocation of resources based on causal insights

One of the most direct outcomes of the Happiness Agenda has been the enhancement of residents' daily lives. By embedding well-being metrics into policy design and service delivery, Dubai has succeeded in creating more livable urban environments. This includes expanded access to green spaces, upgraded public transportation, better healthcare infrastructure, and greater availability of cultural and recreational amenities. These developments have contributed to a noticeable increase in citizen satisfaction and have strengthened the connection between urban governance and human experience (Dani et al. 2023). To model how these improvements translate into perceived well-being, we propose a Multimodal Satisfaction Prediction Network (MSPN)—an advanced AI system that integrates heterogeneous urban data sources to predict population-level satisfaction scores across neighborhoods. Let the predicted satisfaction score $\hat{S}_i$ for area ii be modeled as:

$$\hat{S}_i = \phi\left(g_{\text{env}}\left(x_i^{\text{env}}\right) + g_{\text{mob}}\left(x_i^{\text{mob}}\right) + g_{\text{hlth}}\left(x_i^{\text{hlth}}\right) + g_{\text{cult}}\left(x_i^{\text{cult}}\right)\right)$$

where $\hat{S}_i$ is the predicted satisfaction score in district i, $x_i^{\text{env}}, x_i^{\text{mob}}, x_i^{\text{hlth}}, x_i^{\text{cult}}$ represent feature vectors related to environment (e.g., air quality, green space density), mobility (e.g., travel time, transit reliability), healthcare (e.g., access scores, health service ratings), and culture (e.g., event frequency, cultural venue density), respectively, $g(\cdot)$ are neural sub-networks encoding domain-specific features, and $\phi(\cdot)$ is a fusion function such as a weighted attention mechanism.

To ensure that predictions remain aligned with citizen-reported outcomes, the model is trained using a loss function that penalizes deviation from actual satisfaction survey data:

$$\mathcal{L}(\theta) = \frac{1}{n}\sum_{i=1}^{n}\left(\hat{S}_i - S_i\right)^2 + \lambda\sum_{k}\|\theta_k\|^2$$

where $\left(\hat{S}_i - S_i\right)^2$ is the actual satisfaction score from surveys in region i, θ denotes the set of trainable model parameters, and λ is a regularization coefficient to prevent overfitting.

This multimodal learning framework enables the city to track which urban investments yield the greatest returns in citizen satisfaction, revealing domain-specific bottlenecks and informing targeted interventions. MSPN not only reflects present quality-of-life perceptions but also forecasts how future infrastructure upgrades or policy shifts will affect public contentment, allowing planners to simulate policy outcomes and refine urban well-being strategies dynamically.

Framework: Multimodal Satisfaction Prediction Network for Urban Well-being (MSPN-UW)
Layered AI Architecture for Happiness-Centered Urban Governance

Layer	Main functionality	AI module(s) and components	Operational scope (Input → Processing → Output)
Layer 1	Data collection and preprocessing	Citizen data integration and feature engineering pipeline	Input → Social media data, surveys, event attendance → Output → Preprocessed data for satisfaction prediction model
Layer 2	Learning: multimodal satisfaction prediction and optimization	Multimodal learning, neural networks	Input → Preprocessed data (sentiment, accessibility, service quality) → Output → Predicted satisfaction scores for neighborhoods
Layer 3	Policy simulation and impact forecasting	Decision scenario simulator and policy outcome engine	Input → Predicted satisfaction scores → Output → Simulation of policy outcomes and infrastructure improvements
Layer 4	Satisfaction monitoring and policy adjustment	Satisfaction analytics and adaptive policy tuning platform	Input → Resident feedback, satisfaction data → Output → Adjusted urban policies and resource allocation strategies

The agenda has also contributed to Dubai's economic dynamism by cultivating a socially and psychologically supportive urban climate. A happier population is generally more productive, collaborative, and innovative—traits that are essential for entrepreneurship and sustained economic growth. Dubai's emphasis on happiness has helped position the city as an attractive destination for global talent, startups, and foreign investment. As residents feel more valued and engaged, they are more likely to participate actively in civic life and the local economy, creating a reinforcing cycle of growth and well-being (El-Agamy et al. 2024).

To formally model this feedback loop between collective happiness and economic performance, we propose an AI-driven Urban Happiness–Productivity Feedback Model (UHP-FM) based on deep recurrent econometrics. This model predicts changes in regional economic indicators as a function of lagged happiness metrics, social engagement scores, and entrepreneurial activity using Long Short-Term Memory (LSTM) networks.

Let the economic performance index E_t at time t be predicted as:

$$E_t = f_{\text{LSTM}}([H_{t-\tau}, P_{t-\tau}, C_{t-\tau}, G_{t-\tau}], \theta)$$

where E_t is the economic performance score at time t, $H_{t-\tau}$ denotes the lagged happiness index, $P_{t-\tau}$ is the productivity metric (e.g., GDP per capita or labor efficiency), $C_{t-\tau}$ captures civic engagement (e.g., voting participation, volunteer hours), $G_{t-\tau}$ represents entrepreneurial activity (e.g., new business registrations), τ is the look-back horizon, f_{LSTM} is the recurrent prediction function, and θ are the model parameters trained via backpropagation through time.

To evaluate the contribution of happiness to economic dynamism while accounting for nonlinear interactions and potential mediating variables, we introduce a decomposition of total economic growth impact via attention-based feature attribution:

$$E_t = \sum_j \alpha_j \cdot \psi_j(H_{t-\tau}, P_{t-\tau}, C_{t-\tau}, G_{t-\tau})$$

where α_j is the learned attention weight of factor j, and ψ_j is a nonlinear transformation capturing its marginal effect on E_t. This formulation enables interpretability by quantifying which happiness-driven components most strongly influence urban economic growth at any given time.

The model supports city planners by highlighting the latent economic value of sustained well-being initiatives, informing policy not only by traditional economic metrics but also through psycho-social determinants. As Dubai continues to attract global talent, the UHP-FM reinforces the notion that civic happiness is not peripheral to economic development—it is foundational.

Framework: Urban Happiness–Productivity Feedback Model for Economic Growth (UHP-FM-EG)

Layered AI Architecture for Happiness-Driven Urban Economic Performance

Layer	Main functionality	AI module(s) and components	Operational scope (Input → Processing → Output)
Layer 1	Data collection and preprocessing	Socio-economic and civic signals integration and feature engineering pipeline	Input → Social media data, surveys, civic engagement metrics → Output → Preprocessed data for happiness-productivity model
Layer 2	Learning: deep recurrent econometrics for happiness-driven economic modeling	Long Short-Term Memory (LSTM), attention mechanisms	Input → Preprocessed data (happiness, productivity, civic engagement) → Output → Predicted economic performance scores based on happiness metrics
Layer 3	Impact prediction and policy simulation	Counterfactual economic scenario generator and intervention evaluation engine	Input → Predicted economic performance scores → Output → Simulation of policy interventions and infrastructure impacts
Layer 4	Economic policy adjustment and feedback loop	Adaptive economic governance analytics and resource reallocation platform	Input → Resident feedback, satisfaction scores → Output → Policy adjustment recommendations and resource reallocation

Beyond individual and economic benefits, the Happiness Agenda supports Dubai's broader commitment to sustainability. By prioritizing policies that address environmental quality, social inclusion, and community resilience, the agenda ensures that happiness is aligned with long-term urban health. Initiatives that reduce pollution, promote sustainable mobility, and enhance resource efficiency also contribute to residents' overall well-being. In this way, the city's focus on happiness intersects with its environmental objectives, offering a holistic model for sustainable urban development that is both people-centered and future-oriented (Dani et al. 2023).

To formalize this intersection, we propose an AI-Driven Sustainability–Happiness Alignment Model (SHAM) using Multi-Objective Reinforcement Learning (MORL) to balance and optimize environmental, social, and well-being goals within urban governance systems.

The agent operates over a composite state $s_t = [e_t, h_t, r_t]$, representing the environment, happiness, and resource states at time t, and seeks to maximize a vectorized reward function across three dimensions:

$$\vec{R}(s_t, a_t) = \left[\omega_1 \cdot Q_{\text{env}}(s_t, a_t), \omega_2 \cdot Q_{\text{happiness}}(s_t, a_t), \omega_3 \cdot Q_{\text{resilience}}(s_t, a_t)\right]$$

where $Q_{\text{env}}(s_t, a_t)$ measures environmental performance (e.g., air quality, CO_2 reduction), $Q_{\text{happiness}}(s_t, a_t)$ quantifies well-being based on affective surveys and life satisfaction indices, $Q_{resilience}(s_t, a_t)$ evaluates urban robustness to stressors (e.g., energy security, food systems), and ω_1, ω_2, ω_3, are adjustable weights reflecting the city's current strategic priorities.

To resolve conflicting objectives in decision-making, we compute the Pareto-optimal policy $\pi * \pi *$ that satisfies:

$$\pi^* = \arg \underset{\pi}{max} \left(\text{ParetoFront} \left\{ \vec{R}^{\pi} (s_t, a_t) \right\} \right)$$

This ensures that no single goal—happiness, sustainability, or resilience—is pursued at the expense of others. The MORL framework allows dynamic reweighting of priorities through civic feedback loops or environmental alerts, enabling the system to adapt to changing public values or emergent ecological threats.

This integrated SHAM model empowers Dubai to pursue equitable urban futures, wherein socio-environmental sustainability and psychological well-being are optimized together through AI-guided governance. As the city evolves, this model becomes instrumental in safeguarding quality of life not just for present residents, but for future generations—thus operationalizing sustainability through the lens of collective happiness.

Framework: Sustainability–Happiness Alignment Model for Urban Governance (SHAM-UG)

Layered AI Architecture for Multi-Objective Urban Optimization

Layer	Main functionality	AI module(s) and components	Operational scope (Input → Processing → Output)
Layer 1	Data collection and preprocessing	Environmental and social data integration and normalization pipeline	Input → Environmental data, happiness metrics, civic engagement scores → Output → Preprocessed data for multi-objective optimization
Layer 2	Learning: multi-objective reinforcement learning for sustainability and happiness	Multi-Objective Reinforcement Learning (MORL), Deep Reinforcement Learning (DRL)	Input → Preprocessed data (happiness, environmental metrics, social engagement) → Output → Optimized policy for balancing happiness, sustainability, and resilience
Layer 3	Policy simulation and impact prediction	Multi-domain policy simulation and impact forecasting engine	Input → Optimized policy, environmental and happiness predictions → Output → Simulated policy outcomes and urban impacts
Layer 4	Feedback and dynamic adjustment	Adaptive governance feedback and policy realignment platform	Input → Feedback loops, public satisfaction data → Output → Adjusted policies and rebalanced priorities

These outcomes reveal that when happiness becomes a central principle of urban policy, it generates ripple effects across infrastructure, economy, and environment. Dubai's approach offers an important case study in how cities can adopt a more integrated and ethical model of development—one that values prosperity not only in financial terms, but in human terms as well.

3.2.4 Challenges and Future Directions

While the Happiness Agenda in Dubai has delivered significant progress in integrating well-being into urban governance, its continued success depends on overcoming several key challenges. These issues highlight the evolving complexity of aligning smart city strategies with human-centered values, especially in a context where technological innovation intersects with ethical, social, and institutional considerations.

A central component of the Happiness Agenda is the collection and analysis of data to measure citizen satisfaction and optimize public services. However, this data-driven approach introduces significant challenges related to privacy, consent, and digital security. As the city increasingly relies on real-time data from sensors, mobile devices, and digital platforms, safeguarding the confidentiality of personal information becomes imperative. Balancing the need for actionable insights with the protection of individual privacy requires robust governance frameworks, ethical data protocols, and transparent communication with the public to maintain trust (El-Agamy et al. 2024).

To address the profound ethical and technical challenges embedded in the data-driven nature of Dubai's Happiness Agenda, we introduce a far superior AI-driven framework—the Federated Differential Privacy Governance Model (FDPGM). This model is specifically designed to support large-scale citizen data collection and analysis while maintaining rigorous privacy standards, minimizing data centralization risks, and maximizing transparency in public service optimization. At its foundation, the model leverages federated learning, which enables AI models to be trained across decentralized data sources (e.g., mobile devices or municipal sensors) without transferring raw data to a central server. To guarantee privacy throughout the training process, a differential privacy mechanism is integrated at the local update level. The local model update from device i at round t is:

$$\tilde{g}_i^t = g_i^t + \mathcal{N}\left(0, \sigma^2 I\right)$$

where $\tilde{g}_i^t$ is the privatized gradient sent to the server, g_i^t is the original gradient based on local citizen satisfaction data, $\mathcal{N}(0, \sigma^2 I)$ is Gaussian noise added to guarantee differential privacy, and σ is the noise scale parameter controlling privacy strength. This function ensures no individual's data point disproportionately influences the model, enabling GDPR-compliant analysis.

The aggregated model on the server, used for policy optimization, is calculated as:

$$\theta^{t+1} = \theta^t - \eta \cdot \frac{1}{N} \sum_{i=1}^{N} \tilde{g}_i^t$$

where θ^t is the global model at time t, η is the learning rate, N is the number of participating data nodes, and $\tilde{g}_i^t$ are the noise-protected gradients. This federated update mechanism ensures that learning is both secure and distributed.

To ensure transparency and citizen trust, the system incorporates a consent-aware information disclosure function, defined as:

$$I_t = \sum_{i=1}^{N} \mu_i \cdot \log\left(\frac{P(r_i|c_i)}{P(r_i)}\right)$$

where I_t quantifies information disclosure at time t, μ_i is the user's consent weight (ranging from 0 to 1), $P(r_i| c_i)$ is the probability of revealing user report r_i given consent c_i, and $P(r_i)$ is the unconditional probability. This allows the system to modulate transparency based on informed consent levels, dynamically adjusting data visibility without compromising trust.

Altogether, this AI framework ensures that the Happiness Agenda's data analytics are not only functionally effective but also aligned with international standards of digital ethics, privacy by design, and citizen autonomy. Through the integration of privacy-preserving techniques into each level of the data pipeline—from collection to aggregation and decision-making—the FDPGM positions Dubai as a global benchmark for trustworthy AI in citizen-centric governance.

Framework: Federated Differential Privacy Governance Model for Smart Cities (FDPGM-SC)

Layered AI Architecture for Privacy-Preserving Data Analysis in Urban Governance

Layer	Main functionality	AI module(s) and components	Operational scope (Input → Processing → Output)
Layer 1	Data collection and preprocessing	Decentralized data ingestion and privacy-preserving preprocessing pipeline	Input → Real-time citizen data (e.g., mobile devices, municipal sensors) → Output → Preprocessed data for federated learning
Layer 2	Learning: federated learning with differential privacy for privacy-preserving analysis	Federated learning, differential privacy	Input → Decentralized data (e.g., mobile sensors, citizen data) → Output → Privacy-preserved model updates and policy optimization
Layer 3	Privacy-preserving policy optimization and decision-making	Secure policy optimization and risk-aware decision-support platform	Input → Aggregated privacy-preserved model → Output → Optimized policies for urban governance and public services
Layer 4	Consent-aware transparency and feedback mechanism	Consent management, transparency reporting, and citizen feedback analytics portal	Input → Citizen feedback, consent levels → Output → Adjusted data visibility and transparency for public trust

Another critical challenge is ensuring that the benefits of the Happiness Agenda are distributed fairly across all social groups. While technological advancements and service improvements may uplift the city as a whole, disparities in access can persist, particularly among marginalized or underrepresented communities. Addressing this requires proactive policies that focus on social equity, including targeted investments in underserved neighborhoods, multilingual and inclusive platforms, and the removal of barriers to participation. Only by prioritizing equity can the agenda fulfill its vision of happiness as a universal right rather than a selective privilege (Dani et al. 2023).

To ensure that the Happiness Agenda fulfills its vision of universal well-being rather than benefiting only privileged segments, we propose a Fairness-Constrained Adaptive Policy Optimization Model (FCAPOM). This advanced AI-driven model is designed to dynamically adjust service delivery strategies to promote social equity across all demographic and geographic groups within the city. It integrates real-time fairness constraints into both policy learning and deployment, ensuring that improvements in urban services reach underrepresented and marginalized populations. We begin with the adaptive policy function defined for each region i as:

$$\pi^*(x_i) = \arg \max_{\pi} \mathbb{E}_{x_i \sim \mathcal{D}}[R(x_i, \pi(x_i))] - \lambda \cdot \Delta_{\text{eq}}(\pi)$$

where $\pi^*(x_i)$ is the optimal policy for population subgroup or region i, x_i is the feature vector including socio-economic status, service access level, and geographic data, $\mathcal{D}$ is the data distribution over all city zones, $R(x_i, \pi(x_i))$ is the reward for applying policy π (e.g., satisfaction or accessibility increase), λ is a fairness regularization parameter, and $\Delta_{\text{eq}}(\pi)$ is the equity deviation penalty, described next.

To quantify and penalize disparity, we define:

$$\Delta_{\text{eq}}(\pi) = |\, \mathbb{E}_{x \in G_0}[y^\pi] - \mathbb{E}_{x \in G_1}[y^\pi] \,|$$

where G_0 and G_1 represent disadvantaged and advantaged social groups, respectively, and y^π denotes the expected service outcome (e.g., access to healthcare, transport, digital services) under policy π. This equity constraint minimizes discrepancies in service delivery outcomes across social strata.

To operationalize inclusive deployment, the model includes a multilingual, participatory feedback mechanism trained via Cross-Language Semantic Alignment:

$$s_i = \text{cosine}(E_L(m_i), E_{\text{en}}(r_i))$$

where s_i is the semantic alignment score between multilingual citizen message m_i and standard English reference report r_i, E_L and E_{en} are language-specific and English embedding functions, respectively. This ensures fair representation of feedback regardless of linguistic background.

By combining fairness-optimized policy learning with inclusive, real-time feedback integration, FCAPOM ensures that AI interventions uplift all communities—

not just those with better access to digital infrastructure or higher socio-economic status. This model provides a robust, actionable strategy for realizing equity-aware happiness governance, aligning Dubai's Happiness Agenda with the ethical imperatives of digital justice and inclusive urban development.

Framework: Fairness-Constrained Adaptive Policy Optimization Model for Urban Governance (FCAPOM-UG)

Layered AI Architecture for Equitable Urban Service Delivery

Layer	Main functionality	AI module(s) and components	Operational scope (Input → Processing → Output)
Layer 1	Data collection and preprocessing	Unified civic signals ingestion and quality control pipeline	Input → Social and demographic data, service access, geographic data → Output → Preprocessed data for policy optimization
Layer 2	Learning: fairness-constrained adaptive policy optimization	Fairness-constrained optimization, reinforcement learning	Input → Preprocessed data (citizen satisfaction, socio-economic data, service access) → Output → Optimized policy for equitable service delivery
Layer 3	Policy simulation and impact evaluation	Scenario-based policy simulator and outcome attribution engine	Input → Optimized policy, socio-economic impact predictions → Output → Policy simulation outcomes, urban impact analysis
Layer 4	Feedback loop and dynamic adjustment	Continuous feedback analytics and resource reallocation platform	Input → Public feedback, service satisfaction data → Output → Adjusted policies, rebalanced resource allocation

The long-term success of the Happiness Agenda depends heavily on the sustained involvement of citizens in shaping and refining urban policies. While initial public enthusiasm may be strong, maintaining engagement over time—especially as everyday routines normalize or political priorities shift—can be challenging. Ensuring ongoing participation calls for continuous innovation in civic platforms, transparent feedback loops, and educational initiatives that reinforce the value of citizen voices. Ultimately, a resilient happiness strategy must be co-created with the people it aims to serve, requiring adaptive governance structures that evolve alongside community expectations and technological change (El-Agamy et al. 2024). To ensure the sustained success of the Happiness Agenda, it is essential to embed AI-driven participatory governance mechanisms that actively promote continuous, inclusive citizen involvement. We propose a Dynamic Civic Engagement Reinforcement Model (DCERM)—a cutting-edge AI framework that learns, adapts, and optimizes civic participation strategies over time by leveraging Reinforcement Learning (RL), user feedback analytics, and behavior-driven engagement scoring.

The model defines the participatory value function as follows:

$$V^{\pi}(s_t) = \mathbb{E}_{\pi}\left[\sum_{k=0}^{\infty} \gamma^k R(s_{t+k}, a_{t+k})\right]$$

where $V^{\pi}(s_t)$ is the expected cumulative reward for engagement state s_t under policy π, γ is the discount factor for future engagement, and $R(s_{t+k}, a_{t+k})$ is the reward for civic action a_{t+k} taken in state s_{t+k}, such as commenting on a policy draft or attending a local consultation.

The reward function is constructed to capture both immediate and long-term value from each citizen interaction:

$$R(s_t, a_t) = \beta_1 \cdot f_{\text{quality}}(a_t) + \beta_2 \cdot f_{\text{diversity}}(a_t) + \beta_3 \cdot f_{\text{impact}}(a_t)$$

where $f_{\text{quality}}(a_t)$ measures the informativeness or constructiveness of the action (e.g., a suggestion), $f_{\text{diversity}}(a_t)$ measures participation from underrepresented groups or regions, and $f_{\text{impact}}(a_t)$ reflects how the action influences policy decisions. Weights β_1, β_2, β_3, calibrate the importance of these dimensions.

To adapt participation mechanisms dynamically, we incorporate a Citizen Sentiment Tracking Module (CSTM) using Natural Language Processing (NLP):

$$s_i = \tanh(W \cdot E(x_i) + b)$$

where s_i is the inferred engagement sentiment score for citizen input x_i, $E(x_i)$ is the encoded text vector, and W, b are learnable parameters. This sentiment feedback is used to modulate the engagement strategy in real time—such as sending personalized reminders, improving platform UX, or opening new dialogue channels.

Finally, a Policy Adaptation Trigger (PAT) detects systemic disengagement by computing a decayed engagement gradient:

$$G_t = \delta \cdot G_{t-1} + (1 - \delta) \cdot \Delta P_t$$

where G_t captures engagement trends over time, ΔP_t is the change in participation levels at time t, and δ is the decay factor. If G_t falls below a predefined threshold, adaptive mechanisms (e.g., new gamified participation tools or civic hackathons) are activated.

This Dynamic Civic Engagement Reinforcement Model enables continuous co-creation of urban policy by ensuring feedback loops are not static but evolve intelligently with citizen behavior and community sentiment. It transforms citizen participation from a one-time consultation into a living, AI-augmented process—ensuring that the Happiness Agenda remains not only people-centered in design but also people-driven in practice.

Framework: Dynamic Civic Engagement Reinforcement Model (DCERM)
Layered AI Architecture for Sustainable Urban Participation

Layer	Main functionality	AI module(s) and components	Operational scope
Layer 1	Citizen data collection: gather and preprocess engagement data from various civic platforms	Civic data integration and engagement preprocessing pipeline	Input → Citizen participation data (comments, actions) → Output → Preprocessed data for AI models
Layer 2	Engagement strategy learning: use RL to learn optimal participation strategies	Reinforcement Learning (RL)	Input → Preprocessed data, engagement rules → Output → Optimized participation strategies
Layer 3	Reward calculation: compute rewards based on quality, diversity, and impact of actions	Multidimensional engagement reward analytics engine	Input → Engagement action data → Output → Calculated reward (quality, diversity, impact)
Layer 4	Sentiment tracking: analyze sentiment of citizen input using NLP	Natural Language Processing (NLP)	Input → Citizen input text → Output → Engagement sentiment score
Layer 5	Engagement trend monitoring: track and adjust strategies based on changes in citizen participation	Adaptive participation monitoring and strategy refinement platform	Input → Engagement data, participation levels → Output → Adaptive participation strategies

These challenges do not undermine the value of the Happiness Agenda but rather illustrate the complexity of embedding human well-being within the structures of smart governance. As Dubai moves forward, the future of the agenda will depend on its ability to protect privacy, close equity gaps, and cultivate lasting civic engagement. By addressing these issues directly, the city can strengthen the foundation of its human-centric urban model and provide a replicable path for other cities seeking to align technological innovation with holistic quality of life. Within the governance blueprint articulated above, Dubai can be read as a case where legitimacy is pursued through governance-by-design: privacy and fairness are embedded as formal constraints, while civic engagement is sustained through a continuously optimized feedback channel rather than a one-off consultation.

Figure 3.1 illustrates Dubai's "Happiness Agenda," a framework focused on improving residents' quality of life through innovation and citizen engagement. The central node, "Dubai's 'Happiness Agenda,'" branches into four main components: "Conceptual Framework of the Happiness Agenda," "Implementation Strategies," "Impacts," and "Challenges." The "Conceptual Framework" includes metrics for happiness measurement, data-driven decisions using real-time analytics, and active citizen engagement in governance. "Implementation Strategies" involve smart city technologies like IoT and AI, service optimization in areas such as transport and healthcare, and cultural investments in parks and community spaces.

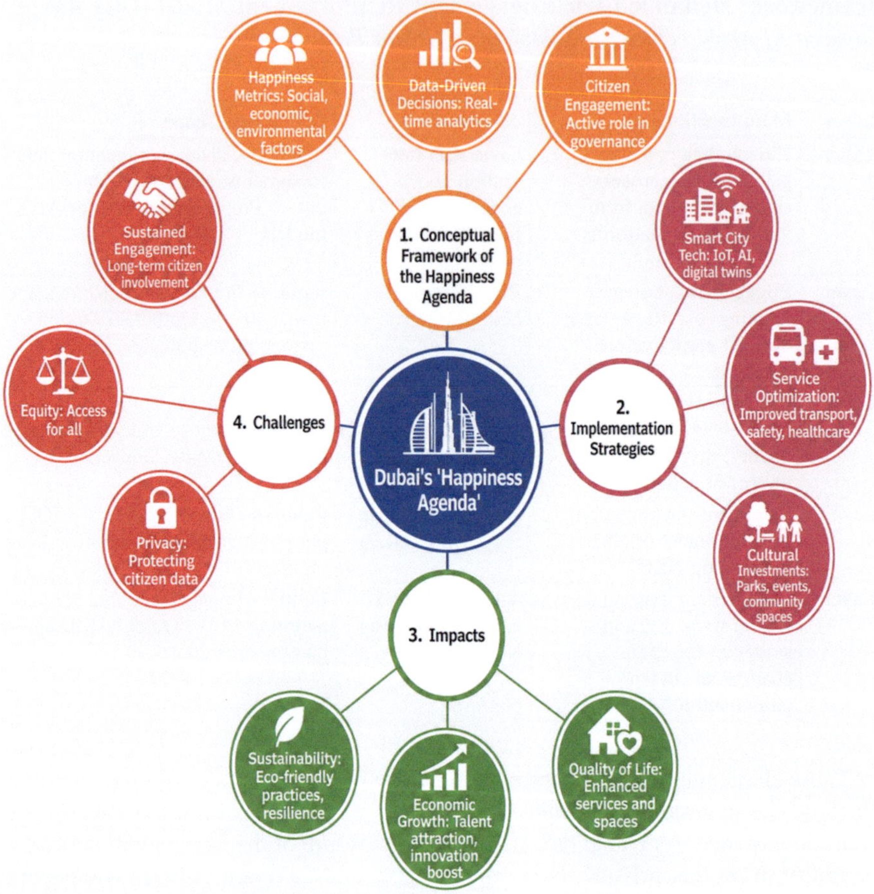

Fig. 3.1 Dubai's 'Happiness Agenda'

The "Impacts" section highlights enhanced quality of life, economic growth through talent attraction, and sustainability with eco-friendly practices. "Challenges" include privacy protection, equitable access for all citizens, and sustaining long-term citizen involvement.

Dubai's "Happiness Agenda" exemplifies a proactive approach to urban governance that prioritizes citizen well-being. Through the integration of advanced technologies, data-driven decision-making, and active citizen engagement, the city is setting a benchmark for how urban centers can foster happiness and satisfaction among residents. As Dubai continues to evolve, addressing the challenges associated with privacy, equity, and engagement will be crucial in ensuring the long-term success of the Happiness Agenda and its contribution to sustainable urban development.

3.3 Herrenberg, Germany

Herrenberg, a small town in Germany, serves as an exemplary case study of how smart city initiatives can be implemented effectively through the integration of digital twin technology. This analysis delves into the various aspects of Herrenberg's smart city strategies, focusing on the role of digital twins in urban planning, citizen engagement, and sustainability. Herrenberg is characterized by its commitment to sustainable urban development and innovative technology integration. The town has adopted a forward-looking approach to urban management, utilizing digital twin technology to create a comprehensive model of its urban environment. This initiative is part of a broader movement within Germany and Europe to enhance urban living through smart technologies, fostering resilience and sustainability (Dembski et al. 2020).

3.3.1 *Digital Twin Technology in Herrenberg*

Digital twin technology refers to the digital replica of physical entities, allowing for real-time data monitoring and simulation. In Herrenberg, the digital twin serves multiple purposes, including urban planning, infrastructure management, and environmental monitoring. For instance, to realize the potential of digital twin technology, we propose an AI-enhanced framework that combines Graph Neural Networks (GNNs) with real-time simulation modules. This approach models the town's physical infrastructure—such as roads, buildings, utilities, and environmental sensors—as a dynamic graph, where each node and edge represents a distinct part of the urban system. The AI model continuously learns from sensor data and citizen inputs to support planning decisions, infrastructure monitoring, and environmental forecasting within the digital twin.

Each infrastructure component is represented as a node in the graph. To simulate interactions among them (e.g., between a traffic signal and road congestion), the GNN updates each node's state by aggregating information from its connected neighbors:

$$h_v^{(t+1)} = \sigma\left(\sum_{u \in \mathcal{N}(v)} W h_u^{(t)} + b\right)$$

where $h_v^{(t+1)}$, updated feature representation of infrastructure node v at time $t + 1$, $\mathcal{N}(v)$, set of nodes directly connected to v (e.g., adjacent utilities or road segments), W, learnable weight matrix defining the influence of neighbor features, $h_u^{(t)}$, feature vector of neighbor node u at time t, b, bias term, and σ, nonlinear activation function introducing flexibility. This function enables each infrastructure node to learn from its surroundings, modeling the dynamic flow of energy, transport, or water across the town's physical layout. These evolving node states are then used to simulate system-level behavior within the digital twin environment.

To assess how the overall urban system behaves, especially under hypothetical planning scenarios or emergencies, the aggregated state of the entire graph is passed through a decision function:

$$\hat{y}_t = \text{MLP}\left(\text{READOUT}\left(\left\{h_v^{(t)} \mid v \in V\right\}\right)\right)$$

where $\hat{y}_t$, predicted outcome for the urban system at time t (e.g., traffic congestion level, energy load, environmental risk), $h_v^{(t)}$, current state of each infrastructure node v, V, full set of nodes in the city graph, READOUT, permutation-invariant aggregation function (such as mean or sum), and MLP, multilayer perceptron used for final prediction. This output can be visualized directly within Herrenberg's digital twin interface, providing city planners with up-to-date, interpretable results that help them evaluate potential interventions, such as adding a bike lane, adjusting waste pickup schedules, or reconfiguring traffic routes.

To adapt the model in real time, especially as weather changes or citizen reports are received through digital platforms, we integrate a temporal learning module based on gated recurrence:

$$h_v^{(t+1)} = f_{\text{GRU}}\left(h_v^{(t)}, x_v^{(t)}\right)$$

where $h_v^{(t+1)}$, predicted future state of node v, $h_v^{(t)}$, current hidden state, $x_v^{(t)}$, new input from environmental or citizen data at time t, and f_{GRU}, a Gated Recurrent Unit that allows the system to learn time-sensitive changes in infrastructure behavior.

This final step ensures the model stays responsive to continuous data input—allowing the city's digital twin not just to reflect the current city state, but to predict and adapt to what will happen next.

The use of Graph Neural Networks with GRU-based forecasting transforms digital twin technology from a static replica into an adaptive intelligence system. It empowers planners to visualize the consequences of urban changes before they are made, prioritize resources more effectively, and engage citizens with transparent, data-backed decision tools.

Framework: AI-Enhanced Urban System Modeling (AIEUSM)
Layered AI Architecture for Real-Time Urban Management

Layer	Main functionality	AI module(s) and components	Operational scope
Layer 1	Urban data collection: gather real-time data from sensors and citizen inputs for urban planning	Real-time urban data ingestion and harmonization platform	Input → Sensor data, citizen inputs → Output → Preprocessed data for AI models

(continued)

Layer	Main functionality	AI module(s) and components	Operational scope
Layer 2	Graph representation: model urban infrastructure as a dynamic graph using GNNs	Graph Neural Networks (GNN)	Input → Preprocessed data → Output → Dynamic graph representation of infrastructure
Layer 3	Node state update: aggregate information from connected nodes to update infrastructure states	Graph Neural Networks (GNN)	Input → Neighbor node data → Output → Updated node state for infrastructure
Layer 4	System prediction: use MLP to aggregate node information and predict system-level outcomes	Multilayer Perceptron (MLP)	Input → Node state data → Output → Predicted system outcomes (e.g., traffic, energy)
Layer 5	Temporal learning: adapt the model in real-time using GRU to learn from time-sensitive data	Gated Recurrent Unit (GRU)	Input → Environmental/citizen data → Output → Real-time model adaptation

Governance lens (Herrenberg). Herrenberg's digital twin, as framed here under AIEUSM and its downstream modules, presupposes a mixed data ecology spanning planning-grade geospatial layers and infrastructural telemetry, alongside civic inputs that become analytically actionable through the participation pathway discussed below. In governance terms, "privacy and data governance" should not be treated as a purely abstract desideratum, but as the operational management of data access, data quality, and data privacy across the lifecycle of the AI-enabled urban system (Papagiannidis et al. 2025). Since digital twin smart city deployments typically integrate heterogeneous data streams and institutional stakeholders (Dani et al. 2023; El-Agamy et al. 2024), we propose that the Herrenberg configuration be strengthened by an explicit data inventory that distinguishes infrastructural telemetry from citizen-contributed inputs, specifies purpose of use and the relevant institutional or legal basis, assigns named stewardship responsibilities per data domain, and defines tiered access regimes so that purpose limitation and provenance remain inspectable as the system scales. Where civic input is computationally consequential rather than merely consultative, consent and transparency should be treated as constitutive design constraints, not optional "add-ons," particularly when citizen-facing channels feed directly into analytics and decision support (Papagiannidis et al. 2025).

The modeling plurality embedded in the stack increases capability, yet it also multiplies the points at which accountability can fail if evidence is not standardized across modules. A pragmatic anchor is therefore to institutionalize documentation as audit evidence: adoption in high-scrutiny settings often turns on whether oversight bodies can inspect human influence and oversight, evaluate data quality through documentation of data collection and preparation, and verify how AI is used in concrete decision processes (Konigstorfer and Thalmann 2022). To make such evidence portable across the AIEUSM stack, structured documentation artifacts can be mandated at system or module level, aligned with compliance-oriented documentation logic that represents intended use, context of use, technical specification, and risk management in both human- and machine-readable form (Golpayegani et al. 2024). In parallel, robustness-oriented monitoring should be framed as an enforceable operational requirement: procedural governance practices that include ongoing evaluation, compliance monitoring, and review processes are specifically motivated by the need to mitigate and correct unintended risks under real operational pressure (Papagiannidis et al. 2025). On that basis, drift and stability SLOs can be introduced here as a concrete engineering instantiation of the review-and-correction requirement, rather than as a claim attributed verbatim to the cited literature.

Because the twin is positioned as an instrument that can shape planning choices, infrastructure interventions, and environmental alerts, decision rights should be made explicit rather than tacit. A minimally defensible arrangement distinguishes (1) who authorizes deployments and parameter changes, (2) who validates high-impact interventions suggested by the twin, and (3) who holds override authority under contestation or emergency. This separation is a precondition for meaningful public accountability, since the participatory promise of civic input becomes credible only when residents can understand how inputs are used, how trade-offs are decided, and how decisions are reviewed and revised over time, especially where automated recommendations become materially consequential (Konigstorfer and Thalmann 2022).

3.3.2 Urban Planning and Development

Herrenberg's digital twin is utilized extensively for urban planning purposes. The digital model integrates various data sources, including geographic information systems (GIS), sensor data, and public feedback, to create a virtual representation of the town. Planners can simulate potential developments and assess their impacts on the community, ensuring that new projects align with the town's sustainability goals and citizens' needs (Dembski et al. 2020). For instance, to fully leverage the digital twin for participatory and sustainable urban planning, we integrate Spatial-Temporal Graph Attention Networks (ST-GATs). These models expand on standard GNNs by adding temporal learning and adaptive attention mechanisms, allowing planners to weigh the importance of spatial and citizen-derived features over time. This AI framework enables planners to evaluate the projected outcomes of future developments, considering not only infrastructure constraints but also dynamic social and environmental signals.

The core mechanism begins by learning the attention-based interaction between urban features in the digital twin. The attention-weighted message passing across the graph is formulated as:

$$h_v^{(l+1)} = \sigma\left(\sum_{u \in \mathcal{N}(v)} \alpha_{vu}^{(l)} W^{(l)} h_u^{(l)}\right)$$

where $h_v^{(l+1)}$, updated node feature of urban unit v at layer $l + 1$, $\mathcal{N}(v)$, neighborhood nodes in the spatial graph (e.g., adjacent buildings or roadways), $\alpha_{vu}^{(l)}$, attention coefficient that determines the influence of node u on node v, $W^{(l)}$, learnable transformation weight matrix at layer l, and σ, activation function. The attention coefficient $\alpha_{vu}^{(l)}$ is itself learned from both spatial proximity (e.g., GIS distance) and citizen-prioritized importance (e.g., high-feedback zones), allowing the model to differentiate between routine and socially sensitive areas.

As urban developments often involve future planning scenarios, the model also includes a temporal component that forecasts future graph states across different planning phases. This is done through a gated temporal aggregation:

$$h_v^{(t+1)} = f_{\text{GRU}}\left(h_v^{(t)}, z_v^{(t)}\right)$$

where $h_v^{(t+1)}$, predicted node state for planning scenario at future time $t + 1$, $h_v^{(t)}$, current embedded feature of spatial unit v, $z_v^{(t)}$, temporal signal vector including citizen feedback scores, sensor readings, and infrastructure load estimates, and f_{GRU}, gated recurrent unit that learns time dependencies and event dynamics.

Once spatial and temporal updates are integrated, the digital twin model outputs projected impacts of proposed urban developments:

$$\hat{y}_{\text{impact}} = \text{MLP}\left(\text{READOUT}\left(\left\{h_v^{(T)} \mid v \in V\right\}\right)\right)$$

where $\hat{y}_{\text{impact}}$, predicted socio-environmental impact score of a proposed plan, $h_v^{(T)}$, final updated node features after simulation time window T, V, full set of nodes (e.g., buildings, roads, parks) in the digital twin graph, READOUT, aggregation function over spatial units, and MLP, multilayer perceptron translating node features into a measurable impact index.

This AI model allows city planners to assess proposals not only in terms of infrastructure capacity but also in terms of equity, livability, and public approval. For example, a proposed housing development can be evaluated for its effects on nearby traffic nodes, green space access, and social feedback from previously underrepresented neighborhoods—all before a single brick is laid.

By embedding Spatial-Temporal Graph Attention Networks into the digital twin, the city's planning ecosystem becomes both predictive and participatory. The city gains a powerful decision-support system that models infrastructure behaviors, anticipates urban dynamics, and aligns future developments with sustainability goals and resident well-being. This ensures that smart planning is not just data-driven, but socially informed and environmentally balanced, an exemplary model for future-ready small cities across Europe.

Framework: Spatial-Temporal Urban Planning AI Model (STUPAM)
Layered AI Architecture for Participatory and Sustainable Urban Development

Layer	Main functionality	AI module(s) and components	Operational scope
Layer 1	Urban data integration: collect GIS, sensor data, and public feedback for urban planning	Urban data ingestion, harmonization, and quality control platform	Input → GIS data, sensor data, citizen feedback → Output → Integrated urban data
Layer 2	Graph representation: model urban infrastructure using Spatial-Temporal Graph Attention Networks (ST-GATs)	Spatial-Temporal Graph Attention Networks (ST-GATs)	Input → Integrated data → Output → Dynamic graph representation with spatial and temporal features
Layer 3	Attention mechanism: aggregate spatial and citizen-derived features using attention coefficients	Spatial-Temporal Graph Attention Networks (ST-GATs)	Input → Graph node features → Output → Attention-weighted node feature updates
Layer 4	Temporal forecasting: learn time dependencies and predict future states using GRUs	Gated Recurrent Unit (GRU)	Input → Node features, citizen feedback → Output → Predicted future states of infrastructure
Layer 5	Impact prediction: evaluate socio-environmental impacts using an MLP-based decision function	Multilayer Perceptron (MLP)	Input → Node feature data → Output → Projected socio-environmental impact score

This modeling capability allows stakeholders to visualize changes in real-time, providing insights into traffic patterns, environmental impacts, and the effectiveness of public services. For instance, by simulating traffic scenarios, planners can optimize road networks and public transport routes before any physical changes are made, thereby reducing potential disruptions and enhancing mobility (Dembski et al. 2020). To transform digital twin outputs into operational improvements—particularly in urban mobility—Reinforcement Learning (RL) is integrated into the city's simulation framework. Once the spatial-temporal graph representation of the city is learned:

$$G = (V, E)$$

where *G*, graph representation of the city's infrastructure, *V*, set of nodes representing urban elements such as roads, buildings, and traffic signals, *E*, set of edges representing interactions between nodes, such as traffic flow or connectivity. This function enables the model to represent the city's infrastructure in a graph format, where each node's state can be dynamically updated based on both spatial and temporal data.

Next, an RL agent is deployed to interact with the environment and evaluate the best traffic management actions. These include route reassignments, signal timing adjustments, and congestion mitigation strategies, all based on forecasted patterns from the digital twin.

The agent operates under the following value function, which estimates the long-term benefits of each urban mobility action in real time:

$$V^{\pi}(s_t) = \mathbb{E}_{\pi}\left[\sum_{k=0}^{\infty} \gamma^k R(s_{t+k}, a_{t+k})\right]$$

where $V^{\pi}(s_t)$, expected value of traffic state s_t under policy π, γ, discount factor weighting future rewards, $R(s_{t+k}, a_{t+k})$, reward from taking action a_{t+k} in state s_{t+k}, and $\mathbb{E}_{\pi}$, expected value under policy π. The function evaluates how routing or service adjustments affect not only current traffic flow but long-term system performance—such as energy efficiency, wait times, and air quality. Actions are learned and updated using a policy optimization process based on Proximal Policy Optimization (PPO) to ensure safe and stable training:

$$\mathcal{L}^{\mathrm{PPO}}(\theta) = \mathbb{E}_t[\min(r_t(\theta)A_t, \mathrm{clip}(r_t(\theta), 1-\epsilon, 1+\epsilon)A_t)]$$

where $\mathcal{L}^{\mathrm{PPO}}(\theta)$, objective function to optimize policy parameters θ, $r_t(\theta)$, policy ratio of new to old probabilities, A_t, advantage function indicating the value of action a_t, and ϵ, clipping threshold to limit policy changes.

Each time the RL agent proposes actions such as rerouting vehicles or modifying a bus line—the digital twin immediately simulates their effects based on the graph representation and real-time data. This integrated loop allows planners to preview the impact of policy changes before they are implemented physically, minimizing disruption and improving public service responsiveness.

This AI-enhanced simulation framework ensures that digital twins are not just descriptive, but prescriptive—actively supporting decision-making across traffic management, environmental policy, and service delivery. The seamless interaction between graph-based planning and reinforcement learning agents provides the town with a practical, adaptable system for real-time optimization of urban life, reinforcing its status as a leading example of small-scale smart city transformation.

Framework: RL-Driven Urban Mobility Optimization (RL-UMO)
Layered AI Architecture for Traffic and Public Service Optimization

Layer	Main functionality	AI module(s) and components	Operational scope
Layer 1	Urban data integration: collect real-time data for urban mobility optimization	Mobility data integration and streaming analytics platform	Input → Real-time data (traffic, sensors, public feedback) → Output → Integrated urban data

(continued)

Layer	Main functionality	AI module(s) and components	Operational scope
Layer 2	Graph representation: model the city's traffic network using spatial-temporal graphs	Spatio-temporal graph construction and embedding service	Input → Integrated data → Output → Dynamic spatial-temporal graph representation
Layer 3	Action evaluation: use RL to optimize traffic management actions such as route reassignments	Reinforcement learning (RL)	Input → Graph representation, traffic scenarios → Output → Evaluated traffic management actions
Layer 4	Policy optimization: apply PPO to optimize traffic management policies and ensure stable training	Proximal Policy Optimization (PPO)	Input → RL agent, policy data → Output → Optimized policy parameters using PPO
Layer 5	Impact simulation: predict the effects of actions based on real-time data and graph representations	Traffic impact simulation and performance forecasting engine	Input → Real-time simulation data, traffic actions → Output → Impact projections for traffic flow and public service

3.3.3 *Infrastructure Management*

The digital twin also plays a critical role in managing Herrenberg's infrastructure. By integrating data from various sources, the town can monitor the condition of its roads, utilities, and public spaces in real-time. This proactive approach enables timely maintenance interventions, reducing the risk of infrastructure failure and ensuring the reliability of urban services (Yigitcanlar and Cugurullo 2020).

For example, sensors embedded in roadways can detect wear and tear, allowing maintenance teams to address issues before they escalate into costly repairs or safety hazards. This data-driven approach enhances the overall efficiency of urban management and promotes sustainable infrastructure practices. To operationalize this real-time infrastructure monitoring, we propose a predictive maintenance framework powered by Extreme Gradient Boosting (XGBoost), trained on historical degradation patterns, current sensor inputs, and digital twin simulation data. Each road segment, utility line, or structural element is modeled as an entity with condition features that evolve over time. These features are learned from embedded sensors that measure vibration, temperature, traffic load, and surface irregularities.

The predicted degradation index at time t is expressed as:

$$\hat{y}_t = \sum_{k=1}^{K} f_k(x_t), f_k \in \mathcal{F}$$

where $\hat{y}_t$, predicted degradation score of a given infrastructure component at time t, x_t, feature vector including current sensor readings, material age, usage load, and environmental conditions, f_k, decision tree at stage k in the XGBoost ensemble, K, total number of trees, and FF, the space of regression functions.

The degradation score $\hat{y}_t$ is then used to estimate time-to-failure and prioritize infrastructure inspections. If the score exceeds a predefined threshold, the system flags the component for preventive action. To enhance reliability and avoid overfitting, the model's objective function includes a regularization term:

$$\mathcal{L}(\phi) = \sum_{t=1}^{T} l(y_t, \hat{y}_t) + \sum_{k=1}^{K} \Omega(f_k)$$

where $\mathcal{L}(\phi)$, total training loss, $l(y_t, \hat{y}_t)$, difference between observed and predicted degradation for time t, and $\Omega(f_k)$, regularization function that penalizes tree complexity to prevent model overfitting across heterogeneous infrastructure types.

In practice, this model is embedded in the digital twin and runs continuously on incoming sensor data. For example, if embedded roadway sensors detect increased vibration frequency and declining traction values, the model can forecast pavement deterioration several weeks in advance. Maintenance teams receive early warnings and can act preventively, minimizing disruption and cost.

Moreover, the model helps identify infrastructure segments with abnormal wear relative to expected baselines, signaling potential systemic issues such as overloaded utility corridors or high-impact transit zones. These insights are visualized directly in the digital twin interface, enabling data-driven prioritization of investment and repairs.

This AI model transforms real-time sensor inputs into actionable intelligence, allowing public service teams to intervene early, extend asset life cycles, and reduce carbon footprints associated with emergency repairs. The outcome is a smarter, safer, and more sustainable infrastructure system—tailored not only to operational efficiency but also to the long-term resilience of the urban environment.

Framework: Predictive Maintenance for Infrastructure Monitoring (PMIM)
Layered AI Architecture for Predictive Infrastructure Maintenance

Layer	Main functionality	AI module(s) and components	Operational scope (Input → Processing → Output)
Layer 1	Data collection: gather real-time sensor data from infrastructure components	Infrastructure sensor ingestion and signal conditioning pipeline	Input → Real-time sensor data (vibration, temperature, load) → Output → Preprocessed input for AI models
Layer 2	Learning: XGBoost for degradation prediction and maintenance decision-making	Extreme Gradient Boosting (XGBoost)	Input → Historical data, sensor streams → Output → Failure prediction and maintenance decisions
Layer 3	Optimization: regularization for model performance and generalization	Structural risk minimization and complexity control service	Input → Model parameters, sensor data → Output → Regularized XGBoost model for optimal prediction

(continued)

Layer	Main functionality	AI module(s) and components	Operational scope (Input → Processing → Output)
Layer 4	Monitoring: continuous assessment of infrastructure condition	Asset health dashboards and anomaly surveillance platform	Input → Real-time sensor data → Output → Continuous monitoring reports and decision recommendations
Layer 5	Reporting: actionable feedback for maintenance teams	Maintenance workorder prioritization and dispatch orchestrator	Input → Degradation score, real-time data → Output → Maintenance alerts and intervention prioritization

3.3.4 Environmental Monitoring

Herrenberg's digital twin also facilitates comprehensive environmental monitoring. By collecting data on air quality, noise levels, and green space usage, the town can assess its environmental performance and implement targeted interventions. For instance, if air quality data indicates high pollution levels in a specific area, the town can explore solutions such as increased greenery or improved traffic management to mitigate these issues (Dembski et al. 2020). For example, to analyze and forecast environmental conditions with high spatial resolution, we integrate Gaussian Process Regression (GPR) into its digital twin framework. GPR is particularly suitable for environmental applications due to its ability to model uncertainty and spatial correlations in data collected from heterogeneous urban sensors. The model operates over a continuous domain, estimating environmental values (e.g., air pollution levels) not just at sensor locations but across all points in urban space.

At its core, GPR predicts the environmental condition $\hat{y}(x_*)$ at an unsampled location x_* using observations from a set of nearby sensor points:

$$\hat{y}(x_*) = k(x_*, X)^{\top} \left[K(X, X) + \sigma_n^2 I\right]^{-1} y$$

where $\hat{y}(x_*)$, predicted environmental value (e.g., air quality index) at location x_*, X, matrix of known sensor locations, y, vector of observed environmental values (e.g., PM2.5 readings), $k(x_*, X)$, covariance vector between test point x_* and known points X, $K(X, X)$, covariance matrix of all training points, $\sigma_n^2 I$, noise variance, and I, identity matrix for regularization. This function enables the model to infer values at locations where no sensor is present, by learning from the spatial relationships of environmental variables. For instance, when air quality data is sparse in a residential zone, the GPR model can still estimate pollution levels with confidence intervals, guiding targeted interventions even in under-monitored areas.

To extend the model's forecasting ability, especially in identifying future risk zones, the GPR formulation is augmented with a temporal kernel that captures changes in environmental variables over time:

$$k((x,t),(x',t')) = k_{\text{space}}(x,x') \cdot k_{\text{time}}(t,t')$$

where $k((x,t),(x',t'))$, combined spatio-temporal covariance function, $k_{\text{space}}(x,x')$, spatial kernel representing similarity between locations, and $k_{\text{time}}(t,t')$, temporal kernel modeling trends across time intervals.

This allows the city's environmental monitoring system to not only understand the current spatial distribution of pollution or noise but also anticipate changes due to weather shifts, seasonal effects, or urban activity patterns.

The predicted results are rendered on the digital twin interface in the form of dynamic heatmaps or risk zones. If air pollution levels in a certain quadrant are predicted to exceed safety thresholds in the next 48 h, city officials can respond in advance—by adjusting traffic flows, increasing vegetation buffers, or issuing early warnings to vulnerable populations. This AI model empowers the town to proactively improve air quality, reduce noise pollution, and optimize green infrastructure planning. Most importantly, it transforms environmental sensing into a forward-looking system that aligns perfectly with the town's sustainability goals and resilience strategies, as envisioned in Europe's broader smart city roadmap.

Framework: Environmental Monitoring and Forecasting for Herrenberg (EMFH)

Layered AI Architecture for Environmental Monitoring and Forecasting

Layer	Main functionality	AI module(s) and components	Operational scope (Input → Processing → Output)
Layer 1	Data collection: gather real-time environmental sensor data	Environmental sensor integration and signal conditioning pipeline	Input → Environmental sensor data (air quality, noise, green space) → Output → Preprocessed input for AI models
Layer 2	Learning: Gaussian Process Regression (GPR) for environmental condition prediction	Gaussian Process Regression (GPR)	Input → Sensor data, spatial data → Output → Predicted environmental conditions at unsampled locations
Layer 3	Optimization: spatio-temporal kernel for forecasting future environmental conditions	Spatio-temporal kernel learning and forecasting engine	Input → GPR predictions, temporal data → Output → Forecasted environmental conditions over time
Layer 4	Monitoring: continuous environmental condition assessment	Environmental state surveillance and anomaly detection platform	Input → Real-time environmental sensor data → Output → Continuous monitoring reports and predictive insights
Layer 5	Reporting: actionable feedback for environmental intervention	Intervention recommendation and regulatory compliance alerting service	Input → Predicted environmental values, safety thresholds → Output → Environmental alerts and intervention prioritization

3.3.5 Citizen Engagement and Participation

A key feature of Herrenberg's approach to smart city development is the emphasis on citizen engagement. The digital twin technology enables a participatory governance model where residents can actively contribute to urban planning and decision-making processes.

Herrenberg has developed platforms that allow citizens to provide feedback on urban projects and policies. By integrating these platforms with the digital twin, residents can visualize proposed changes and express their opinions more effectively. This engagement fosters a sense of community ownership and ensures that urban development reflects the desires and needs of residents (Yigitcanlar and Cugurullo 2020). To turn public feedback into structured, actionable input for the digital twin, we propose an AI framework based on Transformer-based Natural Language Processing (NLP), particularly using Bidirectional Encoder Representations from Transformers (BERT). This model processes unstructured citizen input—such as open comments, policy suggestions, and sentiment-laden feedback—and maps it onto urban planning categories and location-specific concerns. Each citizen input s_i is transformed into a vectorized representation using BERT embeddings:

$$\mathrm{e}_i = \mathrm{BERT}(s_i)$$

where e_i, the semantic embedding of the citizen's statement s_i, capturing both the linguistic context and emotional tone of the message.

These embeddings are then clustered and classified into relevant urban planning themes (e.g., mobility, green space, safety) using a softmax-based classification function:

$$\hat{y}_i = \mathrm{softmax}(W\mathrm{e}_i + b)$$

where $\hat{y}_i$, the probability distribution over planning categories for feedback i, W, learnable weight matrix, and v, bias vector. The output helps identify what aspect of city planning the citizen comment refers to and allows the system to route that input to the corresponding simulation module within the digital twin.

To reflect community sentiment in planning scenarios, a sentiment-aware weighting function is applied to modify citizen engagement scores based on emotional tone and urgency:

$$w_i = \tanh\left(\alpha \cdot s_i^{\mathrm{sentiment}}\right)$$

where w_i, engagement weight for feedback i, $s_i^{\mathrm{sentiment}}$, sentiment score derived from NLP sentiment analysis (ranging from negative to positive), and α, scaling factor controlling sentiment influence.

This engagement weight w_i is then used to prioritize highly relevant or emotionally charged community responses in digital twin simulations. For instance, if feedback from multiple citizens in a specific district expresses strong negative sentiment toward a proposed zoning change, the system amplifies this input in the decision model, prompting planners to explore alternative scenarios.

All processed feedback is visualized directly in the digital twin interface, allowing residents to see how their voices shape the city. This closes the loop between participation and outcome, transforming civic engagement from symbolic involvement to direct co-creation.

This approach ensures that public input is not only collected but meaningfully interpreted and applied in planning simulations. It promotes transparency, accountability, and trust between residents and decision-makers—reinforcing the principle that smart cities must be shaped not only by data but by the lived experiences and aspirations of their communities.

Framework: Citizen Engagement for Urban Planning (CEUP)

Layered AI Architecture for Citizen Feedback and Urban Planning

Layer	Main functionality	AI module(s) and components	Operational scope (Input → Processing → Output)
Layer 1	Data collection: gather citizen feedback through platforms integrated with the digital twin	Multichannel feedback ingestion and text normalization pipeline	Input → Citizen feedback (open comments, suggestions, sentiment) → Output → Raw input for processing
Layer 2	Learning: BERT-based NLP for semantic embedding of citizen input	Bidirectional Encoder Representations from Transformers (BERT)	Input → Raw citizen feedback → Output → Semantic embeddings representing context and emotional tone
Layer 3	Classification: Softmax-based categorization of feedback into urban planning themes	Thematic classification and topic modeling engine	Input → BERT embeddings → Output → Categorized citizen feedback (mobility, green space, safety)
Layer 4	Sentiment-aware weighting: modify citizen engagement scores based on emotional tone	Emotion-weighted engagement scoring and prioritization module	Input → Sentiment score → Output → Engagement weight reflecting emotional urgency and tone
Layer 5	Reporting: prioritize and visualize citizen feedback in the digital twin interface	Interactive feedback visualization and insight generation platform	Input → Engagement weight, categorized feedback → Output → Visualized feedback on digital twin interface, actionable insights for planning

The town also invests in educational initiatives to familiarize citizens with digital twin technology and its benefits. Workshops, community meetings, and online resources help demystify the technology and encourage participation in the smart city agenda. By empowering residents with knowledge, Herrenberg strengthens community ties and builds a collaborative environment for urban governance (Dembski et al. 2020).

3.3.6 Sustainability and Resilience

Herrenberg's digital twin initiatives align closely with broader sustainability goals. By optimizing resource usage and enhancing the efficiency of urban systems, the town aims to create a more resilient urban environment. The digital twin allows for more efficient management of resources such as water, energy, and waste. By analyzing consumption patterns and identifying areas for improvement, Herrenberg can implement strategies that minimize waste and reduce the carbon footprint of urban activities (Yigitcanlar and Cugurullo 2020).

To support these goals with real-time, adaptive decision-making, we propose a Deep Q-Learning (DQL) framework designed for multi-objective resource optimization. The model is embedded in the digital twin and operates across multiple utility sectors (e.g., electricity, water distribution, and waste systems). At each time step, the agent interacts with the simulated environment, making decisions that aim to reduce environmental impact while preserving operational quality. The expected return for a given system state s_t and action a_t is modeled through the Q-function:

$$Q(s_t, a_t) = \mathbb{E}\left[\sum_{k=0}^{\infty} \gamma^k R(s_{t+k}, a_{t+k})\right]$$

where $Q(s_t, a_t)$, the expected cumulative reward of taking action a_t in state s_t, γ, the discount factor balancing immediate versus long-term sustainability outcomes, and $R(s_{t+k}, a_{t+k})$, the reward function reflecting environmental efficiency and resource balance.

The reward function is defined as a multi-objective trade-off among three key sustainability metrics:

$$R(s_t, a_t) = -\alpha \cdot f_{\text{energy}}(s_t, a_t) - \beta \cdot f_{\text{waste}}(s_t, a_t) - \delta \cdot f_{\text{emissions}}(s_t, a_t)$$

where $f_{\text{energy}}(s_t, a_t)$, projected energy usage for a given policy or urban adjustment, $f_{\text{waste}}(s_t, a_t)$, estimated waste production from the action taken, $f_{\text{emissions}}(s_t, a_t)$, anticipated emissions generated, and α, β, δ weighting coefficients that reflect the town's current sustainability priorities.

To learn the optimal action selection policy, the Deep Q-Network is trained using the temporal difference loss:

$$\mathcal{L}(\theta) = \left[R(s_t, a_t) + \gamma \max_{a'} Q(s_{t+1}, a'; \theta^-) - Q(s_t, a_t; \theta)\right]^2$$

where $\mathcal{L}(\theta)$, the loss function used to train the neural network with parameters θ, θ^-, the parameters of the target network used for stability, and $Q(s_{t+1}, a'; \theta^-)$, the next state's maximum predicted value used to update the current estimate.

This approach enables the town to continuously balance multiple sustainability objectives based on real-time data. For example, if sensor data shows elevated energy use in a sector, the DQL agent may recommend redistributing load, adjusting public lighting schedules, or optimizing HVAC systems in public buildings—all while keeping emissions and waste output within acceptable bounds.

The AI model allows the town to pursue resilience through dynamic, data-informed trade-offs—ensuring that every operational decision contributes to long-term environmental goals, resource efficiency, and smart urban stewardship.

Framework: Multi-Objective Resource Optimization for Smart Cities (MORO)
Layered AI Architecture for Resource Optimization and Sustainability

Layer	Main functionality	AI module(s) and components	Operational scope (Input → Processing → Output)
Layer 1	Data collection: gather real-time data from resource consumption systems	Resource telemetry integration and feature engineering pipeline	Input → Real-time resource data (energy, water, waste) → Output → Preprocessed input for AI models
Layer 2	Learning: deep Q-learning for multi-objective resource optimization	Deep Q-Learning (DQL)	Input → Real-time consumption data → Output → Resource optimization decisions
Layer 3	Optimization: multi-objective reward function for balancing energy, waste, and emissions	Multi-objective reward design and policy trade-off engine	Input → Resource consumption data, sustainability priorities → Output → Optimized multi-objective actions
Layer 4	Monitoring: continuous assessment of resource usage and sustainability performance	Sustainability performance analytics and threshold alerting platform	Input → Optimized resource usage actions → Output → Continuous monitoring reports and performance metrics
Layer 5	Reporting: visualization and action recommendation for resource management	Decision-support dashboards and executive action recommendation service	Input → Resource optimization results, sustainability goals → Output → Actionable insights and visual reports for decision-makers

3.3.7 Climate Resilience

The integration of digital twin technology also enhances Herrenberg's ability to respond to climate challenges. By modeling different climate scenarios, the town can assess potential risks and develop adaptation strategies to protect infrastructure and community well-being. This proactive approach ensures that Herrenberg remains resilient in the face of changing environmental conditions (Dembski et al. 2020). For instance, to strengthen climate resilience, we propose a Bayesian Deep Learning framework that supports simulation under uncertainty by generating probabilistic forecasts of climate-related risks. The model operates over a spatio-temporal grid that includes environmental inputs such as temperature trends, rainfall intensity, floodplain elevation, soil moisture, and surface runoff dynamics.

The deep neural network learns predictive functions for each climate impact variable, while Monte Carlo Dropout is applied at inference time to approximate uncertainty in the model's outputs. The expected prediction for a given spatial region x and time t is computed as:

$$\hat{y}_t(x) = \frac{1}{M} \sum_{m=1}^{M} f_{\mathrm{w}_m}(x, t)$$

where $\hat{y}_t(x)$, expected value of the climate impact variable (e.g., surface flooding or heat index) at location x and time t, f_{w_m}, predicted output of the deep network with dropout-applied weights w_m, and M, number of stochastic forward passes used to estimate uncertainty.

This formulation allows the digital twin to simulate how climate stressors may affect different parts of the town under various adaptation strategies, such as increased tree cover, green roofs, permeable surfaces, or stormwater channeling infrastructure. The standard deviation of predictions across the M passes quantifies uncertainty and is given by:

$$\sigma_{\hat{y}}^2(x, t) = \frac{1}{M} \sum_{m=1}^{M} \left(f_{\mathrm{w}_m}(x, t) - \hat{y}_t(x) \right)^2$$

where $\sigma_{\hat{y}}^2(x, t)$, predictive variance indicating the confidence interval of the forecasted climate impact, used by planners to assess risk sensitivity and resource allocation under uncertainty.

These probabilistic maps are integrated into the digital twin interface, allowing city officials to visualize future vulnerabilities and evaluate how different policies would mitigate projected risks. For instance, if high flood risk is forecasted in a low-lying district, planners can compare the simulated effectiveness of green infrastructure versus engineered drainage systems—while considering the reliability of each outcome. This AI-driven model enables the town to simulate, quantify, and respond to future risks with strategic foresight, aligning adaptation measures with both environmental realities and urban resilience planning. In doing so, the city reinforces its commitment to climate resilience, ensuring that smart development is not only efficient and inclusive but also future-proof.

Framework: Climate Resilience and Risk Forecasting for Smart Cities (CRRF)
Layered AI Architecture for Climate Resilience and Adaptation Planning

Layer	Main functionality	AI module(s) and components	Operational scope (Input → Processing → Output)
Layer 1	Data collection: gather real-time environmental data (temperature, rainfall, etc.)	Environmental data integration and quality control pipeline	Input → Environmental data (temperature, rainfall, floodplain elevation) → Output → Preprocessed input for AI models
Layer 2	Learning: Bayesian deep learning for probabilistic forecasting of climate impacts	Bayesian Deep Learning (BDL)	Input → Environmental data, climate impact variables → Output → Probabilistic climate impact predictions
Layer 3	Uncertainty estimation: Monte Carlo Dropout for approximating model uncertainty	Stochastic reliability assessment and predictive variance engine	Input → Deep learning model, stochastic forward passes → Output → Uncertainty estimates and predictive variance
Layer 4	Simulation: climate risk simulation under various adaptation strategies	Scenario-based climate impact orchestrator and adaptation strategy evaluator	Input → Climate impact predictions, adaptation strategies (green roofs, tree cover) → Output → Simulated climate stressors and impact scenarios
Layer 5	Reporting: visualization and decision support for planners based on simulated risks	Vulnerability mapping workspace and strategic decision-support portal	Input → Simulated climate risks, policy effectiveness data → Output → Visualized vulnerability maps and actionable planning recommendations

Figure 3.2 presents the structure of Herrenberg's smart city initiatives centered around digital twin technology. The central node, "Herrenberg, Germany: Digital Twin Technology in Smart City Initiatives," connects to six main areas: "Digital Twin Technology in Herrenberg," "Urban Planning and Development," "Infrastructure Management," "Environmental Monitoring," "Citizen Engagement and Participation," and "Sustainability and Resilience." "Digital Twin Technology" supports real-time monitoring and simulation for urban management. "Urban Planning and Development" uses digital twins to integrate GIS, sensor data, and public feedback for simulations in traffic and environmental impact. "Infrastructure Management" enables real-time monitoring of roads and utilities, facilitating proactive maintenance. "Environmental Monitoring" focuses on tracking air quality, noise, and green space, with targeted interventions for pollution. "Citizen Engagement and Participation" involves public platforms for feedback and educational initiatives to foster digital twin awareness. "Sustainability and Resilience" enhances resource optimization for water and energy, as well as climate resilience through scenario modeling.

Herrenberg, Germany, exemplifies how digital twin technology can be effectively leveraged to enhance urban planning, infrastructure management, and citizen engagement within the framework of smart city development. By integrating real-time data and fostering community participation, Herrenberg not only optimizes its urban services but also promotes sustainability and resilience. As cities globally face

Fig. 3.2 Herrenberg, Germany: digital twin technology in smart city initiatives

increasing challenges related to urbanization and climate change, the lessons learned from Herrenberg's innovative approach provide valuable insights for other urban centers aiming to adopt smart technologies for improved governance and quality of life.

3.4 West Cambridge, UK

West Cambridge serves as a significant case study in the integration of smart city technologies and urban planning. This area, which includes various research institutions, academic facilities, and residential developments, exemplifies how digital technologies can be harnessed to enhance urban living and foster innovation. This analysis explores the smart city initiatives in West Cambridge, focusing on the implementation of digital twins, data-driven urbanism, and the resulting impacts on sustainability and community engagement. West Cambridge is characterized by its focus on research and innovation, housing institutions like the University of Cambridge and numerous technology companies. As part of a broader initiative to develop smart city concepts, West Cambridge has sought to integrate advanced technologies into its urban planning framework, enhancing both functionality and livability. Digital twin technology is central to the smart city initiatives in West Cambridge. This technology creates a virtual representation of physical spaces, allowing for real-time monitoring, data analysis, and simulation of urban dynamics (Pitt et al. 2022).

3.4.1 Urban Planning and Development

The digital twin model in West Cambridge facilitates sophisticated urban planning processes. By integrating various data sources—such as geographic information systems (GIS), traffic patterns, and environmental data—planners can simulate urban scenarios and assess the implications of potential developments before implementation (Pitt et al. 2022). This capacity for simulation enables more informed decision-making, ensuring that new projects align with sustainability goals and community needs. For instance, to support such predictive simulation in urban planning, we propose the integration of a Multilayer Spatio-Temporal Attention Model (STAM). This AI architecture captures both the spatial relationships between urban elements (roads, parcels, districts) and the temporal dynamics of environmental and socio-economic factors (e.g., traffic volume, noise, or population flux). The model processes multi-source inputs and delivers a layered simulation of future scenarios, allowing planners to anticipate system-wide effects before physical implementation.

The spatial attention mechanism first computes the context-aware interaction between urban zones:

$$h_v^{(l+1)} = \sigma\left(\sum_{u \in \mathcal{N}(v)} \alpha_{vu}^{(l)} W^{(l)} h_u^{(l)}\right)$$

where $h_v^{(l+1)}$, updated representation of urban unit v at layer $l + 1$, $\alpha_{vu}^{(l)}$, spatial attention coefficient indicating the influence of neighboring unit u on v, $W^{(l)}$, learnable transformation weights, $h_u^{(l)}$, feature vector of u at layer l, and σ, nonlinear activation function. This attention mechanism enables the model to dynamically prioritize critical spatial interactions—for example, how a new development may increase traffic congestion in adjacent neighborhoods or affect air quality in nearby green zones.

The temporal attention layer is applied next, modeling how historical and forecasted urban metrics evolve:

$$h_t^{(v)} = \text{Softmax}\left(Q_t K_t^{\top}\right) V_t$$

where $h_t^{(v)}$, attention-weighted temporal representation for urban unit v at time t, Q_t, K_t, V_t, query, key, and value matrices generated from past time-series data such as daily vehicle counts, air pollution levels, or citizen engagement metrics.

These spatial and temporal features are fused across multiple layers, producing a comprehensive representation of the urban system under different future scenarios. The final simulation output—such as the projected environmental impact or traffic load—is given by:

$$\hat{y}_{\text{scenario}} = \text{MLP}\left(\text{READOUT}\left(\left\{h_v^{(T)} \mid v \in V\right\}\right)\right)$$

where $\hat{y}_{\text{scenario}}$, predicted system-wide outcome of a simulated development plan (e.g., congestion level, CO_2 footprint, accessibility index), $h_v^{(T)}$, final-layer representation of unit v after T simulation steps, V, set of all urban units in the planning region, READOUT, aggregation function (e.g., weighted sum or average), and MLP, multilayer perceptron that maps the fused representation into interpretable urban impact metrics.

With this model embedded in the digital twin interface, planners can evaluate not only how a proposed infrastructure project will function physically, but also how it interacts with ongoing urban dynamics. For example, a new residential block can be assessed for its downstream effects on public transportation, green space pressure, and neighborhood carbon intensity—all with temporal depth and spatial granularity.

This AI-enhanced simulation allows decision-makers to visualize and optimize future urban states before construction begins, supporting a planning strategy that is data-informed, resilient, and deeply connected to the lived reality of urban systems.

Framework: Spatio-Temporal Urban Planning Simulation (STUPS)
Layered AI Architecture for Predictive Urban Planning and Development

Layer	Main functionality	AI module(s) and components	Operational scope (Input → Processing → Output)
Layer 1	Data collection: integrating GIS, traffic, and environmental data for simulation	Urban data harmonization and feature engineering pipeline	Input → Multi-source urban data (GIS, traffic patterns, environmental data) → Output → Preprocessed data for AI models
Layer 2	Learning: spatio-temporal attention mechanism for context-aware urban interactions	Spatio-temporal attention Model (STAM)	Input → Spatial data, temporal dynamics → Output → Context-aware interaction representations for urban units

(continued)

Layer	Main functionality	AI module(s) and components	Operational scope (Input → Processing → Output)
Layer 3	Attention: temporal attention for forecasting and evolving urban metrics	Time-series context modeling and sequence feature weighting engine	Input → Temporal urban data (e.g., traffic volume, air pollution) → Output → Temporal attention-weighted features
Layer 4	Simulation: final simulation output for urban development scenarios	Urban scenario synthesis and impact forecasting engine	Input → Fused spatial and temporal features → Output → Predicted system-wide urban outcomes (e.g., congestion, CO_2 footprint)
Layer 5	Reporting: visualizing simulation results and evaluating potential urban interventions	Interactive planning insights dashboard and decision support portal	Input → Simulation outcomes (congestion, environmental impacts) → Output → Visualized insights for urban planners and decision-makers

3.4.2 Infrastructure Management

Digital twins also support efficient infrastructure management within West Cambridge. Real-time monitoring of utilities, transportation systems, and public spaces allows for proactive maintenance and optimized resource allocation (Pitt et al. 2022). For example, sensors embedded in roadways can provide data on wear and tear, enabling timely interventions and reducing overall maintenance costs. To support such predictive maintenance and decentralized infrastructure monitoring, we propose the integration of a Federated Edge-Based Learning Framework powered by Gradient-Boosted Predictive Models (XGBoost). This system operates across distributed edge devices embedded in infrastructure components—such as roads, pipelines, or street lighting—where real-time sensor data is analyzed locally while contributing to a shared global model. This preserves privacy, reduces communication latency, and enhances fault prediction accuracy in smart urban systems. Each local node ii maintains its own predictive model for infrastructure degradation:

$$\hat{y}_i^{(t)} = \sum_{k=1}^{K} f_{i,k}\left(x_i^{(t)}\right), f_{i,k} \in \mathcal{F}$$

where $\hat{y}_i^{(t)}$, predicted degradation score of component i at time t, $x_i^{(t)}$, sensor-derived feature vector (e.g., vibration, temperature, load), $f_{i,\ k}$, local gradient-boosted tree function for node i, and K, total number of boosting rounds.

Local models are periodically synchronized with a global federated model using an aggregation step that averages shared model updates:

$$\theta_t^{\text{global}} = \frac{1}{N} \sum_{i=1}^{N} \theta_t^{(i)}$$

where $\theta_t^{(i)}$, parameter vector of the local model at edge node i at time t, and θ_t^{global}, updated global model parameters shared across the network of N edge nodes.

The predicted degradation scores $\hat{y}_i^{(t)}$ are continuously evaluated against predefined thresholds. If the forecasted value surpasses a critical level, the digital twin system flags the component for inspection or preventative action—long before visible failure occurs.

This framework allows urban managers to identify early signals of structural wear, utility stress, or road degradation while ensuring data sovereignty and responsiveness at the edge. For example, a series of anomalous pressure readings from a water pipeline can trigger a targeted maintenance operation without centralizing raw sensor data, preserving both efficiency and privacy.

By embedding a Federated Edge-Based Predictive Maintenance Model within the digital twin, the infrastructure system becomes self-aware, continuously adaptive, and capable of learning across distributed sources. This AI-driven approach ensures early detection of wear, reduces operational costs, and extends asset longevity—all while preserving the scalability, resilience, and responsiveness essential for future-ready cities.

Framework: Federated Edge-Based Predictive Maintenance (FEBPM)

Layered AI Architecture for Decentralized Infrastructure Monitoring and Maintenance

Layer	Main functionality	AI module(s) and components	Operational scope (Input → Processing → Output)
Layer 1	Data collection: real-time sensor data collection from infrastructure components	Edge sensor ingestion and feature extraction pipeline	Input → Real-time sensor data (e.g., vibration, temperature, load) → Output → Local feature vectors for prediction
Layer 2	Learning: XGBoost-based local model for infrastructure degradation prediction	XGBoost	Input → Local sensor data → Output → Predicted degradation score for local component
Layer 3	Federated aggregation: synchronizing local models into a global model	Federated parameter aggregation and model synchronization service	Input → Local model parameters → Output → Aggregated global model parameters
Layer 4	Prediction: continuous monitoring and predictive maintenance for urban infrastructure	Predictive maintenance orchestration and alerting engine	Input → Global model predictions, local degradation scores → Output → Maintenance actions or alerts for infrastructure components
Layer 5	Reporting: visualization of infrastructure health and maintenance schedules	Infrastructure health visualization and maintenance scheduling portal	Input → Predicted degradation scores, maintenance alerts → Output → Visualized reports for decision-making

3.4.3 Environmental Monitoring

The digital twin framework enhances environmental monitoring in West Cambridge. By collecting data on air quality, noise levels, and green space utilization, the area can assess its ecological performance and implement targeted interventions to improve sustainability (Pitt et al. 2022). This proactive approach supports efforts to mitigate pollution and enhance urban green spaces, contributing to residents' overall well-being. For instance, to support such real-time environmental evaluation and spatially adaptive intervention planning, we propose the integration of a Spatio-Temporal Kriging Model enhanced by Variational Autoencoders (VAEs). This AI-driven approach processes distributed environmental sensor data—such as particulate matter, CO_2 concentrations, acoustic profiles, and green space usage—and enables predictive interpolation of ecological conditions across unsensed regions, while also learning latent environmental signatures for anomaly detection.

The spatial interpolation of air quality at an unsampled location x_* is first estimated through spatio-temporal kriging:

$$\hat{y}_t(x_*) = \sum_{i=1}^{n} \lambda_i(x_*) y_t(x_i)$$

where $\hat{y}_t(x_*)$, predicted environmental value (e.g., air quality index) at location x_* and time t, $y_t(x_i)$, observed value at sensor location x_i, and $\lambda_i(x_*)$, kriging weights derived from spatial covariance functions ensuring minimum interpolation variance.

To capture nonlinear patterns and detect irregular ecological behavior, the kriged values are encoded into a latent representation using a variational autoencoder:

$$z_t = f_{\text{enc}}(\hat{y}_t), \tilde{y}_t = f_{\text{dec}}(z_t)$$

where z_t, latent environmental vector summarizing complex spatial-temporal dynamics at time t, f_{enc}, encoder neural network mapping inputs to latent space, f_{dec}, decoder reconstructing environmental patterns, and $\tilde{y}_t$, reconstructed prediction used to identify anomalies or ecological trends.

The reconstruction error $\|\hat{y}^t - \tilde{y}^t\|^2$ serves as an environmental anomaly score, highlighting areas where air quality, noise, or ecological usage deviates from historical baselines. When thresholds are crossed, targeted interventions—such as traffic redistribution, planting vegetation buffers, or urban cooling measures—can be simulated and deployed via the digital twin.

These predictions are visualized on geospatial dashboards that guide sustainability teams in prioritizing areas for remediation or policy response. For example, consistently high reconstruction errors in a school zone could trigger air filtration measures or traffic rerouting to protect public health.

By embedding a Spatio-Temporal Kriging-VAE Hybrid Model into the digital twin infrastructure, cities gain a predictive, adaptive, and spatially granular tool for ecological monitoring. This AI-driven framework enables proactive management of environmental quality, supports sustainable urban policy, and directly contributes to healthier, greener, and more resilient urban ecosystems—aligning technological innovation with ecological responsibility and human well-being.

Framework: Spatio-Temporal Environmental Monitoring and Intervention (STEMI)
Layered AI Architecture for Environmental Quality Monitoring and Adaptive Interventions

Layer	Main functionality	AI module(s) and components	Operational scope (Input → Processing → Output)
h	Data collection: collect distributed environmental sensor data (air quality, noise, etc.)	Environmental sensor integration and data preprocessing pipeline	Input → Environmental sensor data (e.g., PM, CO_2, noise levels) → Output → Raw input for spatial and temporal analysis
Layer 2	Learning: spatio-temporal kriging for spatial interpolation of environmental conditions	Spatio-temporal kriging	Input → Sensor data from multiple locations → Output → Interpolated environmental conditions at unsensed locations
Layer 3	Anomaly detection: variational autoencoder for detecting irregular ecological behavior	Variational Autoencoders (VAE)	Input → Kriged data (spatio-temporal) → Output → Latent representations and anomaly detection
Layer 4	Simulation: predictive simulation for ecological trends and intervention planning	Predictive environmental simulation and trend analysis engine	Input → Latent representations and historical data → Output → Simulated future environmental outcomes under various interventions
Layer 5	Reporting: visualizing anomalies and simulating interventions for urban planning	Ecological outcome visualization and intervention strategy platform	Input → Predicted ecological outcomes, anomaly scores → Output → Visualized reports and actionable insights for decision-making

3.4.4 Citizen Engagement and Community Involvement

A hallmark of the smart city initiatives in West Cambridge is the emphasis on citizen engagement. The implementation of digital twin technology facilitates participatory governance, allowing residents to contribute actively to urban development processes.

West Cambridge has developed various platforms that enable citizens to provide feedback on urban projects and policies. These platforms leverage the digital twin model to visualize proposed changes, allowing residents to understand potential impacts and express their opinions effectively (Yigitcanlar and Cugurullo 2020). This participatory approach fosters a sense of community ownership and ensures that urban development aligns with the desires of residents. For instance, to support such participatory urban development within the digital twin, we propose the integration of a Transformer-Based Dynamic Feedback Analysis System, powered by Bidirectional Encoder Representations from Transformers (BERT) and a Topic-Aware Attention Layer. This AI-driven model allows cities to transform open-ended citizen input into structured insights that influence simulation outcomes and planning decisions.

Each citizen-provided comment s_i is first encoded into a dense semantic vector using the BERT model:

$$\mathrm{e}_i = \mathrm{BERT}(s_i)$$

where e_i, contextual embedding of statement s_i, representing both linguistic content and emotional nuance. These embeddings are sensitive to local dialects, tone, and domain-specific terminology (e.g., "bike lane," "tree coverage," and "zoning").

To identify core concerns and assign input to relevant policy categories (mobility, environment, housing, etc.), a topic-aware attention mechanism refines the embedding via weighted projections:

$$\hat{y}_i = \mathrm{softmax}(W \cdot \mathrm{Attention}(\mathrm{e}_i, T) + b)$$

where $\hat{y}_i$, predicted relevance distribution over planning topics, T, learned topic matrix representing predefined thematic areas, Attention(·), a dot-product attention function that scores topic alignment with input e_i, and W, b, learnable parameters of the classification layer. This function determines not only the subject of the citizen feedback, but also the degree to which it aligns with ongoing digital twin simulation modules. For instance, feedback about traffic congestion near schools will be routed to the transport optimization engine within the twin, while concerns about noise pollution will influence spatial-environmental monitoring nodes.

To further prioritize community input in planning, the model calculates a participation influence score based on topic relevance and public consensus:

$$I_i = \tanh\left(\alpha \cdot r_i \cdot s_i^{\mathrm{sentiment}}\right)$$

where I_i, the influence weight of the feedback i, r_i, relevance score based on topic attention, $s_i^{\mathrm{sentiment}}$, sentiment polarity score of the feedback (positive or negative), and α, scaling factor for controlling sensitivity. Feedback with strong sentiment and topic alignment receives greater weight in scenario simulations and planning decisions.

These influence scores are mapped onto the digital twin's simulation outcomes, ensuring that public opinion is not only recorded but actively used to reshape proposed developments, zoning plans, or service adjustments. The digital twin becomes a participatory interface where each policy is tested not only for its physical viability but also for its social acceptability.

This AI framework ensures that citizen input is quantified, contextualized, and embedded into decision processes—transforming residents from passive observers into active urban co-creators. As such, it redefines the role of digital twins in modern urbanism: not merely as technical tools, but as instruments of democratic, data-informed design.

Framework: Transformer-Based Dynamic Feedback Analysis for Urban Planning (TFDAUP)

Layered AI Architecture for Citizen Feedback Integration and Participatory Urban Planning

Layer	Main functionality	AI module(s) and components	Operational scope (Input → Processing → Output)
Layer 1	Data collection: collect citizen feedback through platforms integrated with digital twin	Citizen feedback aggregation and preprocessing engine	Input → Citizen feedback (open comments, suggestions) → Output → Raw input for AI models
Layer 2	Learning: BERT-based encoding of citizen feedback into semantic embeddings	Bidirectional Encoder Representations from Transformers (BERT)	Input → Citizen feedback → Output → Contextual semantic vector representations
Layer 3	Attention: topic-aware attention mechanism for categorizing feedback into relevant urban themes	Topic classification and urban theme prioritization engine	Input → BERT embeddings → Output → Relevance distribution across urban planning topics
Layer 4	Prioritization: calculate participation influence scores based on relevance and sentiment	Sentiment-weighted influence scoring and feedback optimization platform	Input → Topic relevance, sentiment polarity → Output → Influence weight for feedback prioritization
Layer 5	Reporting: integrating citizen feedback into digital twin simulations and planning decisions	Feedback integration and urban planning optimization platform	Input → Prioritized feedback, simulation models → Output → Actionable insights and adjusted urban planning scenarios

3.4.5 Educational Initiatives

To enhance public understanding of digital twin technology and its benefits, West Cambridge invests in educational initiatives. Workshops, seminars, and online resources help demystify the technology, encouraging residents to engage with the smart city agenda. By empowering citizens with knowledge, West Cambridge strengthens community ties and builds a collaborative environment for urban governance (Dani et al. 2023).

3.4.6 Sustainability and Resilience

The integration of digital twin technology in West Cambridge aligns closely with broader sustainability goals. By optimizing resource use and enhancing urban resilience, the area aims to create a more sustainable environment.

The digital twin framework allows for more efficient management of resources, including energy, water, and waste. By analyzing consumption patterns and identifying areas for improvement, West Cambridge can implement strategies that minimize waste and reduce the carbon footprint of urban activities (Dani et al. 2023). To support such adaptive optimization of urban resource systems, we propose the integration of a multi-objective optimization framework based on Deep Deterministic Policy Gradient (DDPG) within the digital twin. This AI-driven approach enables continuous decision-making over resource flows such as electricity, water, and solid waste—balancing efficiency with sustainability goals in real time. The model is particularly suitable for smart urban systems where actions must be chosen from continuous spaces (e.g., adjusting water flow rates or energy load distributions), and where trade-offs exist among competing objectives.

The policy network μ_θ defines a deterministic mapping from the current urban system state s_t to the optimal resource control action a_t:

$$a_t = \mu_\theta(s_t)$$

where a_t, action vector representing infrastructure-level adjustments (e.g., energy reallocation or waste routing), s_t, current state of the city's resource environment as encoded in the digital twin, and θ, parameters of the actor network optimized during learning.

The critic network $Q_\phi(s_t, a_t)$ evaluates the expected multi-objective utility of performing action a_t in state s_t:

$$Q_\phi(s_t, a_t) = \mathbb{E}\left[\sum_{k=0}^{\infty} \gamma^k R(s_{t+k}, a_{t+k})\right]$$

where $Q_\phi(s_t, a_t)$, expected long-term benefit of the action, $\gamma^k R(s_{t+k}, a_{t+k})$, reward signal incorporating sustainability criteria, γ, discount factor weighing future versus immediate outcomes, and ϕ, parameters of the critic network.

The reward function itself is composed of multiple sustainability sub-goals:

$$R(s_t, a_t) = -\alpha \cdot f_{\text{energy}} - \beta \cdot f_{\text{water}} - \delta \cdot f_{\text{waste}}$$

where $f_{\text{energy}}, f_{\text{water}}, f_{\text{waste}}$, forecasted consumption or inefficiency measures in each domain, and α, β, δ, weighting factors reflecting local sustainability priorities.

Through iterative simulation and policy refinement, the model converges toward actions that simultaneously reduce environmental impact and improve urban resource allocation. For example, if water demand is projected to exceed sustainable thresholds in one district, the model may reduce irrigation or reschedule non-critical usage while preserving essential services. Similar logic applies to optimizing grid-level energy loads or dynamically adjusting waste collection schedules.

This AI system continuously learns from real-time conditions and historical usage patterns, generating policy adjustments that reduce waste, balance supply and demand, and minimize carbon intensity. In doing so, it transforms sustainability from a static design principle into a dynamic, operational goal—positioning cities to meet long-term environmental targets with adaptive intelligence.

Framework: Multi-Objective Resource Optimization for Smart Cities (MORO)
Layered AI Architecture for Sustainable Resource Management and Urban Resilience

Layer	Main functionality	AI module(s) and components	Operational scope (Input → Processing → Output)
Layer 1	Data collection: collect real-time data on energy, water, and waste systems	Sensor data aggregation and preprocessing system	Input → Real-time sensor data on energy, water, waste → Output → Processed data for optimization models
Layer 2	Learning: Deep Deterministic Policy Gradient (DDPG) for resource flow optimization	Deep Deterministic Policy Gradient (DDPG)	Input → Sensor data, resource system states → Output → Optimal resource control actions
Layer 3	Continuous learning: critic network evaluating long-term utility of actions	Long-term impact assessment and utility evaluation engine	Input → Action predictions, reward signals → Output → Critic's evaluation of long-term rewards
Layer 4	Optimization: multi-objective reward function for balancing sustainability goals	Multi-objective resource allocation and sustainability optimization engine	Input → Resource control actions, sustainability objectives → Output → Optimized resource allocation actions
Layer 5	Reporting: visualization of optimized resource flows and sustainability impact	Resource flow visualization and decision-support dashboard	Input → Resource optimization results, simulation outputs → Output → Visualized reports and actionable insights for urban planners

3.4.7 Climate Resilience

The use of digital twin technology enhances West Cambridge's ability to respond to climate challenges. By modeling different climate scenarios, planners can assess potential risks and develop adaptation strategies to protect infrastructure and community well-being (Yigitcanlar and Cugurullo 2020). This proactive approach ensures that West Cambridge remains resilient in the face of changing environmental conditions. To support such climate-responsive scenario modeling and adaptive urban resilience planning, we propose the integration of a Bayesian Deep Spatio-Temporal Forecasting Framework using Monte Carlo Dropout in Convolutional Long Short-Term Memory Networks (ConvLSTM). This AI-driven model forecasts the spatio-temporal evolution of climate impacts—such as heatwaves, flooding, or urban drought—and quantifies uncertainty, enabling planners to simulate multiple adaptation strategies and assess their effectiveness under different risk conditions.

The model takes a sequence of historical spatio-temporal climate inputs and forecasts the next-step environmental condition using ConvLSTM layers, which preserve spatial structure while modeling temporal dynamics:

$$\hat{Y}_{t+1} = \mathrm{ConvLSTM}(X_{t-n:t})$$

where $\hat{Y}_{t+1}$, predicted spatial climate map (e.g., flood extent or surface temperature) for time $t + 1$, and $X_{t-n:t}$, sequence of past observed climate maps over the last n time steps.

To capture model uncertainty and simulate multiple plausible future scenarios, Monte Carlo Dropout is applied at inference:

$$\mathbb{E}\left[\hat{Y}_{t+1}\right] = \frac{1}{M}\sum_{m=1}^{M}\hat{Y}_{t+1}^{(m)},\ \mathrm{Var}\left[\hat{Y}_{t+1}\right] = \frac{1}{M}\sum_{m=1}^{M}\left(\hat{Y}_{t+1}^{(m)} - \mathbb{E}\left[\hat{Y}_{t+1}\right]\right)^2$$

where $\hat{Y}_{t+1}^{(m)}$, predicted map from the m-th stochastic forward pass, $\mathbb{E}\left[\hat{Y}_{t+1}\right]$, expected (mean) forecast, and $\mathrm{Var}\left[\hat{Y}_{t+1}\right]$, predictive variance that highlights uncertainty in high-risk regions.

This probabilistic forecast allows planners to compare adaptation strategies—such as flood-resilient road surfacing, heat-mitigating urban greening, or cooling corridor installations—across a range of possible futures. Scenario outputs are visualized within the digital twin's geospatial interface, enabling risk-sensitive planning decisions that are both data-driven and uncertainty-aware.

For example, if the model forecasts with high variance a 40% probability of flash flooding in a transit corridor, planners can preemptively simulate and test infrastructure interventions such as elevated walkways, stormwater retention tanks, or pervious surfaces, adjusting strategies as new data updates the model.

This AI-driven architecture empowers planners to prepare for environmental uncertainty, evaluate trade-offs among adaptation options, and design cities that remain safe, functional, and livable in the face of accelerating climate variability.

Framework: Bayesian Deep Spatio-Temporal Forecasting Framework (BDSTFF)

Layered AI Architecture for Climate Resilience and Urban Adaptation

Layer	Main functionality	AI module(s) and components	Operational scope
Layer 1	Data collection: collect historical spatio-temporal climate data	Historical climate data integration and preprocessing pipeline	Input → Historical climate maps → Output → Preprocessed input for ConvLSTM
Layer 2	Learning: spatio-temporal forecasting of climate impacts	ConvLSTM	Input → Historical climate maps → Output → Predicted climate map for future climate impact
Layer 3	Uncertainty estimation: simulate uncertainty using Monte Carlo Dropout	Monte Carlo Dropout	Input → Predicted climate maps → Output → Mean and variance of predictions
Layer 4	Scenario simulation: simulate adaptation strategies for climate impacts	Adaptation strategy simulation and scenario evaluation platform	Input → Probabilistic forecasts, adaptation strategies → Output → Simulated outcomes for adaptation planning
Layer 5	Visualization and decision support: support urban resilience planning through digital twin interface	Risk visualization and urban resilience decision support interface	Input → Simulated outcomes → Output → Visualized adaptation scenarios and risk-sensitive planning decisions

3.4.8 Challenges and Future Directions

While West Cambridge has made significant progress in implementing smart city initiatives, several challenges must still be addressed to ensure the long-term effectiveness and inclusivity of these efforts. One of the foremost concerns is data privacy and security. As smart systems increasingly rely on the continuous collection and analysis of personal and environmental data, ensuring the protection of citizen information becomes critical. Balancing the need for real-time data to enhance public services with the ethical obligation to safeguard individual privacy remains an ongoing challenge for policymakers and system designers (Dani et al. 2023).

Within the West Cambridge programme, privacy should be treated as a constitutive governance design condition rather than as an external constraint, because the technical vision explicitly anticipates pervasive sensing, real-time processing, and increasingly intelligent inference in the built environment, while foregrounding the privacy implications of environments "knowing" occupant presence, location, and behavior and the requirement that occupants become first-class participants rather than passive subjects of a building manager (Pitt et al. 2022). This stance is operationalized through an architecture in which the client-facing layer provides APIs and user interfaces as the point of contact for accessing sensor data, making it a natural locus for enforceable access decisions (Pitt et al. 2022). Pitt et al. implement a privacy framework in which access is specified through permissions treated as first-class data objects with temporal historical logging and governed by declarative rules, including practical policies such as "occupants can see sensor information and derived analysis relevant to the space they occupy" and a propagation logic in which permission on a parent object can imply permissions on its child objects (Pitt et al. 2022). Read as governance, we therefore propose making the permissions regime institutionally accountable through an explicit data inventory that distinguishes environmental telemetry from person-referable streams and aligns access tiers with stated purpose limitation, so that runtime enforcement remains traceable to declared norms rather than being left implicit (Papagiannidis et al. 2025). Where analytic or predictive models are constructed atop these streams, model governance is strengthened by standardized, machine-readable documentation of intended use, context of use, and identified risks, for example, through AI Cards, so that assurance evidence is comparable across modules (Golpayegani et al. 2024). In parallel, procedural controls such as compliance monitoring, enforcement, and incident-response mechanisms remain necessary to manage drift, misuse, or emergent harms under operational pressure (Papagiannidis et al. 2025), while audit-oriented documentation should keep human influence, oversight points, and data quality inspectable to external reviewers (Konigstorfer and Thalmann 2022).

To ensure both real-time data utility and citizen privacy in the evolving smart city infrastructure of West Cambridge, we propose a cutting-edge AI-driven framework anchored in Differential Privacy with Federated Learning—an approach that allows decentralized model training across distributed edge devices without transmitting sensitive data. At the core of this privacy-preserving architecture lies the mechanism of ε-Differential Privacy, which provides formal mathematical guarantees that individual user data remains indistinguishable within analytical outputs. This framework is further augmented by an adaptive noise calibration module and a secure gradient aggregation protocol, which collectively maintain utility while enforcing legal and ethical safeguards aligned with regulations such as GDPR.

The foundation of the model begins with the canonical differential privacy guarantee, defined as:

$$\forall S \subseteq \text{Range}(M), \quad \Pr[M(D_1) \in S] \leq e^{\varepsilon} \cdot \Pr[M(D_2) \in S]$$

where D_1 and D_2 are neighboring datasets differing by only one individual's data, M is the randomized algorithm (mechanism) used for analysis, S is any subset of possible outputs, and ε is the privacy loss parameter controlling disclosure risk. This equation ensures that the inclusion or exclusion of a single individual's data does not significantly affect the output of the AI model, thus rendering individual re-identification infeasible even to adversaries with auxiliary knowledge. The parameter ε represents a trade-off between data utility and privacy protection, with smaller values offering stronger guarantees.

To enable collaborative training across multiple smart infrastructure nodes without centralizing raw data, we incorporate a Federated Averaging mechanism. The secure global model update at each round t is computed as:

$$\theta_t = \sum_{i=1}^{N} \frac{n_i}{n} \cdot \theta_t^{(i)}$$

where θ_t is the aggregated global model at time t, $\theta_t^{(i)}$ is the locally trained model at node i, n_i is the number of data points at node i, and $n = \sum_{i=1}^{N} n_i$ is the total data size across all participating nodes. This function allows the central aggregator to compute a weighted average of local updates without accessing the raw data, ensuring that all personal information remains confined to edge locations such as citizen devices, smart sensors, or local urban service hubs. Each local model is differentially privatized before aggregation to prevent leakage through gradients.

To formally integrate privacy into the training gradients, we introduce a Gaussian noise mechanism into each local update:

$$\tilde{g}_t^{(i)} = g_t^{(i)} + \mathcal{N}\left(0, \sigma^2 I\right)$$

where $\tilde{g}_t^{(i)}$ is the privatized gradient for node i at time t, $g_t^{(i)}$ is the true gradient, $\mathcal{N}(0, \sigma^2 I)$ is zero-mean Gaussian noise with variance σ^2, and I is the identity matrix. This mechanism ensures that even if gradients are intercepted, they cannot be reverse-engineered to recover sensitive data. The noise scale σ is calibrated to the sensitivity of the data and the desired privacy budget ε, with higher noise levels offering stronger anonymity guarantees.

Finally, to quantify the cumulative privacy loss over multiple training rounds or data accesses, we adopt the Moments Accountant framework, defined recursively by:

$$\alpha(\lambda) = \sum_{t=1}^{T} \log \mathbb{E}\left[e^{\lambda Z_t}\right]$$

where $\alpha(\lambda)$ is the moment generating function of privacy loss for moment order λ, T is the number of iterations or accesses, and Z_t is the privacy loss random variable at time t. This cumulative accounting allows system designers to monitor and control the total privacy risk incurred during prolonged smart city operations, ensuring compliance with evolving legal standards and reinforcing ethical AI governance.

Taken together, this AI-driven model—uniting Differential Privacy, Federated Learning, and adaptive noise calibration—establishes a mathematically rigorous and ethically robust framework for secure data utilization in smart urban infrastructures. It allows cities like West Cambridge to retain the power of real-time analytics while guaranteeing that no individual's privacy is sacrificed in the process of innovation.

This balance of utility and confidentiality is essential in fostering citizen trust and ensuring the longevity and inclusivity of intelligent city ecosystems.

Framework: Differential Privacy with Federated Learning for Smart City Privacy (DPFL-SCP)

Layered AI Architecture for Privacy-Preserving Smart City Data Utilization

Layer	Main functionality	AI module(s) and components	Operational scope
Layer 1	Data collection: collect local data from smart city infrastructure nodes	Privacy-preserving data collection and preprocessing pipeline	Input → Raw data from citizen devices, smart sensors → Output → Local data for privacy-preserving analysis
Layer 2	Privacy mechanism: apply differential privacy to individual data points	Differential privacy	Input → Personal data → Output → Privacy-preserving data with ε-Differential Privacy guarantees
Layer 3	Collaborative learning: perform decentralized learning using Federated Averaging	Federated learning	Input → Local model updates from edge devices → Output → Aggregated global model without centralizing raw data
Layer 4	Gradient privacy: add noise to gradients to ensure privacy during training	Privacy-enhanced gradient computation and noise injection system	Input → Training gradients → Output → Privatized gradients with Gaussian noise to prevent data leakage
Layer 5	Privacy accounting: track cumulative privacy loss during training	Privacy loss accounting	Input → Multiple training rounds → Output → Cumulative privacy loss over time
Layer 6	Model deployment and monitoring: deploy the trained model and monitor privacy compliance	Privacy-compliant model deployment and continuous monitoring platform	Input → Trained model → Output → Model deployed with privacy safeguards and ongoing monitoring for compliance

Another important issue is equity in access. For smart city initiatives to truly benefit all residents, it is essential to address disparities in digital literacy, connectivity, and access to services. Without deliberate efforts to ensure inclusivity, vulnerable populations may be excluded from the advantages of technological transformation, thus reinforcing existing social inequalities. Bridging these gaps will be vital to ensure that smart urban development promotes fairness and cohesion across all segments of society (Yigitcanlar and Cugurullo 2020). To bridge the equity divide in smart urban systems, we introduce a Fairness-Constrained Multi-Objective Reinforcement Learning (FC-MORL) framework that dynamically allocates digital services while ensuring equitable access for all demographic segments. This model captures disparities in access by introducing fairness constraints directly into the reward structure, thereby allowing AI agents to optimize not only efficiency but also social equity across distributed smart city infrastructure. By incorporating demographic-aware fairness metrics, the model avoids reinforcing existing inequalities and actively works to reduce digital exclusion in real time.

At the core of this model lies the Fairness-Constrained Reward Function, which is formally defined as:

$$R_t = U_t - \lambda \cdot \mathcal{F}_t$$

where R_t is the fairness-adjusted reward at time t, U_t is the utility derived from service delivery (e.g., bandwidth, access time, or service quality), $\mathcal{F}_t$ is a dynamic fairness penalty reflecting inequitable access across population groups, and λ is a hyperparameter controlling the trade-off between system utility and fairness. This formulation allows the AI system to learn optimal service strategies while actively penalizing decisions that disproportionately favor digitally advantaged groups. The fairness penalty $\mathcal{F}_t$ is derived using the statistical disparity across protected attributes (such as income level, neighborhood, or digital literacy class), computed as:

$$\mathcal{F}_t = \sum_{g=1}^{G} \left| \frac{A_t^{(g)}}{P^{(g)}} - \frac{\sum_{h=1}^{G} A_t^{(h)}}{\sum_{h=1}^{G} P^{(h)}} \right|$$

where $A_t^{(g)}$ is the amount of digital services allocated to group gg at time t, $P^{(g)}$ is the population of group g, G is the total number of demographic groups, and the expression computes the absolute deviation of each group's per capita access from the overall population average. This term enforces proportional fairness, ensuring that all demographic groups receive digital services in accordance with their population share. If a certain group is underserved, its deviation increases the penalty, prompting the learning agent to reallocate future resources more equitably.

To ensure long-term fairness adherence while maintaining adaptability, the policy update is governed by the Multi-Objective Proximal Policy Optimization (MO-PPO) function:

$$L_{\text{MO}-\text{PPO}}(\theta) = \mathbb{E}_t \left[min \left(r_t(\theta)\hat{A}_t^{U}, \text{clip}(r_t(\theta), 1-\epsilon, 1+\epsilon)\hat{A}_t^{U} \right) - \beta \cdot \hat{A}_t^{\mathcal{F}} \right]$$

where $L_{\text{MO}-\text{PPO}}(\theta)$ is the loss function used to update the policy parameters θ, $r_t(\theta)$ is the policy ratio at time t, $\hat{A}_t^{U}$ is the utility-based advantage function, $\hat{A}_t^{\mathcal{F}}$ is the fairness-based advantage function, ϵ is the clipping parameter used to ensure stability in policy updates, and β is a weight that adjusts the relative importance of fairness during training. This formulation preserves the benefits of Proximal Policy Optimization in maintaining stable updates while embedding fairness as an explicit optimization objective. The fairness advantage term $\hat{A}_t^{\mathcal{F}}$ ensures that the model prioritizes actions that correct historical disparities without degrading service quality for all.

Together, these components form a high-performance AI architecture that enables urban systems to learn inclusive, data-driven policies that dynamically correct digital access disparities. By modeling both the utility and fairness of each allocation decision, this AI-driven model ensures that smart services are distributed in a socially equitable manner, thereby upholding the principles of digital justice and inclusive urban transformation. This framework is particularly impactful in heterogeneous urban regions like West Cambridge, where socio-economic diversity demands sensitivity to access equity and technological inclusion.

Framework: Fairness-Constrained Multi-Objective Reinforcement Learning for Smart City Equity (FC-MORL-SCE)

Layered AI Architecture for Equity-Preserving Smart City Service Allocation

Layer	Main functionality	AI module(s) and components	Operational scope
Layer 1	Data collection: gather demographic data from smart city infrastructure	Demographic data aggregation and preprocessing system	Input → Demographic data (e.g., income level, digital literacy) → Output → Local data for equitable service allocation
Layer 2	Fairness mechanism: apply fairness constraints to reward structure	Fairness-constrained reward function	Input → Service utility, fairness metrics → Output → Fairness-adjusted rewards to penalize inequities
Layer 3	Learning: perform decentralized learning with fairness constraints	Fairness-Constrained Multi-Objective Reinforcement Learning (FC-MORL)	Input → Local model updates, fairness rewards → Output → Optimized service allocation based on fairness and efficiency
Layer 4	Policy update: update policy using MO-PPO with fairness objectives	Multi-Objective Proximal Policy Optimization (MO-PPO)	Input → Utility and fairness-based advantage functions → Output → Updated policy that prioritizes equity while ensuring service quality
Layer 5	Monitoring and adaptation: track long-term fairness adherence and adaptability	Long-term fairness monitoring and adaptive policy adjustment platform	Input → Long-term fairness adherence, service allocations → Output → Adjusted policy to ensure equity and inclusion over time

Finally, sustaining citizen engagement over time presents both a logistical and cultural challenge. While initial participation may be strong, maintaining long-term involvement in governance and decision-making processes requires ongoing outreach, transparent communication, and the development of trust-based civic platforms. The success of smart city initiatives depends not only on technological sophistication but also on the active, continuous participation of the community they are designed to serve. Without this sustained engagement, the momentum and responsiveness of smart governance may weaken over time (Dani et al. 2023).

To address the long-term dynamics of civic involvement in smart governance platforms, we propose a state-of-the-art AI-driven model based on Temporal Graph

Neural Networks (TGNN) combined with Sequential Survival Analysis. This hybrid model captures both the temporal patterns of individual engagement behavior and the relational structure of civic networks (e.g., neighborhood councils, digital forums, social influence chains). It allows city administrators to predict disengagement risk, tailor re-engagement strategies, and maintain trust-based communication channels by identifying key influencers and timing outreach based on evolving engagement probabilities.

At the heart of this framework lies a temporal attention-based graph representation function that captures the embedded state of each citizen node over time:

$$h_v^{(t)} = \mathrm{GRU}\left(h_v^{(t-1)}, \sum_{u \in \mathcal{N}(v)} \alpha_{vu}^{(t)} W x_u^{(t)}\right)$$

where $h_v^{(t)}$ is the temporal embedding of citizen v at time t, $h_v^{(t-1)}$ is the previous hidden state, $\mathcal{N}(v)$ is the set of neighboring nodes (citizens or civic actors), $\alpha_{vu}^{(t)}$ is the temporal attention weight between node v and neighbor u at time t, W is the learnable transformation matrix, and $x_u^{(t)}$ is the feature vector of neighbor u at time t. This function allows the model to incorporate both temporal activity and social influence into the representation of each individual, ensuring that patterns of participation, peer effects, and responsiveness to public events are fully captured. The attention weight $\alpha_{vu}^{(t)}$ is derived from a softmax over relevance scores, ensuring dynamic prioritization of impactful neighbors. To quantify the probability that a citizen will remain engaged at future time steps, we embed a survival probability function using the Cox Proportional Hazards Model:

$$\log \lambda_v\left(t \mid h_v^{(t)}\right) = \beta^\top h_v^{(t)}$$

where $\lambda_v\left(t \mid h_v^{(t)}\right)$ is the hazard rate—i.e., the instantaneous risk of disengagement for citizen v at time t, conditioned on the embedding $h_v^{(t)}$, and β is the trainable coefficient vector encoding how latent features influence the risk of disengagement. This formulation models the engagement lifecycle of each citizen, allowing the system to predict when disengagement is likely to occur and which features—such as communication gaps, event fatigue, or network isolation—contribute most to it.

To forecast the cumulative probability that a citizen remains engaged through time TT, we define the survival function as:

$$S_v(T) = \exp\left(-\int_0^T \lambda_v\left(t \mid h_v^{(t)}\right) dt\right)$$

where $S_v(T)$ is the survival probability of continued engagement for citizen v up to time T, based on the cumulative hazard derived from $\lambda_v(t)$. This probabilistic function allows the city's civic systems to target outreach to those citizens most at risk of disengagement, optimizing resources for civic inclusion strategies. When aggregated over the population, these survival curves inform policy designers on how to evolve platform design, outreach cadence, and participatory feedback loops.

Together, the integration of TGNN embeddings and survival modeling creates a predictive, adaptive AI system that monitors long-term civic engagement across the smart city landscape. It enables the identification of at-risk individuals and community clusters, empowers proactive retention interventions, and ultimately reinforces the cultural and logistical sustainability of participatory governance platforms. This approach ensures that smart governance in West Cambridge not only begins with strong civic enthusiasm but maintains it as a structural feature of inclusive urban innovation.

Framework: Temporal Graph Neural Networks and Sequential Survival Analysis for Civic Engagement (TGNN-SSA)

Layered AI Architecture for Long-Term Civic Engagement in Smart Governance

Layer	Main functionality	AI module (s) and components	Operational scope
Layer 1	Data collection: gather civic engagement data from smart city platforms	Data preprocessing module	Input → Citizen participation data (e.g., forum activity, meeting attendance) → Output → Processed data for engagement modeling
Layer 2	Graph representation: model civic networks as temporal graphs with attention weights	Temporal Graph Neural Networks (TGNN)	Input → Citizen interaction data, social network structure → Output → Temporal embedding of citizen nodes considering social influence
Layer 3	Engagement risk prediction: predict disengagement risk using survival analysis	Cox proportional hazards model	Input → Temporal embeddings of citizens → Output → Predicted disengagement risk for each citizen
Layer 4	Survival probability forecasting: compute the probability of continued engagement	Survival function (engagement forecast)	Input → Hazard rate, engagement embeddings → Output → Cumulative engagement probability for each citizen
Layer 5	Monitoring and adaptation: monitor engagement and adjust strategies for re-engagement	Re-engagement strategy optimizer	Input → Disengagement predictions, participation data → Output → Adjusted outreach strategies based on predicted engagement trends

Figure 3.3 illustrates the smart city initiatives and digital twin integration in West Cambridge, UK. At the center is the primary node "West Cambridge, UK: Smart City Initiatives and Digital Twin Integration," connected to seven main areas: "Digital Twin Technology in West Cambridge," "Urban Planning and Development," "Infrastructure Management," "Environmental Monitoring," "Citizen Engagement and Community Involvement," "Sustainability and Resilience," and "Challenges and Future Directions." "Digital Twin Technology" enables real-time monitoring, data analysis, and urban simulation. "Urban Planning and Development" uses simulations with GIS and environmental data for informed decision-

Fig. 3.3 West Cambridge, UK: smart city initiatives and digital twin integration

making. "Infrastructure Management" supports real-time utility monitoring and proactive maintenance. "Environmental Monitoring" collects data on air quality, noise, and green spaces for targeted ecological improvements. "Citizen Engagement and Community Involvement" involves public platforms for feedback and educational initiatives. "Sustainability and Resilience" focuses on resource optimization and climate resilience strategies. "Challenges and Future Directions" address data privacy, equity in access, and sustaining long-term citizen engagement.

West Cambridge exemplifies how digital twin technology can be effectively leveraged to enhance urban planning, infrastructure management, and citizen engagement within the framework of smart city development. By integrating real-time data and fostering community participation, West Cambridge not only optimizes its urban services but also promotes sustainability and resilience. As cities globally face increasing challenges related to urbanization and climate change, the lessons learned from West Cambridge's innovative approach provide valuable insights for other urban centers aiming to adopt smart technologies for improved governance and quality of life.

3.5 Málaga City, Spain

Málaga, a vibrant city located on the southern coast of Spain, serves as a notable example of smart city development through the integration of advanced technologies and data-driven solutions. This analysis delves into the various initiatives undertaken by Málaga to enhance urban living, improve sustainability, and foster citizen engagement, with a particular focus on the application of digital twin technology and other smart city strategies.

Málaga has embraced the smart city concept as part of its urban development strategy, aiming to position itself as a leader in innovation and technology in Spain and Europe. The city's strategic approach to smart city initiatives is driven by the need to address urban challenges, such as traffic congestion, environmental sustainability, and citizen engagement, while enhancing the overall quality of life for residents.

Digital twin technology plays a crucial role in Málaga's smart city framework. By creating virtual replicas of physical assets and systems, digital twins facilitate real-time monitoring, data analysis, and simulation of urban dynamics (Dani et al. 2023).

3.5.1 Urban Planning and Infrastructure Management

Málaga utilizes digital twin technology to support urban planning and infrastructure management. In the configuration described in this case study, the digital twin integrates geographic information systems (GIS), sensor-network telemetry, and public feedback to construct a comprehensive virtual representation of the city, which can then be used to simulate alternative urban scenarios, assess the impacts of proposed developments, and optimize infrastructure investments. Such DT-enabled planning logics, grounded in multi-source integration and scenario-based decision support, have been discussed in the wider smart city digital twin literature, particularly in relation to how heterogeneous streams are assembled into actionable representations for policy and operations (Dani et al. 2023). For example, planners can visualize the likely effects of new construction on traffic patterns, green-space availability, and public services, thereby encouraging a more proactive alignment between project appraisal, sustainability objectives, and articulated community needs.

Governance lens (Málaga). Because the downstream framework stack introduced below operationalizes these planning ambitions through data-intensive analytics, we propose an explicit governance spine that renders the data-to-decision chain institutionally defensible rather than merely technically plausible. In data governance terms, this entails a living data register that enumerates each stream feeding the twin (GIS basemaps, infrastructure and environmental telemetry, and citizen-contributed feedback), specifies purpose of use and the relevant institutional or legal basis, assigns stewardship responsibility for provenance and quality, and defines tiered access regimes, treating baseline safeguards such as robust encryption and access controls as design-time requirements rather than retrofitted hardening (El-Agamy et al. 2024). This positioning is consistent with governance accounts that frame privacy and data governance as lifecycle management of access, quality, and protection, sustained through operational practices such as classification, encryption, and systematic checks that limit the propagation of bias and error under real deployment conditions (Papagiannidis et al. 2025). At the level of model governance, the plurality of modeling components across the Málaga stack increases capability but also multiplies accountability surfaces; accordingly, assurance evidence should be standardized through structured, machine-readable documentation of intended use, context of use, technical characteristics, and risk management, for example, via AI Cards, so that claims remain comparable across modules and versions (Golpayegani et al. 2024), while continuous evaluation, compliance monitoring, and incident-response procedures should be specified as enforceable operating controls when performance degrades or harms emerge (Papagiannidis et al. 2025). Finally, because outputs from the twin can shape planning choices and resource allocations, institutional governance should make decision rights explicit, including who authorizes deployments and parameter changes, who validates high-impact interventions, and who holds override authority under contestation, with audit-oriented documentation preserving the inspectability of human influence, oversight points, data quality, and the concrete role of AI in particular decisions (Konigstorfer and Thalmann 2022).

To address such predictive urban simulation and infrastructure optimization, we propose the integration of an AI-driven Spatio-Temporal Graph Neural Network (ST-GNN) within the digital twin architecture. This model transforms the city's infrastructure into a dynamic graph structure, where nodes represent spatial entities—such as intersections, parks, utility lines, or buildings—and edges encode physical, functional, or geographic relationships among them. By learning both the spatial interactions and temporal patterns from historical and real-time data, the model enables planners to forecast the outcomes of proposed urban developments with precision.

The system first models the evolving infrastructure state of each urban unit v using a graph convolution mechanism:

$$h_v^{(l+1)} = \sigma\left(\sum_{u \in \mathcal{N}(v)} \alpha_{vu}^{(l)} W^{(l)} h_u^{(l)}\right)$$

where $h_v^{(l+1)}$, updated representation of infrastructure node v at layer $l + 1$, $\mathcal{N}(v)$, set of adjacent spatial nodes (e.g., nearby road segments, linked utility systems), $\alpha_{vu}^{(l)}$, spatial attention coefficient capturing the influence of node u on v, $W^{(l)}$, learnable transformation weights, $h_u^{(l)}$, represents the feature vector (also called the embedding) of neighbor node u at layer l of the graph neural network (GNN). and σ, nonlinear activation function.

To capture temporal trends—such as traffic surges, energy demand peaks, or usage of green spaces over time—the model incorporates a temporal gated mechanism using a Gated Recurrent Unit (GRU):

$$h_v^{(t+1)} = f_{\text{GRU}}\left(h_v^{(t)}, x_v^{(t)}\right)$$

where $h_v^{(t+1)}$, updated hidden state of node v at time $t + 1$, $h_v^{(t)}$, previous hidden state, $x_v^{(t)}$, new input features from sensor data or planning scenarios, and f_{GRU}, a gated recurrent function that tracks temporal dependencies.

Finally, to simulate the projected effect of proposed infrastructure developments, the learned graph embeddings are aggregated and passed through a forecasting layer:

$$\hat{y}_t = \text{MLP}\left(\text{READOUT}\left(\left\{h_v^{(t)} \mid v \in V\right\}\right)\right)$$

where $\hat{y}_t$, predicted system-level impact at time t (e.g., congestion levels, public service load, green space coverage), V, set of all urban infrastructure nodes, READOUT, aggregation function over the graph (e.g., sum, mean), and MLP, multilayer perceptron that outputs interpretable urban indicators.

This allows planners to interactively test development scenarios—such as introducing a new residential block, expanding public transit routes, or retrofitting drainage infrastructure—and visualize their downstream consequences. The model ensures that planning decisions are not only aligned with sustainability goals but also socially and operationally robust.

The proposed AI framework integrates spatial structure, historical behavior, and real-time feedback into a unified simulation engine—empowering planners to design future infrastructure with precision, resilience, and citizen-centered insight.

Framework: Spatio-Temporal Graph Neural Networks for Urban Infrastructure Simulation (ST-GNN-UIS)

Layered AI Architecture for Predictive Urban Simulation and Infrastructure Optimization

Layer	Main functionality	AI module (s) and components	Operational scope
Layer 1	Data collection: gather spatial and temporal data from urban infrastructure	Data preprocessing and integration module	Input → Data from GIS, sensors, public feedback → Output → Processed data for model training
Layer 2	Graph representation: model urban infrastructure as a dynamic graph	Spatio-Temporal Graph Neural Network (ST-GNN)	Input → Spatial entities and relationships → Output → Temporal graph embeddings representing infrastructure dynamics
Layer 3	Temporal dynamics: capture temporal trends using GRU	Gated recurrent unit (GRU)	Input → Temporal data (e.g., traffic, energy usage) → Output → Temporal embeddings of infrastructure node behavior
Layer 4	Impact forecasting: simulate projected effects of urban developments	Multilayer Perceptron (MLP)	Input → Aggregated graph embeddings → Output → Predicted system-level impacts (e.g., congestion, green space)
Layer 5	Simulation and optimization: visualize urban planning scenarios and optimize infrastructure investments	Urban simulation and optimization engine	Input → Predicted impacts, urban planning scenarios → Output → Optimized urban development strategies based on simulation results

3.5.2 Environmental Monitoring and Sustainability

The digital twin framework in Málaga enhances the city's ability to monitor and manage environmental conditions. By collecting data on air quality, noise levels, and energy usage, the city can assess its ecological performance and implement targeted interventions to improve sustainability (Mazzetto 2024). This data-driven approach supports efforts to mitigate pollution, enhance energy efficiency, and promote the use of renewable resources. To support such high-resolution environmental monitoring and proactive sustainability interventions, we propose the integration of a Hybrid Graph-Temporal Forecasting Model based on Graph Convolutional Networks (GCNs) and Temporal Convolutional Networks (TCNs) within the digital twin infrastructure. This model processes real-time environmental sensor data—such as CO_2 levels, sound intensity, and electrical load—and predicts spatially distributed ecological outcomes over future time intervals. It enables city officials to detect anomalies, forecast pollution peaks, and identify energy inefficiencies with both spatial precision and temporal foresight.

The spatial dependencies among environmental sensing nodes are first encoded using a graph convolution layer:

$$H^{(l+1)} = \sigma\left(\tilde{D}^{-1/2}\tilde{A}\tilde{D}^{-1/2}H^{(l)}W^{(l)}\right)$$

where $H^{(l)}$, matrix of node feature representations at layer l, $\tilde{A}$, adjacency matrix of the environmental sensor graph with self-loops, $\tilde{D}$, corresponding degree matrix, $W^{(l)}$, learnable weight matrix for layer l, and σ, activation function (e.g., ReLU). This operation propagates information across the environmental network, allowing the model to learn how localized conditions (e.g., pollution hotspots) influence surrounding zones.

To capture temporal trends—such as daily traffic-related pollution cycles or energy consumption waves—the spatial features are passed into a Temporal Convolutional Network:

$$Z_t = \text{TCN}(H_t)$$

where H_t, sequence of graph-encoded features over time, and Z_t, latent temporal features representing evolution patterns of environmental variables.

The final prediction of environmental indicators (e.g., PM2.5 concentration, noise amplitude, or grid load) is generated by:

$$\hat{y}_{t+1} = \text{MLP}(Z_t)$$

where $\hat{y}_{t+1}$, predicted environmental metric for the next time step, and MLP, a multilayer perceptron producing interpretable output from latent features.

This forecast enables the digital twin to simulate and evaluate potential mitigation strategies. For example, if the model predicts a spike in CO_2 levels in a commercial district during peak hours, the system can simulate the effectiveness of alternative interventions—such as traffic rerouting, increasing EV-only zones, or scheduling green corridor misting—before actions are deployed.

This AI-driven system enhances ecological awareness, supports data-informed environmental governance, and provides city planners with precise, actionable insights for reducing pollution, improving energy efficiency, and aligning resource use with climate-resilient urban goals.

Framework: Hybrid Graph-Temporal Forecasting Model for Environmental Monitoring (HG-TFM-EM)

Layered AI Architecture for Predictive Environmental Simulation and Sustainability Optimization

Layer	Main functionality	AI module(s) and components	Operational scope
Layer 1	Data collection: gather environmental sensor data (e.g., CO_2 levels, noise, energy usage)	Environmental data integration and preprocessing module	Input → Environmental sensor data → Output → Processed data for spatial and temporal modeling
Layer 2	Spatial representation: encode spatial		Input → Environmental sensor graph → Output → Spatially

(continued)

Layer	Main functionality	AI module(s) and components	Operational scope
	dependencies between environmental nodes	Graph Convolutional Network (GCN)	encoded node feature representations
Layer 3	Temporal dynamics: capture temporal trends and dependencies in environmental data	Temporal Convolutional Network (TCN)	Input → Temporal sequence of graph features → Output → Latent temporal features representing environmental trends
Layer 4	Impact forecasting: predict environmental outcomes based on learned features	Multilayer Perceptron (MLP)	Input → Latent temporal features → Output → Predicted environmental indicators (e.g., pollution levels, energy consumption)
Layer 5	Scenario simulation: simulate and optimize interventions based on forecasted outcomes	Scenario simulation: intervention simulation and optimization engine	Input → Predicted environmental outcomes → Output → Simulated interventions and optimization strategies (e.g., pollution control, energy efficiency)

3.5.3 Citizen Engagement and Participation

Málaga places a strong emphasis on citizen engagement in its smart city initiatives. By leveraging digital twin technology and other tools, the city encourages residents to participate actively in governance and urban development processes.

The city has developed platforms that enable citizens to provide feedback on urban projects, policies, and services. By integrating these platforms with the digital twin model, residents can visualize proposed changes and express their opinions more effectively. This participatory approach fosters a sense of community ownership and ensures that urban development reflects the desires and needs of residents (Yigitcanlar and Cugurullo 2020). For instance, to support such participatory governance within the digital twin framework, we propose the integration of a Transformer-Based Feedback Analysis Model using RoBERTa embeddings combined with Latent Dirichlet Allocation (LDA) and a sentiment-aware weighting function. This AI-driven system processes citizen feedback—whether through surveys, open comment platforms, or community discussion threads—and transforms it into structured inputs that can influence urban development simulations. It enables a scalable, consistent method for incorporating resident voices into data-driven planning.

Each citizen statement s_i is first converted into a semantic vector using the RoBERTa model:

$$\mathrm{e}_i = \mathrm{RoBERTa}(s_i)$$

where e_i, is the contextual embedding of the feedback text s_i, capturing its linguistic meaning, emotional tone, and topical context.

These embeddings are then passed through a topic assignment layer using a probabilistic LDA mechanism:

$$P(z_k \mid s_i) = \frac{n_{z_k}^{(s_i)} + \beta}{\sum_{k'} n_{z_{k'}}^{(s_i)} + K\beta}$$

where $P(z_k \mid s_i)$, the probability that feedback sisi belongs to topic z_k, $n_{z_k}^{(s_i)}$, the count of terms in s_i associated with topic z_k, K, total number of topics, and β, Dirichlet prior smoothing parameter.

To prioritize which feedback items exert the greatest influence in the digital twin's planning simulations, a sentiment-aware influence score is computed:

$$I_i = \tanh(\alpha \cdot P(z_k \mid s_i) \cdot \text{Sent}(s_i))$$

where I_i, the influence score of feedback i, $P(z_k \mid s_i)$, the topic relevance probability, $\text{Sent}(s_i)$, sentiment polarity score of the text, and α, scaling parameter adjusting influence sensitivity.

Feedback items with high topic relevance and strong sentiment—positive or negative—are used to dynamically adjust urban policy parameters within the digital twin simulation. For example, if many citizens express negative sentiment regarding traffic around schools and the topic model aligns this concern with "Mobility & Safety," the digital twin can prioritize pedestrianization scenarios and simulate their system-wide impacts in real time.

This AI model ensures that resident feedback is not merely symbolic but structurally integrated into simulation logic and infrastructure policy. As such, it elevates citizen engagement from passive consultation to active co-design—bridging democratic values with computational intelligence in shaping the urban future.

Framework: Transformer-Based Feedback Analysis for Citizen Engagement (TFBA-CE)

Layered AI Architecture for Participatory Urban Planning and Governance

Layer	Main functionality	AI module (s) and components	Operational scope
Layer 1	Data collection: collect citizen feedback from various platforms	Feedback data aggregation and cleansing module	Input → Survey responses, open comments, community discussions → Output → Raw feedback data for processing
Layer 2		RoBERTa	Input → Citizen feedback text → Output → Semantic

(continued)

Layer	Main functionality	AI module (s) and components	Operational scope
	Semantic representation: convert citizen feedback into contextual embeddings		embeddings capturing meaning, tone, and context of feedback
Layer 3	Topic assignment: classify feedback into relevant topics using LDA	Latent Dirichlet Allocation (LDA)	Input → Semantic embeddings of feedback → Output → Topic assignments indicating feedback themes
Layer 4	Influence scoring: prioritize feedback based on sentiment and topic relevance	Sentiment-aware influence scoring	Input → Topic relevance, sentiment polarity → Output → Influence scores for each feedback item
Layer 5	Simulation and policy adjustment: adjust urban planning simulations based on feedback	Adaptive urban policy simulation engine	Input → Feedback influence scores, simulation data → Output → Adjusted urban development policies and scenarios based on community input

3.5.4 Educational and Outreach Initiatives

Málaga invests in educational initiatives to familiarize citizens with smart city technologies, particularly digital twins. Workshops, seminars, and online resources help demystify these technologies and encourage residents to engage with the smart city agenda. By empowering citizens with knowledge, Málaga strengthens community ties and promotes a collaborative environment for urban governance (Dani et al. 2023).

3.5.5 Impacts of Smart City Initiatives

The smart city initiatives in Málaga have produced a range of tangible benefits that contribute to urban livability, economic vitality, and environmental sustainability. One of the most evident outcomes is the improvement in the overall quality of life for residents. Through the integration of digital twin technology and data-driven decision-making, the city has enhanced public services, expanded access to green spaces, and fostered a more sustainable urban environment. These advancements have led to increased citizen satisfaction and a stronger sense of urban well-being (Dani et al. 2023). To quantitatively assess and enhance urban well-being within Málaga's digital twin framework, we propose the implementation of a Multi-Objective Deep Reinforcement Learning (MO-DRL) model embedded in a livability optimization loop. This architecture dynamically balances multiple urban indicators—such as access to green space, walkability, air quality, noise levels, and proximity to public services—to guide AI agents in continuously refining urban service delivery. Unlike static policy frameworks, this AI model adapts in real time to citizen feedback, sensor readings, and digital twin simulations, optimizing both objective and subjective dimensions of livability.

At the core of the model lies the multi-objective reward function, which evaluates actions based on their impact across multiple livability dimensions:

$$R_t = \sum_{j=1}^{m} \omega_j \cdot r_t^{(j)}$$

where R_t is the aggregated reward at time t, $r_t^{(j)}$ is the individual reward for livability dimension j (e.g., green access, noise reduction), ω_j is the priority weight assigned to that dimension based on policy targets and citizen input, and m is the total number of livability indicators considered. This reward structure allows the AI agent to evaluate trade-offs among competing objectives—for example, reducing vehicular traffic to improve air quality while ensuring transit accessibility. The policy that guides urban interventions is updated using the Proximal Policy Optimization (PPO) framework to ensure stability during real-time livability refinement:

$$L_{\text{PPO}}(\theta) = \mathbb{E}_t\left[min\left(r_t(\theta)\hat{A}_t, \text{clip}(r_t(\theta), 1-\epsilon, 1+\epsilon)\hat{A}_t\right)\right]$$

where $L_{\text{PPO}}(\theta)$ is the loss function for updating the policy parameters θ, $r_t(\theta) = \dfrac{\pi_\theta(a_t \mid s_t)}{\pi_{\theta_{\text{old}}}(a_t \mid s_t)}$ is the probability ratio between the new and old policies at time t, $\hat{A}_t$ is the advantage function estimating the benefit of action taken, and ϵ is the clipping threshold that prevents large, destabilizing updates to the policy network.

To continuously improve livability outcomes based on the simulated effects of decisions within the digital twin, the model incorporates environmental feedback from real-time IoT sensors and digital urban simulations. This data enters a state encoder, which represents the current urban condition as:

$$s_t = \phi(x_t, g_t, e_t)$$

where s_t is the encoded state of the urban environment at time t, x_t represents current sensor inputs (e.g., air quality, noise), g_t includes citizen-generated goals (e.g., increase walkability), and e_t represents environmental constraints (e.g., zoning, energy load). The function $\phi(\cdot)$ is a deep neural encoder that integrates physical and social variables into a compact latent state used for policy learning. By optimizing livability through direct interaction with environmental sensors, citizen goals, and urban simulation feedback, this AI model not only predicts but actively improves resident quality of life over time. For instance, if simulation results within the digital twin indicate declining satisfaction in a neighborhood due to noise pollution and reduced green space, the AI agent can prioritize green corridor construction or traffic calming strategies, simulate their outcomes, and deploy the most promising intervention.

In this way, the Multi-Objective DRL model, governed by PPO optimization and integrated into Málaga's digital twin, enables continuous, citizen-centered, evidence-based improvement of urban well-being. It transforms smart city systems from reactive monitoring tools into proactive well-being engines—supporting policy agility, public trust, and the adaptive governance essential for next-generation urban sustainability.

Framework: Multi-Objective Deep Reinforcement Learning for Urban Well-being Optimization (MO-DRL-UWO)
Layered AI Architecture for Livability Optimization and Sustainable Urban Development

Layer	Main functionality	AI module (s) and components	Operational scope
Layer 1	Data collection: gather sensor data, citizen feedback, and digital twin simulations	Urban data aggregation and preprocessing module	Input → Sensor data, citizen feedback → Output → Raw data for urban optimization
Layer 2	Reward function: balance multiple urban livability objectives	Multi-objective reward function	Input → Livability indicators (e.g., green space, air quality) → Output → Aggregated reward based on policy targets
Layer 3	Policy update: optimize urban policies using reinforcement learning	Proximal Policy Optimization (PPO)	Input → Reward signal, state data → Output → Updated policies for urban intervention
Layer 4	State encoding: integrate environmental data and citizen goals for policy learning	State encoder for urban environment	Input → Sensor data, citizen goals → Output → Encoded state representing current urban conditions
Layer 5	Simulation and optimization: simulate interventions based on predictions and refine strategies	Urban intervention simulation and strategy refinement engine	Input → Predicted outcomes, interventions → Output → Simulated results for policy refinement and urban service delivery

Moreover, Málaga's transformation into a smart and responsive city has stimulated economic development by attracting new investments and skilled professionals. The combination of intelligent infrastructure, transparent governance, and a high standard of urban services makes the city an appealing destination for businesses and entrepreneurs. This has driven innovation, encouraged new enterprise formation, and contributed to job creation, reinforcing the city's position as a hub of smart economic activity (Mazzetto 2024). To model the economic momentum and entrepreneurial ecosystem generated by Málaga's smart infrastructure, we propose a Graph Neural Network–based Entrepreneurial Activation Model (GNN-EAM), extended with Bayesian Optimization for adaptive investment strategy calibration. This hybrid AI framework captures the structural dynamics of economic agents—startups, investors, accelerators, knowledge institutions—while learning optimal incentive configurations to attract talent, catalyze innovation, and maximize job creation within spatial and policy constraints.

The foundation of the model begins with the structural learning of the entrepreneurial network using a Graph Neural Network (GNN), defined as:

$$H^{(l+1)} = \sigma\left(\tilde{D}^{-1/2}\tilde{A}\tilde{D}^{-1/2}H^{(l)}W^{(l)}\right)$$

where $H^{(l)}$ is the matrix of node embeddings at layer l, $\tilde{A}$ is the adjacency matrix of the economic interaction graph (e.g., partnerships, investments), $\tilde{D}$ is the degree matrix, $W^{(l)}$ is the learnable weight matrix, and $\sigma(\cdot)$ is a nonlinear activation function (e.g., ReLU). This GNN models relational influence and resource flow among entities, enabling the system to identify central actors, innovation clusters, and underutilized nodes.

To adaptively optimize the policy and incentive strategies that activate entrepreneurship and attract investment, the model employs Bayesian Optimization, which iteratively selects optimal hyperparameters (e.g., tax credits, subsidy levels, public–private partnership ratios) to maximize an economic vitality function:

$$\theta^* = \arg \max_{\theta \in \Theta} \mathbb{E}[f(\theta)]$$

where θ represents the set of investment and policy parameters, $f(\theta)$ is the objective function measuring economic performance (e.g., firm creation rate, job generation, innovation output), and $\mathbb{E}[f(\theta)]$ is the expected performance under a probabilistic model learned from historical data and current economic simulations.

To balance exploration (testing new strategies) and exploitation (scaling effective ones), the acquisition function $a(\theta)$ guides the next best configuration to evaluate:

$$\theta_{t+1} = \arg \max_{\theta \in \Theta} a(\theta)$$

where $a(\theta)$ may represent an upper confidence bound (UCB), expected improvement (EI), or probability of improvement (PI), depending on the risk tolerance and strategic objectives of the city's economic office.

This hybrid AI model enables Málaga's digital twin to simulate entrepreneurial scenarios, adjust policy levers in real time, and identify the most effective actions for fostering sustainable economic growth. For example, if the model detects low network centrality among startups in the mobility sector, it can recommend targeted accelerator programs or R&D grants. Alternatively, if job growth plateaus in outer districts, the model can propose zoning incentives or infrastructure enhancements to attract firms.

By harnessing relational economic intelligence (via GNN) with adaptive policy optimization (via Bayesian learning), this far superior AI-driven system turns

Málaga's digital infrastructure into a predictive engine for inclusive and innovation-driven economic development. It transforms economic planning from reactive analysis into proactive orchestration—positioning the city as a model for data-informed, AI-optimized urban prosperity.

Framework: Graph Neural Network-Based Entrepreneurial Activation Model (GNN-EAM)

Layered AI Architecture for Adaptive Economic Development and Investment Strategy Optimization

Layer	Main functionality	AI module (s) and components	Operational scope
Layer 1	Data collection: gather data from entrepreneurial network, investments, and economic agents	Entrepreneurial data aggregation and integration module	Input → Data from startups, investors, accelerators → Output → Raw data for network modeling and optimization
Layer 2	Network representation: learn structural relationships in the entrepreneurial ecosystem	Graph Neural Network (GNN)	Input → Entrepreneurial network data (e.g., partnerships, investments) → Output → Node embeddings representing economic agents and their relationships
Layer 3	Optimization: adaptively optimize investment strategies and policy incentives	Bayesian optimization	Input → Policy parameters (e.g., tax credits, subsidies) → Output → Optimized investment strategies and policy configurations
Layer 4	Exploration and exploitation: balance exploration of new strategies and exploitation of effective ones	Bayesian optimization—acquisition function	Input → Current strategies and performance data → Output → Acquisition function guiding the next strategy evaluation
Layer 5	Simulation and adjustment: simulate economic scenarios and refine strategies based on results	Economic scenario simulation and policy refinement engine	Input → Simulated economic outcomes, interventions → Output → Adjusted policies and strategies based on dynamic simulations and optimization results

In parallel, the initiatives undertaken are deeply aligned with Málaga's broader sustainability goals. By promoting environmentally conscious practices and integrating resilience into urban planning, the city addresses key social and ecological determinants of happiness. This commitment not only reduces the environmental footprint of urban operations but also supports long-term sustainability and the equitable distribution of environmental benefits (Dani et al. 2023). As a result, Málaga demonstrates how the convergence of technology, sustainability, and citizen-centered design can shape a more inclusive and forward-looking urban future. To operationalize the convergence of ecological sustainability, social equity, and citizen well-being in urban planning, we propose the integration of a Multi-Agent Deep Reinforcement Learning model with Fairness Constraints

(MA-DRL-FC). This advanced AI system coordinates the actions of decentralized agents—each representing different urban sustainability domains such as energy distribution, green space access, waste management, and air quality regulation—while explicitly optimizing for both ecological efficiency and equitable benefit allocation across neighborhoods.

At the heart of this system is the multi-agent reward function that integrates both performance and fairness dimensions for each agent i:

$$R_t^{(i)} = E_t^{(i)} - \lambda \cdot I_t^{(i)}$$

where $R_t^{(i)}$ is the adjusted reward for agent i at time t, $E_t^{(i)}$ is the ecological efficiency score (e.g., CO_2 reduction, energy saved, water reused), $I_t^{(i)}$ is an inequality index measuring disparities in access to environmental benefits (e.g., green areas, clean air), and λ is a fairness regularization coefficient that adjusts the trade-off between sustainability maximization and equity.

The inequality term $I_t^{(i)}$ is computed using a spatially normalized Gini-like index across the population distribution of the benefit managed by agent i:

$$I_t^{(i)} = \frac{1}{2\mu^{(i)} n^2} \sum_{j=1}^{n} \sum_{k=1}^{n} | b_j^{(i)} - b_k^{(i)} |$$

where $b_j^{(i)}$ is the benefit level delivered to resident j by agent i, $\mu^{(i)}$ is the mean benefit across the population, and n is the number of residents. This fairness penalty increases when certain groups receive disproportionately lower environmental quality or access.

Each agent updates its action policy using a Multi-Agent Proximal Policy Optimization (MA-PPO) strategy:

$$L_{\text{PPO}}^{(i)}(\theta_i) = \mathbb{E}_t \left[min \left(r_t^{(i)}(\theta_i) \hat{A}_t^{(i)}, \text{clip} \left(r_t^{(i)}(\theta_i), 1 - \epsilon, 1 + \epsilon \right) \hat{A}_t^{(i)} \right) \right]$$

where $L_{\text{PPO}}^{(i)}(\theta_i)$ is the PPO loss for updating agent i's policy θ_i, $r_t^{(i)}(\theta_i)$ is the probability ratio of policy improvement, $\hat{A}_t^{(i)}$ is the advantage function, and ϵ is the policy update threshold for stability. This structure allows each urban agent to learn not only how to improve environmental outcomes but also how to do so equitably across spatial and social demographics.

Embedding this fairness-aware reinforcement learning into the digital twin ecosystem, Málaga's smart governance model can simulate the distributive effects of sustainability interventions—ensuring that policies like green infrastructure expansion or electric mobility subsidies benefit both central and peripheral neighborhoods equitably. As such, the MA-DRL-FC model represents a far superior AI architecture that harmonizes environmental resilience with social inclusion. It demonstrates how equity is not a constraint on innovation but a functional component of adaptive urban intelligence—empowering cities like Málaga to build truly sustainable, citizen-centered futures.

Framework: Multi-Agent Deep Reinforcement Learning with Fairness Constraints (MA-DRL-FC)
Layered AI Architecture for Sustainable and Equitable Urban Development

Layer	Main functionality	AI module (s) and components	Operational scope
Layer 1	Data collection: gather data on environmental indicators and citizen feedback	Environmental and feedback data integration module	Input → Environmental data, citizen input → Output → Raw data for sustainability modeling
Layer 2	Reward function: balance ecological efficiency and fairness across urban sectors	Multi-agent reward function	Input → Ecological efficiency, inequality index → Output → Aggregated reward for each agent with trade-offs between sustainability and equity
Layer 3	Fairness calculation: measure disparities in the distribution of environmental benefits	Inequality INDEX (Spatially Normalized Gini-like)	Input → Benefit levels for residents, population distribution → Output → Fairness penalty to ensure equity in benefits
Layer 4	Policy update: update agent policies using reinforcement learning for optimal actions	Multi-Agent Proximal Policy Optimization (MA-PPO)	Input → Reward signal, agent actions → Output → Updated policy for sustainable and equitable urban interventions
Layer 5	Simulation and adjustment: simulate interventions and refine policies based on outcomes	Intervention simulation and strategy refinement engine	Input → Simulated intervention outcomes → Output → Adjusted urban planning strategies based on data-driven insights

3.5.6 Challenges and Future Directions

While Málaga has made considerable progress in the deployment of smart city initiatives, several critical challenges must still be addressed to ensure the long-term effectiveness, inclusivity, and resilience of its digital transformation. One of the foremost concerns is the issue of data privacy and security. As the city increasingly relies on continuous data collection to inform real-time decision-making and improve public services, safeguarding citizen data becomes both a technical and ethical imperative. Balancing the demand for rich urban data with the protection of personal privacy remains a complex challenge that requires robust cybersecurity protocols, transparent data governance, and public trust (Dani et al. 2023). To address the tension between pervasive urban sensing and the ethical obligation to protect citizen data, we introduce a Differentially Private Federated Learning Framework with Encrypted Aggregation (DP-FL-EA)—an AI model designed to support privacy-preserving real-time analytics in distributed smart city environments. This architecture enables decentralized AI training across edge nodes—such as public infrastructure sensors, mobile devices, and IoT platforms—without transferring raw personal data, while ensuring formal privacy guarantees and resistance to cyber intrusion.

The foundation of this privacy-preserving mechanism is the ε-differential privacy guarantee, which ensures that outputs of urban analytics are statistically indistinguishable with or without any single individual's data:

$$\forall S \subseteq \text{Range}(M), \quad \Pr[M(D_1) \in S] \leq e^{\varepsilon} \cdot \Pr[M(D_2) \in S]$$

where D_1 and D_2 are neighboring datasets differing by one individual's data, M is the privacy-preserving mechanism (e.g., differentially private model), S is any subset of possible outputs, and ε is the privacy loss parameter controlling the allowable data leakage.

Each participating smart city node performs local model training and adds privacy-preserving Gaussian noise to its gradients before transmission:

$$\tilde{g}_t^{(i)} = g_t^{(i)} + \mathcal{N}(0, \sigma^2 I)$$

where $\tilde{g}_t^{(i)}$ is the privatized gradient of node i at time t, $g_t^{(i)}$ is the true gradient, $\mathcal{N}(0, \sigma^2 I)$ is Gaussian noise calibrated to the privacy budget ε, and I is the identity matrix ensuring isotropic noise dispersion.

To secure the aggregation of local updates without revealing individual contributions, the model uses homomorphic encryption for encrypted federated averaging:

$$\theta_t = \text{Decrypt}\left(\sum_{i=1}^{N} \text{Encrypt}\left(\tilde{g}_t^{(i)}\right)\right)$$

where $\tilde{g}_t^{(i)}$ are the encrypted noisy gradients, θ_t is the updated global model, and encryption/decryption operations ensure that the aggregator learns only the collective update—not the individual contents of any local device.

To continuously monitor the cumulative privacy leakage over time, a moment accounting function is used to maintain compliance with the city's privacy governance thresholds:

$$\alpha(\lambda) = \sum_{t=1}^{T} \log \mathbb{E}\left[e^{\lambda Z_t}\right]$$

where $\alpha(\lambda)$ is the λ-th moment of the privacy loss random variable, Z_t is the instantaneous privacy loss at time t, and T is the total number of data access rounds.

This DP-FL-EA framework enables Málaga to support continuous, AI-powered optimization of public services—such as mobility forecasting, energy grid management, or predictive maintenance—without compromising the digital rights of residents. Unlike centralized models that expose raw data and risk re-identification, this system offers a provable privacy shield, real-time analytics, and transparent governance. It transforms data protection from a regulatory burden into an architectural principle of ethical and sustainable urban intelligence.

This AI model ensures that citizen trust remains foundational—not incidental—to the deployment of smart systems. It operationalizes GDPR-compliant data use, fosters algorithmic accountability, and equips the city to scale its digital capabilities while upholding the principles of human-centered governance. In governance terms, the challenge is to convert privacy and security from policy aspirations into enforceable system properties, so that citizen trust becomes a measurable precondition of the digital twin's operation.

Framework: Differentially Private Federated Learning with Encrypted Aggregation (DP-FL-EA)
Layered AI Architecture for Privacy-Preserving Real-Time Analytics in Smart Cities

Layer	Main functionality	AI module (s) and components	Operational scope
Layer 1	Data collection: gather sensor data and citizen feedback	Sensor and feedback data aggregation module	Input → Urban sensor data, citizen input → Output → Raw data for federated learning
Layer 2	Privacy mechanism: ensure privacy preservation through differential privacy	Differential privacy	Input → Data from sensors and citizens → Output → Privacy-preserving data with ε-differential privacy guarantees
Layer 3	Gradient updates: apply Gaussian noise to gradients for privacy-preserving updates	Privacy-preserving gradient update	Input → Local model gradients → Output → Noisy gradients ensuring privacy before transmission
Layer 4	Model aggregation: securely aggregate model updates using homomorphic encryption	Encrypted federated averaging	Input → Noisy gradients from local nodes → Output → Encrypted model updates aggregated into global model
Layer 5	Privacy monitoring: track cumulative privacy loss to ensure compliance	Privacy loss accounting	Input → Data access rounds, privacy loss measurements → Output → Cumulative privacy loss function ensuring compliance with privacy regulations

Another key area of concern is equity in access. The benefits of smart city technologies must be distributed fairly across all segments of the population. Without deliberate policies to reduce digital exclusion, there is a risk that existing social inequalities may be exacerbated. Ensuring equitable access to digital infrastructure, public services, and participatory platforms is essential for building an inclusive smart city that reflects the needs of its entire community (Yigitcanlar and Cugurullo 2020).

To ensure that the deployment of smart city services supports equitable access across all population groups, we propose a Fairness-Constrained Multi-Agent Reinforcement Learning (FC-MARL) framework. This state-of-the-art model consists of a network of intelligent agents—each representing a critical public service or digital infrastructure function—that collaboratively learn optimal service delivery strategies while simultaneously minimizing inequality in resource distribution. The framework explicitly models disparities in access across different demographic and geographic segments, integrating equity as a formal optimization constraint within the AI policy update process.

Each agent i operates under a multi-objective reward function that balances service efficiency and fairness:

$$R_t^{(i)} = U_t^{(i)} - \lambda \cdot \mathcal{D}_t^{(i)}$$

where $R_t^{(i)}$ is the adjusted reward for agent i at time t, $U_t^{(i)}$ represents the utility gained from digital service delivery (e.g., number of users served, latency minimized), $\mathcal{D}_t^{(i)}$ is a fairness disparity metric quantifying the deviation from proportional access, and λ is the fairness sensitivity parameter controlling how aggressively the agent corrects inequality.

The fairness disparity term $\mathcal{D}_t^{(i)}$ is computed based on group-wise proportionality using the Kullback–Leibler divergence between the observed access distribution and the ideal equitable distribution:

$$\mathcal{D}_t^{(i)} = \sum_{g=1}^{G} p_g \cdot \log\left(\frac{p_g}{q_g^{(i)}}\right)$$

where G is the number of demographic groups (e.g., income quintiles, districts, digital literacy levels), p_g is the ideal access proportion for group g, and $q_g^{(i)}$ is the actual proportion of group g served by agent i at time t. This divergence grows when certain groups receive disproportionately less access, prompting the model to reallocate services to reduce imbalance.

Policy updates for each agent are guided by a shared Multi-Agent Proximal Policy Optimization (MA-PPO) formulation to ensure learning stability:

$$L_{\text{PPO}}^{(i)}(\theta_i) = \mathbb{E}_t\left[min\left(r_t^{(i)}(\theta_i)\hat{A}_t^{(i)}, \text{clip}\left(r_t^{(i)}(\theta_i), 1-\epsilon, 1+\epsilon\right)\hat{A}_t^{(i)}\right)\right]$$

where $L_{\text{PPO}}^{(i)}(\theta_i)$ is the PPO loss for agent i's policy parameters θ_i, $r_t^{(i)}(\theta_i)$ is the policy ratio, $\hat{A}_t^{(i)}$ is the advantage function, and ϵ is the clipping threshold to avoid overcorrection in learning.

This FC-MARL framework allows smart city agents to continuously adjust their service strategies—such as Wi-Fi deployment, mobile health unit scheduling, or digital platform access—based not only on demand, but also on demographic fairness metrics. For instance, if certain neighborhoods report lower usage of telemedicine services despite infrastructure availability, the agents will prioritize those zones for improved connectivity, community education, or subsidy adjustment.

Through the integration of fairness as a first-class objective, the FC-MARL system transforms equity from a policy ideal into a functional AI constraint, ensuring that all residents—not only the digitally advantaged—benefit from technological innovation. This framework offers a powerful, future-ready platform for operationalizing inclusive smart city development and correcting structural inequalities in real time.

Framework: Fairness-Constrained Multi-Agent Reinforcement Learning (FC-MARL)

Layered AI Architecture for Equitable Smart City Service Delivery

Layer	Main functionality	AI module(s) and components	Operational scope
Layer 1	Data collection: collect citizen feedback and service usage data	Service usage and feedback integration Module	Input → Citizen feedback, service data → Output → Raw data for processing and analysis
Layer 2	Reward function: balance service efficiency with equity objectives	Multi-agent reward function	Input → Ecological efficiency, fairness disparity → Output → Aggregated reward for each agent, balancing objectives
Layer 3	Fairness disparity calculation: measure disparities in access to services	Fairness disparity calculation using KL divergence	Input → Access data for demographic groups → Output → Fairness penalty metric to guide resource reallocation
Layer 4	Policy update: optimize service strategies with reinforcement learning	Multi-Agent Proximal Policy Optimization (MA-PPO)	Input → Reward signals, agent actions → Output → Updated service delivery policies to improve efficiency and equity
Layer 5	Simulation and strategy adjustment: simulate interventions and optimize service allocation	Service allocation simulation and strategy refinement engine	Input → Simulated service outcomes, fairness metrics → Output → Adjusted strategies for equitable service distribution and urban planning

Equally important is the challenge of sustaining citizen engagement over time. Initial enthusiasm for digital participation platforms often wanes without consistent investment in civic communication, responsiveness, and feedback loops. The long-term success of smart governance depends not only on technical infrastructure but also on the continuous cultivation of a participatory urban culture. To maintain momentum, Málaga must reinforce channels for resident involvement, build civic literacy, and ensure that citizen contributions are visibly integrated into urban decision-making (Dani et al. 2023).

To ensure long-term civic participation in Málaga's smart governance system, we propose a Temporal Graph Neural Network (TGNN) integrated with Neural Survival

Analysis, a far superior AI-driven framework that models the temporal evolution and social influence structure of citizen engagement. This system enables the digital twin to forecast the probability of resident dropout from participation platforms, simulate the impacts of engagement strategies, and prioritize interventions for communities at high risk of disengagement.

At the core of this framework lies a temporal message-passing architecture that learns how each resident's participation behavior is influenced by peers, history, and civic interactions:

$$h_v^{(t)} = \mathrm{GRU}\left(h_v^{(t-1)}, \sum_{u \in \mathcal{N}(v)} \alpha_{vu}^{(t)} W x_u^{(t)}\right)$$

where $h_v^{(t)}$ is the hidden state (engagement embedding) of resident v at time t, $\mathcal{N}(v)$ is the set of socially or civically connected peers, $\alpha_{vu}^{(t)}$ is the temporal attention weight reflecting the influence of peer u on v, $x_u^{(t)}$ is the activity vector of peer u, and W is a learnable weight matrix. This representation integrates past engagement, influence dynamics, and interaction context into each user's engagement trajectory.

To forecast the likelihood of sustained engagement, we apply a Neural Cox Proportional Hazards model to the TGNN-generated embeddings:

$$\log \lambda_v\left(t \mid h_v^{(t)}\right) = \beta^\top h_v^{(t)}$$

where $\lambda_v\left(t \mid h_v^{(t)}\right)$ is the hazard function representing the instantaneous risk of disengagement for resident v, conditioned on their learned temporal-social embedding $h_v^{(t)}$, and β is a trainable coefficient vector that governs the sensitivity of hazard predictions to engagement signals.

The cumulative probability that a resident remains engaged through time T is then computed via the survival function:

$$S_v(T) = \exp\left(-\int_0^T \lambda_v\left(t \mid h_v^{(t)}\right) dt\right)$$

where $S_v(T)$ is the predicted survival probability—i.e., the likelihood that resident v remains civically engaged up to time T. The system prioritizes engagement campaigns, incentives, or civic literacy programs in areas or groups with steep predicted declines.

This AI-driven model empowers Málaga to move beyond reactive engagement efforts by proactively sustaining participation through data-informed interventions. If, for example, youth involvement is forecasted to wane in certain districts, the model can simulate the effect of youth-led urban forums, gamified civic tools, or school-based co-design projects and identify which strategies are most likely to reverse the disengagement trend.

By continuously learning from participation behaviors, peer dynamics, and civic interactions, the TGNN–Survival framework transforms citizen engagement from a volatile phase into a managed lifecycle. It allows Málaga's smart governance system to cultivate enduring participation, reinforce democratic legitimacy, and evolve as a transparent, equitable, and human-centered urban platform.

Framework: Temporal Graph Neural Network-Embedded Survival Model (TGNN-ESM)

Layered AI Architecture for Sustaining Citizen Engagement in Smart Governance

Layer	Main functionality	AI module (s) and components	Operational scope
Layer 1	Data collection: gather real-time citizen engagement data	Citizen engagement data aggregation and preprocessing module	Input → Citizen feedback, platform usage data → Output → Raw data for engagement modeling
Layer 2	Engagement embedding: learn citizen participation behavior over time	Temporal message passing—GRU	Input → Historical participation data, social interactions → Output → Engagement embedding representing behavior trajectory
Layer 3	Engagement risk forecasting: predict likelihood of disengagement using survival analysis	Neural cox proportional hazards model	Input → Engagement embeddings → Output → Predicted risk of disengagement for each citizen
Layer 4	Survival analysis: calculate long-term engagement probability	Survival function	Input → Hazard function → Output → Probability of sustained engagement over time
Layer 5	Intervention prioritization: optimize engagement strategies based on predictions	Engagement strategy optimization engine	Input → Predicted disengagement risks → Output → Prioritized interventions for at-risk communities

Figure 3.4 presents the smart city initiatives and digital twin integration in Málaga, Spain. The central node, "Málaga City, Spain: Smart City Initiatives and Digital Twin Integration," connects to six main areas: "Digital Twin Technology in Málaga," "Urban Planning and Infrastructure Management," "Environmental Monitoring and Sustainability," "Citizen Engagement and Participation," "Impacts of Smart City Initiatives," and "Challenges and Future Directions." "Digital Twin Technology" enables real-time monitoring, data analysis, and urban simulation. "Urban Planning and Infrastructure Management" involves GIS integration and scenario simulation to optimize city planning. "Environmental Monitoring and Sustainability" includes data collection on air quality and climate resilience strategies. "Citizen Engagement and Participation" highlights public platforms for

Fig. 3.4 Málaga City, Spain: smart city initiatives and digital twin integration

feedback and educational outreach. "Impacts of Smart City Initiatives" focuses on improved quality of life, economic development, and sustainability goals. "Challenges and Future Directions" address data privacy, equitable access, and sustaining long-term citizen engagement.

Málaga City exemplifies how digital twin technology and smart city initiatives can be effectively leveraged to enhance urban planning, infrastructure management, and citizen engagement. By integrating real-time data and fostering community participation, Málaga not only optimizes its urban services but also promotes sustainability and resilience. As cities globally face increasing challenges related to urbanization and climate change, the lessons learned from Málaga's innovative approach provide valuable insights for other urban centers aiming to adopt smart technologies for improved governance and quality of life.

3.6 Ålesund, Norway

Ålesund, a picturesque coastal town in Norway, is recognized for its innovative approach to urban management and sustainability. This analysis explores the smart city initiatives undertaken in Ålesund, focusing on the integration of digital twin technology, community engagement, and environmental sustainability. By leveraging advanced technologies, Ålesund aims to enhance the quality of life for its residents while addressing the challenges posed by urbanization and climate change. Ålesund is renowned for its unique architecture and scenic beauty, situated among islands and fjords. As a small yet vibrant urban center, the town has embraced the smart city concept as part of its strategy to optimize urban services, promote sustainability, and engage citizens actively in governance. The city's initiatives align with Norway's broader goals of achieving sustainable development and enhancing urban resilience.

Digital twin technology is central to Ålesund's smart city framework. This technology creates a virtual representation of physical entities, allowing for real-time monitoring and simulation of urban dynamics.

3.6.1 Urban Planning and Development

Ålesund utilizes digital twin technology to support urban planning processes. By integrating data from various sources, including geographic information systems (GIS), sensor networks, and public feedback, the digital twin provides a comprehensive model of the urban environment (Major et al. 2021). Planners can simulate different scenarios, such as traffic flow and land use changes, to assess the potential impacts of new developments before implementation.

Because Ålesund's planning-oriented digital twin is framed as an integrative decision instrument that consolidates heterogeneous urban data, including GIS layers, sensor telemetry, and public feedback, governance pressure accumulates at the interface between civic intelligibility and privacy protection (Major et al. 2021). Major et al. identify privacy concerns and the difficulty that non-expert publics face when interpreting large-scale heterogeneous urban data as salient barriers to citizen engagement, and they correspondingly motivate 3D graphical digital twins and 4D visualizations of geo-localized time series as mechanisms for rendering such complexity tractable in participatory settings (Major et al. 2021). Yet the same spatio-temporal richness that enables fine-grained planning simulations can heighten re-identification risk, which is why privacy protection must be operationalized through concrete design choices such as spatial aggregation for GDPR-aligned, privacy-compliant data handling (Major et al. 2021) and, where sensitive streams are processed, secure-by-design controls including encryption and access governance (El-Agamy et al. 2024). From a broader urban-digital-twin perspective, governance frameworks can then be strengthened by codifying data usage, sharing, and security guidelines, and by institutionalizing participatory oversight arrangements such as public advisory boards and citizen representation, which are discussed as levers for legitimacy and societal acceptance (Mazzetto 2024). In operational terms, this legitimating layer is usefully coupled to lifecycle data governance practices that foreground documented decision rights over data access, sharing, and compliance obligations, so that "public feedback" becomes not merely a data source but a governed input with traceable provenance, quality controls, and authorized uses (Papagiannidis et al. 2025). If the subsequent optimisation and simulation stack introduced below is treated as model-mediated decision support, model governance can further be made inspectable through structured documentation such as AI Cards, which standardize intended use, context of use, and technical specifications, and through audit-oriented documentation that enables external scrutiny of data preparation, human oversight points, and the role of AI in concrete decisions (Golpayegani et al. 2024; Konigstorfer and Thalmann 2022).

For example, by modeling traffic patterns, urban planners can optimize transportation routes and public transport options, reducing congestion and enhancing mobility for residents. This proactive approach helps ensure that urban growth aligns with the community's sustainability goals and quality of life. For instance, to support such scenario-based simulations in urban development, we propose the integration of a Multi-Agent Graph Reinforcement Learning (MAGRL) Framework within the digital twin model. This AI-driven system simulates the complex interplay between heterogeneous urban agents—such as vehicles, land parcels, mobility nodes, and infrastructure components—within a dynamic graph structure. It enables planners to evaluate the system-wide consequences of proposed developments by learning agent behaviors over time and optimizing actions across competing objectives like traffic flow, land efficiency, and public accessibility.

Each urban entity i is modeled as an agent operating on a dynamic graph $G_t = (V_t, E_t)$, where nodes V_t represent spatial features (e.g., intersections, land plots), and edges E_t define functional or physical interactions (e.g., road connectivity or zoning transitions). At each time step t, agent i selects an action a_i^t based on its observed state s_i^t and local graph context:

$$a_i^t = \pi_\theta\left(s_i^t, \mathcal{N}(i)\right)$$

where π_θ, is the shared policy network parameterized by θ, s_i^t, is the local observation of agent i (e.g., congestion level, land use type), and $\mathcal{N}(i)$, denotes the neighborhood of agent i in the graph (adjacent nodes and their features).

The collective goal is to learn optimal agent behaviors that maximize the long-term sustainability and efficiency of the urban system. This is expressed through a cumulative reward function:

$$R = \sum_{t=0}^{T} \sum_{i=1}^{N} \gamma^t r_i^t$$

where r_i^t, is the local reward for agent i at time t (e.g., reduction in travel time, balanced land use allocation), γ, is the discount factor controlling long-term planning sensitivity, and T, is the simulation horizon.

To model decision feedback from multiple scenarios, the digital twin simulates each proposed development as a unique policy trajectory, allowing planners to compare and select the strategy that achieves optimal multi-agent equilibrium under the city's sustainability constraints.

For example, when planning a new residential zone, the model can simulate multiple infrastructure layouts and transit strategies, analyzing their cascading effects on congestion, green space distribution, and pedestrian accessibility. By learning from spatial interactions and historical behaviors, the MAGRL system enables the digital twin to recommend design scenarios that are operationally efficient, environmentally balanced, and socially inclusive.

This AI-driven simulation system empowers decision-makers to anticipate the systemic consequences of planning choices, dynamically optimize land and mobility strategies, and ensure that the city's growth trajectory reflects both sustainability objectives and the lived needs of its residents.

Framework: Multi-Agent Graph Reinforcement Learning (MAGRL)
Layered AI Architecture for Sustainable Urban Planning and Smart City Development

Layer	Main functionality	AI module (s) and components	Operational scope
Layer 1	Data collection: gather urban data from sensors, feedback, and geographic systems	Urban data aggregation and preprocessing system	Input → Sensor data, GIS, public feedback → Output → Raw data for model input and analysis
Layer 2	Agent modeling: represent urban entities as agents in a dynamic graph structure	Multi-agent system (graph representation)	Input → Urban infrastructure, vehicles, land parcels → Output → Representation of agents in a dynamic graph with nodes and edges
Layer 3	Action selection: agents select actions based on local state and neighborhood context	Graph reinforcement learning (policy network)	Input → Agent state, neighborhood context → Output → Selected action based on learned policy
Layer 4	Optimization: maximize long-term sustainability and system efficiency through reward maximization	Cumulative reward function	Input → Agent actions, sustainability objectives → Output → Cumulative reward reflecting system-wide goals
Layer 5	Simulation and strategy adjustment: simulate various urban development strategies and optimize them	Urban development simulation and optimization framework	Input → Simulated outcomes, development strategies → Output → Optimized urban planning recommendations based on system-wide simulations

3.6.2 Infrastructure Management

The digital twin framework in Ålesund also supports efficient infrastructure management. Real-time monitoring of utilities, transportation systems, and public spaces enables the city to identify maintenance needs and optimize resource allocation (Major et al. 2021). For instance, sensors in public infrastructure can detect issues such as wear and tear, allowing maintenance teams to address problems before they escalate, ensuring reliable services for residents. For instance, to support such proactive and distributed infrastructure supervision, we propose the integration of a Federated Predictive Maintenance Architecture using Extreme Gradient Boosting (XGBoost) in combination with Edge-Based Anomaly Detection. This AI-driven framework allows individual edge nodes—such as roadway sensors, bridge strain gauges, or pipeline monitors—to train local models while preserving data privacy and reducing latency. These models predict infrastructure degradation and communicate summarized patterns to a central digital twin, where shared knowledge is aggregated to inform citywide maintenance planning.

Each edge device *ii* trains a local predictive function based on sensor inputs:

$$\hat{y}_i^{(t)} = \sum_{k=1}^{K} f_{i,k}\left(x_i^{(t)}\right), f_{i,k} \in \mathcal{F}$$

where $\hat{y}_i^{(t)}$, predicted degradation level for component i at time t, $x_i^{(t)}$, vector of real-time features (e.g., vibration frequency, temperature, load pressure), $f_{i,\ k}$, decision tree in the ensemble for node i, and $\mathcal{F}$, function space of regression trees.

These local models are periodically synchronized through federated averaging:

$$\theta^{\text{global}} = \frac{1}{N} \sum_{i=1}^{N} \theta^{(i)}$$

where $\theta^{(i)}$, the local parameter set of node i, and θ^{global}, the aggregated global model used to update predictive strategies within the digital twin across all monitored assets.

To enhance the detection of early anomalies that may not yet affect global metrics, an edge-level anomaly scoring function is introduced:

$$\mathcal{A}_i^{(t)} = \left\| x_i^{(t)} - \mu_i \right\|^2 / \sigma_i^2$$

where $\mathcal{A}_i^{(t)}$, is the anomaly score at node i at time t, μ_i, is the rolling average of normal behavior, and σ_i^2, is the variance estimate. If $\mathcal{A}_i^{(t)}$ exceeds a threshold, the system flags the asset for inspection or predictive maintenance.

This distributed architecture enables the city to respond swiftly to micro-level disruptions—such as roadbed subsidence, signal latency in transit nodes, or fluctuating water pressure—while ensuring holistic oversight across its digital twin interface. For example, early detection of wear in a high-traffic pedestrian bridge allows the maintenance unit to schedule intervention during off-peak hours, avoiding safety risks and operational downtime.

This AI framework empowers the city to identify subtle failure patterns, optimize inspection schedules, and extend asset lifespans—all while maintaining system-wide awareness and service reliability. It exemplifies how modern cities can shift from reactive maintenance to intelligent resilience.

Framework: Federated Predictive Maintenance Architecture (FPMA)
Layered AI Architecture for Proactive Infrastructure Management in Smart Cities

Layer	Main functionality	AI module(s) and components	Operational scope
Layer 1	Data collection: gather real-time sensor data from public infrastructure	Infrastructure sensor data aggregation and preprocessing module	Input → Sensor data (e.g., vibration, pressure, temperature) → Output → Raw data for predictive maintenance models
Layer 2	Degradation prediction: train local models to predict infrastructure degradation	XGBoost (Extreme Gradient Boosting)	Input → Sensor data features → Output → Predicted degradation level for each infrastructure component
Layer 3	Model synchronization: aggregate local models into a global model using federated averaging	Federated averaging	Input → Local models from edge devices → Output → Global model for citywide predictive strategies
Layer 4	Anomaly detection: detect early anomalies in infrastructure behavior	Edge-based anomaly detection	Input → Real-time sensor data → Output → Anomaly score indicating abnormal behavior or degradation risk
Layer 5	Maintenance optimization: prioritize maintenance tasks based on predicted degradation and anomalies	Predictive maintenance prioritization and resource allocation system	Input → Anomaly scores, predicted degradation levels → Output → Optimized maintenance schedules and resource allocation

3.6.3 Environmental Monitoring

Ålesund's digital twin technology enhances the city's ability to monitor and manage environmental conditions. By collecting data on air quality, noise levels, and water quality, the city can assess its ecological performance and implement targeted interventions to improve sustainability (Major et al. 2021). This data-driven approach supports efforts to mitigate pollution and enhance the natural environment, contributing to the overall well-being of the community. To advance such adaptive environmental governance, we recommend the deployment of a Variational Spatio-Temporal Gaussian Process (VSTGP) Model within the city's digital twin framework. This AI-driven model enables continuous prediction of environmental variables across both space and time, leveraging sparse sensor data to construct high-resolution pollution and ecological health maps. It allows planners to detect emerging environmental risks and simulate the effects of mitigation strategies before they are physically implemented.

The environmental variable at any location x and time t is modeled as a spatio-temporal Gaussian process:

$$f(x,t) \sim \mathcal{GP}(0, k((x,t),(x',t')))$$

where $f(x, t)$ is the latent environmental function (e.g., pollution concentration), and $k((x, t), (x', t'))$ is a kernel function encoding spatial and temporal correlation between observations at different locations and times.

Because exact inference over large urban datasets is computationally intractable, the model employs variational inference using a set of inducing points $Z = \{z_m\}_{m=1}^{M}$ to approximate the posterior:

$$q(f) = \int p(f \mid \mathrm{u})q(\mathrm{u})d\mathrm{u}$$

where u are latent function values at inducing points (the inducing inputs are denoted by the set $Z = \{z_m\}_{m=1}^{M}$, which are points in the input space (spatio-temporal coordinates) chosen to summarize the behavior of the Gaussian Process across the full dataset), and $q(\mathrm{u})$ is a variational distribution optimized to closely match the true posterior. This formulation allows the system to efficiently produce spatial interpolations and time-forward predictions of pollutant concentration, noise propagation, or water purity.

Each predicted environmental condition is associated with uncertainty bounds:

$$\text{Uncertainty Bound} = \mu(x_t) \pm 2\sigma(x_t)$$

where $\mu(x_t)$ is the predicted mean (or expected value) of the environmental variable at time t and location x_t, $\sigma(x_t)$ is the predicted standard deviation (or uncertainty) at time t and location x_t. The factor of 2 represents the typical confidence interval (95%) for the predictions under a Gaussian assumption. This formula provides a range of possible values for the environmental variable, indicating where the model has high confidence (low uncertainty) or low confidence (high uncertainty), allowing urban managers to assess where confidence is low and where further data collection or citizen science efforts might be deployed. For example, if elevated nitrate levels in a section of the local watershed are detected with high predictive confidence, the system can recommend upstream land use audits, while areas with low confidence may prompt sensor deployment to reduce blind spots in the ecological model.

This AI framework provides planners with an intelligent map of ecological dynamics, offering early warnings of environmental degradation and guiding targeted policy interventions. Ultimately, it reinforces the city's commitment to sustainability, resilience, and data-informed stewardship of its natural assets.

Framework: Variational Spatio-Temporal Gaussian Process (VSTGP)
Layered AI Architecture for Adaptive Environmental Governance and Sustainability in Smart Cities

Layer	Main functionality	AI module(s) and components	Operational scope
Layer 1	Data collection: gather real-time environmental sensor data	Environmental sensor data integration and preprocessing module	Input → Air quality, noise levels, water quality → Output → Raw data for predictive modeling and analysis
Layer 2	Prediction: predict environmental conditions using spatio-temporal Gaussian processes	Spatio-temporal Gaussian process	Input → Environmental sensor data → Output → Predicted pollution concentration, noise levels, or water purity over time
Layer 3	Posterior approximation: approximate the posterior distribution using variational inference	Variational inference	Input → Sensor data, spatio-temporal observations → Output → Posterior approximation for environmental predictions
Layer 4	Uncertainty estimation: provide uncertainty bounds for predictions	Uncertainty quantification	Input → Predictions from GP model → Output → Uncertainty bounds indicating areas with low confidence in predictions
Layer 5	Intervention simulation: simulate and evaluate the impact of mitigation strategies	Mitigation strategy impact simulation engine	Input → Predicted environmental risks, mitigation strategies → Output → Simulated outcomes for decision-making and strategy optimization

3.6.4 Community Engagement and Participation

A significant aspect of Ålesund's smart city initiatives is the emphasis on citizen engagement. The city recognizes that active participation from residents is crucial for effective governance and successful implementation of urban projects.

Ålesund has developed platforms that enable citizens to provide feedback on urban projects and services. By integrating these platforms with the digital twin model, residents can visualize proposed changes and express their opinions more effectively. This participatory approach fosters a sense of community ownership and ensures that urban development reflects the desires and needs of residents (Major et al. 2021). To reinforce this participatory infrastructure within the digital twin, we advocate for the integration of a Dual-Stage Semantic Mapping Model, combining the linguistic depth of RoBERTa embeddings with the hierarchical interpretability of Hierarchical Latent Dirichlet Allocation (hLDA). This AI framework empowers the city to process multilingual citizen input—collected through feedback forms, public consultations, or real-time comment systems—and link resident sentiment to urban development priorities across multiple scales.

In the first stage, each citizen submission s_i is semantically encoded using the RoBERTa model:

$$\mathrm{e}_i = \mathrm{RoBERTa}(s_i)$$

where e_i is the embedding of the input sentence s_i, capturing its grammatical structure, semantic nuance, and emotional valence across languages and dialects.

In the second stage, the model applies hierarchical topic modeling to extract structured thematic layers from aggregated input:

$$P(z_l \mid \mathrm{e}_i) = \frac{n_{z_l}^{(\mathrm{e}_i)} + \eta}{\sum_{z' \in Z_l} n_{z'}^{(\mathrm{e}_i)} + |Z_l|\eta}$$

where $P(z_l \mid \mathrm{e}_i)$ is the probability that e_i pertains to topic z_l at hierarchy level l, $n_{z_l}^{(\mathrm{e}_i)}$ is the count of words in e_i assigned to z_l, Z_l is the topic set at level l, and η is the Dirichlet prior controlling topic sparsity.

This two-stage process allows the digital twin to map feedback hierarchically—for instance, categorizing input into broad themes like "Urban Mobility" or "Green Infrastructure" before refining subtopics such as "bicycle safety" or "urban tree density." These structured insights are then linked to policy levers and simulation parameters within the twin, creating a loop where public discourse directly informs technical planning simulations.

In addition, the system tracks sentiment orientation using an auxiliary scoring function:

$$S(s_i) = \tanh(w^{\top}\mathrm{e}_i + b)$$

where $S(s_i)$ is the sentiment polarity score of citizen input s_i, typically scaled between -1 (very negative) and +1 (very positive), e_i is the RoBERTa embedding of the text s_i, w and b are learnable parameters from a linear classifier (or pre-trained sentiment layer), and tanh ensures bounded sentiment output) guiding planners in identifying areas of high community concern or satisfaction. For example, if feedback clustered under "transport accessibility" consistently exhibits negative sentiment, the digital twin may prioritize scenario testing for improved transit coverage in underserved areas.

This AI-enhanced interface ensures that community voices are not only heard but directly translated into simulation logic, enabling an urban planning process that is democratic, transparent, and continuously responsive to the evolving aspirations of its residents.

Framework: Dual-Stage Semantic Mapping Model (DSSM)
Layered AI Architecture for Citizen Engagement and Urban Planning in Smart Cities

Layer	Main functionality	AI module (s) and components	Operational scope
Layer 1	Data collection: gather multilingual citizen feedback from various platforms	Multilingual feedback aggregation and preprocessing module	Input → Citizen feedback, public consultations → Output → Raw data for processing and analysis
Layer 2	Semantic encoding: encode citizen feedback using RoBERTa embeddings	RoBERTa embedding	Input → Citizen feedback → Output → Semantic embeddings capturing grammatical structure, nuance, and emotional tone
Layer 3	Topic modeling: extract hierarchical themes from feedback using hLDA	Hierarchical Latent Dirichlet Allocation (hLDA)	Input → Semantic embeddings → Output → Hierarchical topic structure categorizing feedback into themes and subtopics
Layer 4	Sentiment analysis: measure sentiment orientation in citizen input	Sentiment scoring function	Input → Semantic embeddings → Output → Sentiment polarity score indicating the emotional tone of feedback
Layer 5	Engagement integration: link feedback insights to urban planning and simulation parameters	Urban planning feedback integration engine	Input → Feedback insights, sentiment scores → Output → Actionable insights for planners to inform decision-making and urban development simulations

3.6.5 *Educational Initiatives*

To enhance public understanding of smart city technologies, Ålesund invests in educational initiatives aimed at familiarizing citizens with the benefits of digital twin technology and data-driven decision-making. Workshops, community meetings, and online resources help demystify these technologies and encourage residents to engage with the smart city agenda. By empowering citizens with knowledge, Ålesund strengthens community ties and promotes collaborative governance (Dani et al. 2023).

3.6.6 Sustainability and Resilience

The smart city initiatives in Ålesund align closely with broader sustainability goals. By optimizing resource usage and enhancing urban resilience, the city aims to create a more sustainable environment for current and future generations.

The digital twin framework allows for more efficient management of resources, including energy, water, and waste. By analyzing consumption patterns and identifying areas for improvement, Ålesund can implement strategies that minimize waste and reduce the carbon footprint of urban activities (Mazzetto 2024). To realize these objectives through adaptive control and predictive optimization, we propose embedding a Multi-Objective Deep Reinforcement Learning Model based on the Soft Actor-Critic (SAC) algorithm within Ålesund's digital twin system. This model supports continuous learning and real-time decision-making over urban resource systems, where the goal is to reduce carbon emissions, improve efficiency, and meet sustainability targets simultaneously. The SAC framework is particularly well-suited for this application, as it enables stochastic policy learning under uncertainty—critical in fluctuating urban conditions such as energy load variability or water demand.

Let s_t be the observed state of the resource environment at time t, including metrics such as energy consumption, water levels, and waste generation. The policy selects a continuous action a_t—representing resource allocation or control measures—according to a stochastic policy π_θ:

$$a_t \sim \pi_\theta(a_t \mid s_t)$$

The model learns to optimize a soft value function that balances cumulative reward with policy entropy:

$$J(\pi) = \sum_{t=0}^{T} \mathbb{E}_{(s_t, a_t)}[r(s_t, a_t) + \alpha \mathcal{H}(\pi(\cdot \mid s_t))]$$

where $r(s_t, a_t)$ is the reward function evaluating sustainability performance (e.g., energy saved, emissions reduced), $\mathcal{H}(\pi)$ is the entropy of the policy, promoting exploration and robustness, and α is the entropy temperature parameter controlling exploration versus exploitation trade-off.

The reward signal incorporates multiple environmental objectives:

$$r(s_t, a_t) = -\lambda_1 f_{\text{energy}}(s_t, a_t) - \lambda_2 f_{\text{water}}(s_t, a_t) - \lambda_3 f_{\text{water}}(s_t, a_t)$$

where each function $f_{\text{energy}}(s_t, a_t)$, $f_{\text{water}}(s_t, a_t)$, and $f_{\text{water}}(s_t, a_t)$ quantifies the inefficiency or environmental cost associated with energy use, water consumption, and waste production, respectively, under the given system state s_t and action a_t; the coefficients λ_1, λ_2, λ_3 are weighting parameters that encode policy preferences for environmental prioritization across domains.

This formulation enables the digital twin to learn optimal policies that dynamically reduce environmental impact. For example, during peak energy demand periods, the system might proactively lower non-essential lighting, balance battery storage across public infrastructure, and reroute waste collection to minimize fuel usage—without requiring predefined schedules.

The model adapts to the city's changing needs, learns from operational feedback, and makes intelligent decisions across interconnected systems—ultimately transforming sustainability from a static policy target into a living, responsive function embedded within the urban fabric.

Framework: Multi-Objective Deep Reinforcement Learning Model (MO-DRL)
Layered AI Architecture for Adaptive Resource Management in Smart Cities

Layer	Main functionality	AI module(s) and components	Operational scope
Layer 1	Data collection: gather real-time environmental data from urban systems	Environmental sensor data aggregation and preprocessing system	Input → Sensor data (energy, water, waste) → Output → Raw data for prediction and optimization models
Layer 2	Action selection: select resource allocation actions based on observed state	Stochastic Policy (SAC)	Input → Observed state (resource metrics) → Output → Continuous action (resource allocation, control measures)
Layer 3	Optimization: learn to maximize reward by optimizing multiple sustainability objectives	Soft Value Function (SAC)	Input → Actions, rewards → Output → Optimized action policy balancing multiple environmental objectives
Layer 4	Reward calculation: evaluate sustainability performance with multi-objective reward function	Reward function (multi-objective)	Input → Sustainability metrics (energy, water, waste) → Output → Reward signal incorporating environmental cost and benefit
Layer 5	Simulation: simulate the impact of resource management decisions on sustainability	Resource management impact simulation engine	Input → Simulated actions, environmental data → Output → Optimized decision-making for long-term sustainability

3.6.7 Climate Resilience

The integration of digital twin technology enhances Ålesund's ability to respond to climate challenges. By modeling various climate scenarios, the city can assess potential risks and develop adaptation strategies to protect infrastructure and community well-being. This proactive approach ensures that Ålesund remains resilient in the face of changing environmental conditions (Yigitcanlar and Cugurullo 2020). In pursuit of a robust and adaptive climate resilience strategy, we recommend embedding a Bayesian Deep Learning Architecture based on Long Short-Term Memory Networks (Bayesian LSTM) with Monte Carlo Dropout into Ålesund's digital twin system. This AI-driven model enables the city to simulate extreme weather scenarios, predict long-term climatic trends, and quantify uncertainty in model forecasts—critical for infrastructure planning and disaster preparedness under environmental volatility.

Let $X_{t-n:t}$ represent a multivariate time series of historical climate indicators (e.g., rainfall, temperature, sea level) observed over the past n time steps. The LSTM-based forecast for the next time step is defined as:

$$\hat{Y}_{t+1} = f_{\text{LSTM}}(X_{t-n:t})$$

where $f_{\text{LSTM}}(\cdot)$ denotes the learned temporal dynamics from the sequence input, and $\hat{Y}_{t+1}$ is the predicted environmental variable (e.g., flood risk level) at time $t + 1$.

To enable probabilistic forecasting and capture model uncertainty, Monte Carlo Dropout is applied during inference:

$$\mathbb{E}\left[\hat{Y}_{t+1}\right] = \frac{1}{M}\sum_{m=1}^{M}\hat{Y}_{t+1}^{(m)}, \text{Var}\left[\hat{Y}_{t+1}\right] = \frac{1}{M}\sum_{m=1}^{M}\left(\hat{Y}_{t+1}^{(m)} - \mathbb{E}\left[\hat{Y}_{t+1}\right]\right)^2$$

where $\hat{Y}_{t+1}^{(m)}$ is the prediction from the m-th forward pass with dropout applied, $\mathbb{E}\left[\hat{Y}_{t+1}\right]$ is the expected forecast, and $\text{Var}\left[\hat{Y}_{t+1}\right]$ captures uncertainty, identifying areas of high climatic risk or model sensitivity.

The resulting probabilistic forecasts are embedded into the digital twin's geospatial interface, enabling scenario comparisons and resource prioritization. For example, if the model identifies a high-probability coastal flooding event with large variance in the city's southern zone, urban managers can simulate adaptation actions such as elevating vulnerable infrastructure, rerouting drainage flows, or activating emergency response simulations.

The city gains not only predictive insights but also probabilistic understanding of climate futures—empowering officials to prepare for disruptions with precision, and to design infrastructure that is both responsive to local risk and robust against global variability.

Framework: Bayesian Deep Learning Architecture (Bayesian LSTM)
Layered AI Architecture for Adaptive Climate Resilience and Urban Planning in Smart Cities

Layer	Main functionality	AI module(s) and components	Operational scope
Layer 1	Data collection: gather historical climate data from sensors	Climate sensor data aggregation and preprocessing module	Input → Sensor data (e.g., rainfall, temperature, sea level) → Output → Raw data for modeling and prediction
Layer 2	Prediction: forecast environmental conditions using LSTM	LSTM (Long Short-Term Memory)	Input → Historical climate data → Output → Predicted environmental variable (e.g., flood risk, temperature)
Layer 3	Uncertainty estimation: capture uncertainty using Monte Carlo Dropout	Probabilistic uncertainty estimation via Monte Carlo Dropout	Input → LSTM predictions → Output → Probabilistic forecast and uncertainty bounds
Layer 4	Scenario simulation: simulate adaptation strategies based on forecasts	Adaptation strategy simulation and evaluation engine	Input → Probabilistic forecasts, adaptation strategies → Output → Simulated outcomes for urban planning and decision-making
Layer 5	Decision support: provide actionable insights for climate resilience planning	Climate resilience decision support and adaptation recommendation system	Input → Simulation results, policy decisions → Output → Adaptation recommendations for infrastructure and disaster preparedness

3.6.8 Challenges and Future Directions

While Ålesund has made commendable progress in deploying smart city technologies, several challenges continue to shape the trajectory of its digital transformation. One of the most pressing concerns is the issue of data privacy and security. As the city increasingly relies on real-time data streams to enhance service delivery and operational efficiency, safeguarding sensitive citizen information becomes an ethical and technical imperative (Major et al. 2021). Ensuring that digital infrastructure adheres to stringent data protection standards while remaining agile enough to support public innovation represents a critical tension that municipal leaders must navigate (Dani et al. 2023).

To safeguard sensitive information while sustaining innovation in smart service delivery, we propose the Privacy-Preserving Federated Graph Learning with Adaptive Differential Privacy (PFG-ADP) framework. This advanced AI-driven architecture allows Ålesund to conduct decentralized learning on dynamic graph-structured smart city data (e.g., mobility, healthcare, energy) while ensuring that citizen identities and behavioral patterns remain untraceable. Each smart node (e.g., IoT sensor cluster, transport terminal, civic database) locally trains graph embeddings using its private subgraph, then securely contributes encrypted updates to a global model without sharing raw data. At the core of the learning framework is the federated graph embedding function:

$$H_v^{(l+1)} = \sigma\left(\sum_{u\in\mathcal{N}(v)} \frac{1}{c_{vu}} W^{(l)} H_u^{(l)}\right)$$

where $H_v^{(l+1)}$ is the node embedding of entity v at layer $l + 1$, $\mathcal{N}(v)$ is the set of neighbors of node v, c_{vu} is a normalization constant (e.g., node degree), $W^{(l)}$ is the trainable graph convolution weight matrix, and σ is a nonlinear activation function (e.g., ReLU). This federated representation is computed locally on each device without centralizing the graph.

To enforce strong privacy guarantees, each local update is perturbed using adaptive differential privacy:

$$\tilde{g}_v = g_v + \mathcal{N}\left(0, \sigma_v^2 I\right)$$

where $\tilde{g}_v$ is the differentially private update vector for node v, g_v is the true gradient from local training, and $\mathcal{N}\left(0, \sigma_v^2 I\right)$ is Gaussian noise calibrated to the local node's sensitivity, dynamically adjusted by privacy demands or group-level exposure risk.

To aggregate local contributions securely, the model uses secure multiparty computation (SMC) or homomorphic encryption:

$$\theta_t = \text{Decrypt}\left(\sum_{i=1}^{N} \text{Encrypt}\left(\tilde{g}_v^{(i)}\right)\right)$$

where θ_t is the updated global model at time t, aggregated from encrypted, noise-perturbed updates $\tilde{g}_v^{(i)}$ from each node ii. This ensures that no entity—central or local—ever sees unprotected data during model convergence.

To quantify and manage privacy leakage over time, a privacy accountant function is used:

$$\epsilon_t = min\{\epsilon : \delta \geq \Pr[\mathcal{M}(D) \neq \mathcal{M}(D')]\}$$

where ϵ_t is the cumulative privacy budget at time t, and δ is the privacy breach threshold (e.g., 1e−5). This budget ensures compliance with GDPR and other legal frameworks.

The PFG-ADP model enables cities to build and deploy graph-based learning systems across their digital infrastructure—such as traffic graphs, utility networks, and social interaction maps—without compromising data privacy. It balances the need for real-time intelligence with the duty to protect citizen data, making it ideal for ethical smart city growth.

This AI-driven model:

- Provides privacy-by-design guarantees aligned with legal and ethical norms.
- Retains full analytical capability across a federated network of urban systems.
- Enables secure digital twin evolution, incorporating behavioral data while shielding personal identities.
- Builds institutional trust by showing mathematically provable protections against surveillance misuse.

Framework: Privacy-Preserving Federated Graph Learning for Smart Cities (PFG-ADP)

Layered AI Architecture for Privacy-Preserving Federated Learning in Smart City Data Systems

Layer	Main functionality	AI module(s) and components	Operational scope (Input → Processing → Output)
Layer 1	Data collection and privacy preparation	Smart city data preprocessing and privacy filtering module	Input → Raw smart city data (mobility, healthcare, energy) → Output → Preprocessed data for federated learning
Layer 2	Graph representation learning for decentralized data processing	Federated Graph Learning (FL), Graph Neural Networks (GNNs)	Input → Local data (subgraph embeddings) → Output → Local graph embeddings
Layer 3	Privacy-preserving local model update	Differential Privacy (DP)	Input → Local model updates (graph embeddings) → Output → Perturbed updates with noise for privacy
Layer 4	Secure aggregation of local updates	Secure Multiparty Computation (SMC), homomorphic encryption	Input → Encrypted local updates from nodes → Output → Aggregated global model updates
Layer 5	Privacy budget management and compliance	Privacy budget monitoring and GDPR compliance engine	Input → Local privacy parameters → Output → Updated privacy budget ensuring GDPR compliance
Layer 6	Deployment and real-time model evolution	Privacy-preserving model deployment and update engine	Input → Updated global model → Output → Deployed privacy-preserving model for real-time smart city data processing
Layer 7	Monitoring and evaluation of privacy protections	Privacy auditing and model performance evaluation framework	Input → Model performance and privacy leakage data → Output → Performance reports and privacy audit results

Another ongoing challenge lies in ensuring equity in access to the benefits of smart city initiatives. While digital technologies can dramatically enhance quality of life, their uneven distribution may exacerbate existing socio-economic disparities if not managed inclusively. Addressing these gaps requires the intentional design of policies and platforms that reach underserved communities and ensure that participation in the smart city ecosystem is not limited by digital literacy, income, or geography (Yigitcanlar and Cugurullo 2020). Equity must be embedded not only in access to services but also in the co-creation of urban solutions. To proactively mitigate inequities in access and participation, we propose the FC-URAM framework—a fairness-aware AI system that learns optimal resource distribution policies under social justice constraints. The model simulates allocation decisions (e.g., public Wi-Fi, mobility subsidies, healthcare units, civic infrastructure) across neighborhoods with heterogeneous needs, digital literacy, and demographic compositions.

At the heart of the framework is a multi-agent policy optimization function that learns to maximize utility while satisfying fairness constraints:

$$\pi^* = \arg\max_{\pi} \mathbb{E}\left[\sum_{t=0}^{T} \gamma^t R(s_t, a_t)\right] \quad \text{s.t.} \quad \mathcal{F}(\pi) \leq \epsilon$$

where π^* is the optimal policy over allocation actions, $R(s_t, a_t)$ is the reward function reflecting urban efficiency (e.g., reduced service delay, access coverage), γ is the discount factor for long-term equity planning, $\mathcal{F}(\pi)$ is a fairness constraint (e.g., disparity in service allocation between privileged and underserved groups), and ϵ is a tunable threshold defining acceptable inequality bounds.

To enforce demographic parity in service access, we introduce a regularization term:

$$\mathcal{L}_{\text{fair}} = |P(a \mid z = 1) - P(a \mid z = 0)|$$

where $P(a \mid z = 1)$ and $P(a \mid z = 0)$ are the probabilities of action a (e.g., allocating high-speed Internet or new clinics) for privileged vs. underprivileged groups, defined by a protected attribute z (e.g., income, ethnicity, district). Minimizing this loss encourages equal treatment across social strata. The total policy optimization objective becomes:

$$\mathcal{L}_{\text{total}} = -\mathbb{E}\left[\sum_{t=0}^{T} \gamma^t R(s_t, a_t)\right] + \lambda \mathcal{L}_{\text{fair}}$$

where λ is a regularization coefficient controlling the trade-off between performance and fairness.

This formulation ensures that smart city services are optimized not only for utility but also for social equity, dynamically adapting policy outputs to the evolving needs of marginalized communities.

The FC-URAM system operationalizes equity as a mathematically embedded goal—not just a post-hoc audit metric. It provides cities with the ability to:

- Proactively identify resource imbalances and underserved regions.
- Simulate multiple allocation strategies and evaluate fairness outcomes.

- Adapt urban policy to shifting socio-demographic realities.
- Maintain algorithmic accountability in public infrastructure deployment.

By incorporating fairness directly into the learning loop, the model guarantees that access to smart services and decision-making influence are not limited by digital divide variables such as income, literacy, or geographic remoteness.

Thus, the proposed AI system transforms equity from a normative goal into a computationally actionable mandate—ensuring that the benefits of digital transformation are co-designed, co-distributed, and co-governed by all sectors of society.

Framework: Fairness-Conscious Urban Resource Allocation Model (FC-URAM)

Layered AI Architecture for Fairness-Aware Urban Resource Allocation

Layer	Main functionality	AI module(s) and components	Operational scope (Input → Processing → Output)
Layer 1	Policy design and demographic analysis	Socio-demographic data analysis and community needs assessment module	Input → Socio-demographic data (income, literacy, geography) → Output → Analysis of underserved communities and needs
Layer 2	Multi-agent policy optimization for resource distribution	Multi-Agent Reinforcement Learning (MARL)	Input → Local resource allocation data, community profiles → Output → Optimized policy for resource distribution
Layer 3	Fairness regularization for resource allocation	Fairness-constrained resource allocation system	Input → Allocation actions for different social groups → Output → Fairness enforcement ensuring equal treatment
Layer 4	Total policy optimization	Integrated policy optimization with fairness constraints	Input → Utility function, fairness regularization → Output → Final policy optimizing both performance and fairness
Layer 5	Simulation of allocation strategies	Resource allocation strategy simulation engine	Input → Policy configurations → Output → Simulated outcomes for various allocation strategies
Layer 6	Dynamic adaptation to shifting demographics	Adaptive policy evolution system for demographic changes	Input → Evolving socio-demographic data → Output → Updated policies to address new needs and imbalances

Equally vital is the challenge of sustaining meaningful citizen engagement over time. Initial enthusiasm around participatory platforms often fades without sustained dialogue, feedback mechanisms, and demonstrable impact. For Ålesund's smart city initiatives to remain dynamic and relevant, ongoing efforts will be needed to cultivate trust, encourage deliberation, and integrate public input into core urban functions (Major et al. 2021). This requires more than technical integration—it demands a cultural commitment to transparency, responsiveness, and civic inclusion at every stage of the planning and implementation process (Dani et al. 2023).

To address the long-term complexity of civic engagement within smart city initiatives, we introduce an AI-driven model entitled the Graph-Based Temporal Engagement Optimization Network (GTEON). This framework builds upon Sequential Graph Neural Networks (SGNN) integrated with a Trust-Aware Engagement Attention Mechanism (TEAM) and Longitudinal Participation Modeling. Unlike generic engagement dashboards or static metrics, GTEON dynamically learns from multiscale behavioral inputs—ranging from survey participation, feedback text sentiment, attendance records, and interaction frequency—to forecast participation decline, identify influential community actors, and optimize engagement strategies over time. The model is designed to integrate directly into a city's civic data architecture, ensuring that public input is not only collected but also predictively modeled and structurally embedded into ongoing urban policy cycles.

At the foundation of the GTEON architecture lies the temporal graph convolution function that captures the evolving participatory state of each citizen node:

$$h_v^{(t)} = \text{GRU}\left(h_v^{(t-1)}, \sum_{u \in \mathcal{N}(v)} \alpha_{vu}^{(t)} W x_u^{(t)}\right)$$

where $h_v^{(t)}$ is the updated representation of citizen or community cluster v at time t, encoding the engagement history and social interactions up to that point, $h_v^{(t-1)}$ is the previous engagement state, $\mathcal{N}(v)$ is the set of connected civic nodes, $\alpha_{vu}^{(t)}$ is the attention weight that prioritizes the social influence of peer u on v, W is the learnable transformation matrix that projects input features into latent space, and $x_u^{(t)}$ is the participation feature vector for neighbor u at time t. This formulation enables the system to capture temporal dependencies and peer dynamics, allowing for nuanced modeling of engagement flows across social and institutional networks.

To quantitatively assess engagement vulnerability, a trust-modulated disengagement risk function is defined using survival modeling:

$$\log \lambda_v\left(t \mid h_v^{(t)}\right) = \beta^\top h_v^{(t)}$$

where $\lambda_v\left(t \mid h_v^{(t)}\right)$ is the hazard rate of civic disengagement for node v at time t, conditional on its latent participatory state $h_v^{(t)}$, and β is the parameter vector that quantifies how underlying engagement patterns influence dropout likelihood. This function allows the model to continuously estimate how trust, feedback mechanisms, and responsiveness impact the risk of losing participants.

To forecast sustained civic involvement over a given horizon T, the survival probability is computed:

$$S_v(T) = \exp\left(-\int_0^T \lambda_v\left(t \mid h_v^{(t)}\right) dt\right)$$

where $S_v(T)$ is the probability that citizen v remains actively engaged through time T, $\lambda_v\left(t \mid h_v^{(t)}\right)$ is the instantaneous hazard rate of disengagement at time t, conditioned on the dynamic engagement embedding $h_v^{(t)}$, and $\int_0^T \lambda_v\left(t \mid h_v^{(t)}\right) dt$ computes the cumulative hazard across the specified horizon. The exponential function maps this risk into a bounded survival probability.

The GTEON framework equips cities with an ethically aligned and computationally adaptive mechanism for sustaining civic inclusion. By embedding trust dynamics, feedback responsiveness, and participatory history into a unified graph-temporal model, it ensures that citizen input is not episodic or reactive but a structural determinant of long-term smart governance success.

Framework: Graph-Based Temporal Engagement Optimization Network (GTEON)

Layered AI Architecture for Sustained Civic Engagement in Smart Cities

Layer	Main functionality	AI module(s) and components	Operational scope (Input → Processing → Output)
Layer 1	Policy design and engagement analysis	Socio-demographic and engagement pattern analysis module	Input → Socio-demographic data (income, literacy, geography) → Output → Analysis of civic engagement patterns and community needs
Layer 2	Temporal graph convolution for engagement dynamics	Sequential Graph Neural Networks (SGNN)	Input → Citizen engagement history and social interactions → Output → Evolving participatory state of each citizen node
Layer 3	Trust-modulated disengagement risk estimation	Trust-aware disengagement risk prediction model	Input → Engagement history and trust dynamics → Output → Estimation of disengagement risk for each citizen node
Layer 4	Engagement survival probability forecasting	Civic engagement survival forecasting engine	Input → Hazard rate, participatory state → Output → Forecasted probability of sustained civic engagement over time
Layer 5	Optimization of engagement strategies	Adaptive engagement strategy optimization system	Input → Predicted disengagement risks, community needs → Output → Optimized engagement strategies to sustain participation

Collectively, these challenges underscore the importance of a human-centered and ethically grounded approach to digital urbanism—one that ensures Ålesund's smart city evolution remains inclusive, secure, and democratically accountable. This illustrates a distinct governance pathway for adoption: the digital twin is sustained not only by data pipelines but by inclusion, ongoing civic literacy, and accountability mechanisms that keep "participation" structurally stable over time.

Figure 3.5 illustrates Ålesund's smart city initiatives centered around digital twin integration. At the core, "Ålesund, Norway: Smart City Initiatives and Digital Twin Integration" connects to seven main areas: "Digital Twin Technology in Ålesund,"

Fig. 3.5 Ålesund, Norway: smart city initiatives and digital twin integration

"Urban Planning and Development," "Infrastructure Management," "Environmental Monitoring," "Community Engagement and Participation," "Sustainability and Resilience," and "Challenges and Future Directions." "Digital Twin Technology" enables real-time monitoring, data analysis, and simulation of urban environments. "Urban Planning and Development" incorporates GIS, sensor data, and public feedback for scenario modeling to optimize urban growth. "Infrastructure Management" focuses on real-time utility monitoring and proactive maintenance. "Environmental Monitoring" tracks air, noise, and water quality, allowing targeted sustainability interventions. "Community Engagement and Participation" highlights platforms for public feedback and educational initiatives on digital twins. "Sustainability and Resilience" prioritizes resource optimization and climate resilience strategies. "Challenges and Future Directions" address data privacy, equitable access, and sustaining long-term citizen engagement.

Ålesund, Norway, exemplifies how digital twin technology and smart city initiatives can be effectively leveraged to enhance urban planning, infrastructure management, and citizen engagement. By integrating real-time data and fostering community participation, Ålesund not only optimizes its urban services but also promotes sustainability and resilience. As cities globally face increasing challenges related to urbanization and climate change, the lessons learned from Ålesund's innovative approach provide valuable insights for other urban centers aiming to adopt smart technologies for improved governance and quality of life.

3.7 Conclusion

From an integrative standpoint, this chapter establishes a unified conceptual and empirical framework demonstrating how Artificial Intelligence (AI) and Digital Twin (DT) technologies now constitute the cognitive, ethical, and operational infrastructure of next-generation urban governance. Through the comparative study of Dubai, Herrenberg, West Cambridge, and Ålesund, it reveals how AI-empowered digital twins are transforming cities from static systems of management into adaptive organisms capable of ethical reasoning, predictive foresight, and participatory evolution. Each case represents a distinctive manifestation of the global transition toward intelligent, human-centric, and sustainability-oriented governance architectures.

Viewed analytically, Dubai exemplifies the transformation of subjective well-being into a measurable governance variable. Its "Happiness Agenda" embeds emotional, psychological, and experiential dimensions of life into algorithmic decision-making through models such as the Utility-Regularized Policy Model for Human-Centered Smart Cities, the Real-Time Urban Adaptation Model, and the Fairness-Constrained Adaptive Policy Optimization Model. These formulations, grounded in reinforcement learning and fairness-aware optimization, create a self-regulating governance loop where citizen satisfaction functions as a mathematical constraint within the policy objective itself. Through federated privacy governance and civic reinforcement frameworks, Dubai pioneers a system in which ethical safeguards, inclusivity, and data protection coexist with computational efficiency. The city's digital twin thus becomes a learning entity that aligns happiness, fairness, and sustainability through continuously adaptive optimization.

Herrenberg, operating at the scale of a European small town, provides a complementary paradigm rooted in decentralized intelligence and participatory sustainability. Its AI-Enhanced Urban System Modeling and Spatial-Temporal Urban Planning AI Models employ Graph Neural Networks, Gated Recurrent Units, and Spatial-Temporal Graph Attention Networks to simulate the town as a living relational graph. These models unify infrastructure dynamics, citizen feedback, and environmental signals within a single adaptive topology, allowing planners to anticipate system-level behaviors before interventions occur. Reinforcement learning and predictive maintenance mechanisms convert the digital twin into a preventive rather than reactive governance tool. Herrenberg's participatory platforms, supported by transformer-based language models, ensure that citizen contributions are not peripheral consultations but computationally encoded feedback loops that shape urban evolution in real time.

West Cambridge extends the paradigm into the domain of research-driven interpretability and federated ethics. The region's digital twin functions as a scientific ecosystem linking academic inquiry, municipal management, and environmental stewardship. Through Spatio-Temporal Attention Models, federated edge analytics, and interpretability modules, the twin synthesizes data sovereignty with analytic transparency. Here, AI serves not only predictive accuracy but epistemic accountability—each decision path remains explainable, traceable, and auditable. West Cambridge thus demonstrates that when digital twins are embedded within research infrastructures, cities can institutionalize continuous innovation without sacrificing privacy, explainability, or public trust.

Ålesund introduces a distinctly Nordic interpretation grounded in environmental resilience, adaptive planning, and multilingual inclusion. As a coastal city highly exposed to climate variability, Ålesund integrates multi-agent reinforcement learning, uncertainty-aware climate forecasting, and entropy-regularized sustainability optimization within its digital twin ecosystem. Through Bayesian deep learning and spatio-temporal environmental inference, it predicts and mitigates climate impacts across sectors such as mobility, energy, and water management. Its Soft-Actor-Critic resource optimization and multi-agent graph frameworks enable balanced trade-offs between efficiency, resilience, and ecological integrity. Importantly, Ålesund institutionalizes participatory co-creation through bilingual feedback interfaces and real-time public dashboards, ensuring that citizens remain active agents within the city's cognitive infrastructure.

From a comparative synthesis, the four cities collectively illustrate five governing principles that define the next phase of AI-driven urbanism. First, value internalization: ethical and social objectives such as happiness, equity, and sustainability must be mathematically embedded within optimization targets, not appended as external audits. Second, uncertainty governance: by employing probabilistic and Bayesian methods, cities must learn to act under uncertainty rather than against it. Third, participatory semantics: civic discourse—captured through natural language and sentiment analysis—must be algorithmically translated into actionable variables within the digital twin, closing the loop between voice and policy. Fourth, privacy and sovereignty by design: federated and edge-based learning architectures ensure that data ownership and transparency remain fundamental civic rights. Fifth, operational foresight: predictive maintenance and adaptive resource control redefine urban management from reactive administration to anticipatory stewardship.

Viewed holistically, scale becomes less important than institutional design. Dubai demonstrates that a megacity can integrate emotion and policy at metropolitan scale; Herrenberg proves that small cities can lead through participatory ethics; West Cambridge shows how research ecosystems can secure interpretability and sovereignty; Ålesund confirms that resilience, inclusivity, and uncertainty-aware learning can coexist even in compact, resource-constrained settings. Together, they delineate an epistemic shift from "smart" to "wise" cities—entities capable of self-reflection, moral calibration, and collective reasoning.

From a governance standpoint, these findings reveal that digital twins are evolving into civic epistemic institutions—constitutions encoded in data, algorithms, and adaptive feedback loops. When fairness penalties, privacy gradients, and participatory rewards are embedded directly within their mathematical structure, cities acquire not only predictive power but normative coherence. Each policy update becomes a micro-experiment in ethical optimization, while each feedback loop becomes a rehearsal in civic self-understanding.

In final analysis, the convergence of AI and Digital Twin systems across Dubai, Herrenberg, West Cambridge, and Ålesund marks the emergence of a new ontology of the city as an intelligent moral system—one capable of learning from itself, reasoning about itself, and governing itself through quantifiable yet humanistic intelligence. The chapter concludes that the true measure of an intelligent city lies not in its computational capacity but in its ethical elasticity: its ability to sustain happiness, fairness, and resilience as co-evolving state variables within the grand equation of urban life.

References

Croket, I., Michiels, P., Lefever, S., Hilgers, G. (2025). Unlimited Potential: The Reusable Local Digital Twin. In: Raes, L., Ruston McAleer, S., Croket, I., Kogut, P., Brynskov, M., Lefever, S. (eds) Decide Better. Springer, Cham. https://doi.org/10.1007/978-3-031-81451-8_5

Dani, A.A.H., Supangkat, S.H., Lubis, F.F., Nugraha, I.G.B.B., Kinanda, R., & Rizkia, I. (2023). Development of a Smart City Platform Based on Digital Twin Technology for Monitoring and Supporting Decision-Making. *Sustainability*, 15, 14002. https://doi.org/10.3390/su151814002

Dembski F, Wössner U, Letzgus M, Ruddat M, Yamu C. Urban Digital Twins for Smart Cities and Citizens: The Case Study of Herrenberg, Germany. *Sustainability*. 2020; 12(6):2307. https://doi.org/10.3390/su12062307

El-Agamy, R.F., Sayed, H.A., & Al Akhatatneh, A.M. (2024). Comprehensive Analysis of Digital Twins in Smart Cities: A 4200-paper bibliometric study. *Artificial Intelligence Review*, 57, 154. https://doi.org/10.1007/s10462-024-10781-8

Golpayegani, D., Hupont, I., Panigutti, C., Pandit, H. J., Schade, S., O'Sullivan, D., & Lewis, D. (2024). AI Cards: Towards an Applied Framework for Machine-Readable AI and Risk Documentation Inspired by the EU AI Act. https://doi.org/10.48550/arXiv.2406.18211

Jørgensen, N., & Ma, J. (2025). Digital Twins Under EU Law: A Unified Compliance Framework Across Smart Cities, Industry, Transportation, and Energy Systems. Electronics 2025, 14(24), 4881; https://doi.org/10.3390/electronics14244881

Konigstorfer, F., & Thalmann, S. (2022). AI documentation: A path to accountability. *Journal of Responsible Technology*, *11*, 100043. https://doi.org/10.1016/j.jrt.2022.100043

Major, P., Li, G., Hildre, H. P., & Zhang, H. (2021). The use of a data-driven digital twin of a smart city: A case study of Ålesund, Norway. *IEEE Instrumentation & Measurement Magazine*, 24(7), 39–49. https://doi.org/10.1109/MIM.2021.9549127

Mazzetto, S. (2024). A Review of Urban Digital Twins Integration, Challenges, and Future Directions in Smart City Development. *Sustainability*, 16, 8337. https://doi.org/10.3390/su16198337

McAleer, S.R., Glidden, J. (2025). Force Multiplier: Realising the Benefits of Open and Re-usable Local Digital Twins. In: Raes, L., Ruston McAleer, S., Croket, I., Kogut, P., Brynskov, M., Lefever, S. (eds) Decide Better. Springer, Cham. https://doi.org/10.1007/978-3-031-81451-8_10

Mureddu, F., Paciaroni, C., Pavelka, M., Pemberton, S., & Remotti, A. (2025). Rights and responsibilities: Legal and ethical considerations in adopting local digital twin technology. In L. Raes, M. Beeckman, C. Jacobs, & G. Poels (Eds.), *Decide Better: Open and Interoperable Local Digital Twins* (pp. 291–309). Springer. https://doi.org/10.1007/978-3-031-81451-8_11

Papagiannidis, D., Mikalef, N., & Conboy, J. K. (2025). Responsible artificial intelligence governance: A review and research framework. *Journal of Strategic Information Systems, 34*, 101885. https://doi.org/10.1016/j.jsis.2024.101885

Pitt, J., Ruiz, C., Kumar, K., Vahdatikhaki, F., & Hartmann, T. (2022). *CDBB West Cambridge Digital Twin: Lessons learned* [Preprint]. arXiv. https://doi.org/10.48550/arXiv.2209.15290

Raes, L., De Lathouwer, B., Hilgers, G., Lefever, S. (2025). Public Development: Towards Open Standards and Components for Local Digital Twins. In: Raes, L., Ruston McAleer, S., Croket, I., Kogut, P., Brynskov, M., Lefever, S. (eds) Decide Better. Springer, Cham. https://doi.org/10.1007/978-3-031-81451-8_6

Sargiotis, D. (2024). Data Governance A Guide Springer, Cham. https://doi.org/10.1007/978-3-031-67268-2

Yigitcanlar, T., & Cugurullo, F. (2020). The Sustainability of Artificial Intelligence: An Urbanistic Viewpoint from the Lens of Smart and Sustainable Cities. Sustainability, 12, 8548 https://doi.org/10.3390/su12208548

Chapter 4
Evaluation Frameworks and Challenges in Smart City Development: Lessons from Five UK Cities

This chapter examines how five leading United Kingdom cities—Birmingham, Bristol, Manchester, Milton Keynes, and Peterborough—transform digital aspiration into institutional capability, redefining evaluation as the cognitive foundation of intelligent urbanism. It demonstrates how governance design, data infrastructure, and citizen participation intertwine when measurement becomes an intrinsic part of decision-making, not a post hoc exercise.

Birmingham establishes a comprehensive urban architecture where open data resources, the British Standards Institution Publicly Available Specification 181 Smart City Framework (BSI PAS 181), and inclusion-oriented services converge to position evaluation as a democratic mechanism of accountability and transparency. Here, data-driven governance acts as a catalyst for both civic empowerment and operational coherence.

Bristol advances an experimental, participatory model through initiatives such as the Future City Demonstrator and the Bristol Is Open network, leveraging the Internet of Things (IoT) to enable real-time learning across the urban fabric. The Renaissance of Places with Innovative Citizenship and Technology (REPLICATE) programme, funded under Horizon 2020, expands this experimentation into applied domains such as energy efficiency, mobility, and digital inclusion. The city's evolution lies in consolidating fragmented Key Performance Indicators (KPIs) into an adaptive evaluative framework capable of continuous self-assessment and transformation.

Manchester repositions the role of data platforms as instruments of reflexive governance. Through initiatives like CityVerve, the city integrates information flows from mobility, health, and environmental systems to enhance policy learning and accountability. The emerging challenge is to institutionalize these analytics within a unified evaluative structure that endures beyond individual projects, reinforcing urban knowledge as a shared civic resource.

Milton Keynes epitomizes the mid-sized city as an innovation laboratory. The MK:Smart initiative and the MK Data Hub transform data into strategic intelligence by linking urban sensors, satellite observations, and citizen-generated

D. Sargiotis, *Blueprints for Future Living*,
https://doi.org/10.1007/978-3-032-24178-8_4

information into cohesive decision systems. The alignment with British Standards Institution (BSI) frameworks such as Publicly Available Specification 181 (PAS 181) ensures interoperability and replicability, bridging experimentation with institutional learning.

Peterborough redefines smartness through the lens of the circular economy. The Peterborough DNA initiative and the Smart Fengate project champion inter-organizational collaboration, by-product exchange, and open data as instruments of urban resilience. Rather than prioritizing technological accumulation, Peterborough's model emphasizes ecological interdependence, embedding the principles of fairness, efficiency, and sustainability within its operational framework.

The chapter articulates a comparative grammar of intelligence across diverse city scales and governance models. Standards provide the necessary comparability for cross-urban benchmarking; platforms grant visibility into complex interdependencies, and participatory engagement ensures legitimacy and trust. Together, they reveal how National Health Service (NHS) partnerships, Small and Medium-Sized Enterprises (SMEs), and the United Nations Sustainable Development Goals (SDGs) can be translated into measurable, ethical, and adaptive urban outcomes.

4.1 Comparative Case Studies of Smart City Evaluation Frameworks in the United Kingdom

4.1.1 Birmingham: Smart Governance, Data Infrastructure, and Adaptive Evaluation

Birmingham exemplifies a pragmatic yet visionary model of smart city transformation within the United Kingdom, structured through institutional coordination and participatory governance. At the core of this transformation lies the Birmingham Smart City Commission, a cross-sectoral governance body uniting leadership from municipal authorities, academia, industry, and civic society (Birmingham City Council 2014). This institutional arrangement underscores a deliberate attempt to align innovation ecosystems with social purpose. The city's strategic agenda—articulated through three interdependent pillars, Technology and Place, People, and Economy—reflects a holistic approach to embedding digital intelligence into the everyday metabolism of urban life, ensuring that technological progress translates into social inclusion and economic competitiveness (Caird and Hallett 2019).

From a technological standpoint, Birmingham has consolidated its digital infrastructure through pioneering initiatives such as the Big Data Corridor and the Data Factory Portal, both designed to enable high-speed information exchange, open-data collaboration, and innovation-led entrepreneurship (Birmingham City Council 2014). These platforms form the cognitive and operational backbone of the city's data ecosystem, transforming raw data into actionable intelligence that supports cross-sector coordination in mobility, energy, and environmental management.

Their significance extends beyond technical capacity; they institutionalize data governance as a civic asset, illustrating Birmingham's ability to treat digital infrastructure not as an end in itself but as a strategic enabler of adaptive policy and transparent governance (Caird and Hallett 2019).

The "People" dimension of the city's smart strategy advances a distinctly human-centric orientation. Through partnerships with the National Health Service (NHS) and local community organizations, Birmingham promotes digital inclusion and telehealth solutions that extend healthcare access to marginalized and aging populations (Birmingham City Council 2014). This approach redefines smartness as a matter of equitable connectivity rather than technological density—demonstrating that digital transformation gains legitimacy only when it advances social justice, accessibility, and well-being.

Economically, Birmingham's smart city model is designed to balance innovation-driven growth with ecological and social sustainability. Strategic initiatives in smart mobility, intelligent energy systems, and small- and medium-sized enterprise (SME) digital capacity-building foster a circular relationship between technological advancement, workforce development, and low-carbon transitions (Birmingham City Council 2014; Caird and Hallett 2019). In doing so, Birmingham aligns its innovation agenda with the United Kingdom's broader decarbonization and green-growth objectives, embedding digitalization within a sustainable economic narrative.

Birmingham remains in a transitional yet progressive phase. While a comprehensive suite of Key Performance Indicators (KPIs) is still under development, the city has initiated an evaluation framework guided by the British Standards Institution (BSI) Publicly Available Specification (PAS) 181 Smart City Framework (Caird and Hallett 2019). This framework provides a procedural and ethical architecture for aligning stakeholder engagement, data management, and benefit realization with the city's strategic priorities. Rather than pursuing premature quantification, Birmingham emphasizes methodological coherence and governance adaptability, ensuring that its evaluation practices evolve in tandem with its digital transformation. The city's partial adoption of PAS 181 reveals an institutional awareness that evaluation in smart cities must remain reflexive, transparent, and contextually embedded.

Birmingham's trajectory exposes a productive tension between operational execution and evaluative maturity. The city has prioritized the tangible delivery of innovation projects while progressively building evaluative mechanisms capable of capturing multidimensional outcomes—economic, environmental, and social. This incremental maturation underscores an essential principle for global smart city practice: evaluation must emerge from institutional learning rather than be externally imposed as a compliance exercise.

Birmingham's smart city evolution demonstrates a balanced governance architecture—technologically ambitious, socially grounded, and normatively coherent. By integrating open-data infrastructures, citizen inclusion, and standards-based evaluation within a single strategic vision, Birmingham transcends the instrumental logic of digitalization. It instead models a civic intelligence framework, in which digital transformation becomes a means of reinforcing democracy, transparency, and collective well-being. As such, Birmingham provides a replicable reference point for international cities seeking to institutionalize digital transformation not as an abstract technological project, but as an ongoing exercise in ethical, adaptive, and human-centered urban governance.

4.1.2 Bristol: Human-Centered Innovation and the Evolution of Reflexive Smart Governance

Bristol stands as one of the United Kingdom's most advanced and internationally recognized smart city ecosystems, defined by the deliberate integration of technological experimentation, sustainability objectives, and social innovation. Its overarching governance vision—articulated through the One City Plan and coordinated by the Bristol Futures Directorate—envisions a "city of inclusion, creativity, and resilience" (Lockwood 2020). Operationally, this vision is implemented through the Connecting Bristol strategy, structured around six interrelated pillars: digital connectivity, innovation ecosystems, responsible data use, innovation management, public service innovation, and challenge-led experimentation. Collectively, these themes aim to embed data intelligence and ethical digitalization into every layer of urban policy, infrastructure, and citizen service (Lockwood 2020).

At the technological core of this transformation lie two flagship projects: the Bristol Future City Demonstrator and Bristol Is Open. The Future City Demonstrator functions as a large-scale urban testbed integrating smart data, transport, and energy systems, while Bristol Is Open—a joint venture between the City Council and the University of Bristol—establishes an open, programmable city network enabling experimentation with the Internet of Things (IoT), real-time analytics, and digital twin simulations under live urban conditions (Caird and Hallett 2019). These initiatives embody an experimental governance paradigm, wherein technology, policy, and civic participation co-evolve through continuous feedback, co-design, and institutional learning. Through such "living laboratory" mechanisms, Bristol transforms its urban environment into a site of iterative discovery and governance-by-experimentation.

Distinctively, Bristol's innovation philosophy departs from market-driven, technology-push models, advancing instead a challenge-led framework where social and environmental problems define the technological agenda (Lockwood 2020). This adaptive ethos—rooted in "learning by doing"—is exemplified in the REPLICATE (Renaissance of Places with Innovative Citizenship and Technology) programme, a €7 million Horizon 2020 demonstrator addressing fuel poverty, energy efficiency, and sustainable mobility in disadvantaged neighborhoods such as Ashley, Easton, and Lawrence Hill (Lockwood 2020). REPLICATE operationalizes the Bristol Approach, a participatory methodology that empowers residents to access and interpret urban data, co-design technological interventions, and share responsibility for innovation outcomes (Michalec et al. 2019). Through this co-creative model, digital transformation becomes a collective civic endeavor rather than a technocratic process.

Social inclusion thus forms the normative backbone of Bristol's smart city philosophy. Citizens are not treated as passive data providers but as co-producers of digital futures, aligning technological progress with the principles of fairness, accessibility, and community empowerment. The city's responsible innovation agenda corresponds with the United Nations Sustainable Development Goals (SDGs), ensuring that environmental, economic, and social dimensions of sustainability remain integrated within all digital initiatives (Lockwood 2020). This people-centered framework spans diverse sectors—from telehealth and low-carbon transport to open-data ecosystems and digital literacy programmes—demonstrating that a city's intelligence resides not in its algorithms but in its capacity to mobilize human creativity and institutional empathy.

Nevertheless, despite its international recognition and institutional sophistication, Bristol continues to face a methodological challenge of evaluation. As Caird and Hallett (2019) observe, while specific Key Performance Indicators (KPIs) have been developed for individual projects such as Bristol Energy and smart transport, these remain fragmented and siloed, limiting the city's ability to integrate findings into a unified performance framework. The Council already manages over 150 separate indicators annually, leading to reporting fatigue and the risk of metric inflation (Caird and Hallett 2019). Consequently, Bristol is exploring approaches to embed evaluation within existing governance architectures, aligning measurement practices with strategic priorities and participatory learning rather than imposing additional layers of bureaucratic assessment.

This persistent evaluative fragmentation reveals a structural paradox of innovation-driven governance: the faster a city evolves technologically, the harder it becomes to consolidate methodological coherence across its expanding digital ecosystem. Bristol's experience illustrates that meaningful evaluation must emerge organically from adaptive governance and institutional reflexivity, rather than being externally prescribed through standardized models. In this regard, the city exemplifies a transitional stage in smart city evolution—one shifting from technological experimentation toward reflexive institutionalization, where evaluation itself becomes a mode of civic intelligence.

Bristol represents both an inspiration and a cautionary exemplar. It showcases how open data infrastructures, participatory innovation, and ethical digital practices can collectively reimagine urban governance. Yet, without an integrative evaluation framework capable of measuring systemic outcomes—such as social equity, trust, and long-term sustainability—Bristol risks fragmenting its transformative vision into disconnected experiments. The next frontier of its smart city journey, therefore, is not technological but epistemological to design reflexive evaluation architectures that transform innovation into learning, experimentation into policy wisdom, and data into enduring civic progress.

4.1.3 *Manchester: From Digital Regeneration to Evaluative Intelligence*

Manchester has evolved from its industrial legacy into a leading smart city laboratory within the United Kingdom, distinguished by its long-term commitment to urban regeneration, digital innovation, and knowledge-based development. As Carter (2012) demonstrates, Manchester's early digital development strategies were deeply embedded in its broader regeneration agenda, emphasizing the role of technology in fostering economic restructuring, urban competitiveness, and social inclusion. This historical trajectory—linking physical redevelopment with digital infrastructure—laid the foundation for Manchester's later prominence as an international testbed for smart city experimentation and governance.

At the forefront of this transformation are initiatives such as the CityVerve demonstrator, which exemplifies the city's ambition to integrate data-driven systems across multiple sectors. The project embodies the evolution of Manchester's digital agenda from isolated innovation efforts toward multi-sectoral experimentation, aligning industrial research capacity with public service modernization. As noted by Gledson et al. (2019), CityVerve's technical infrastructure—anchored in a smart city dashboard capable of combining and analyzing multi-source data streams—demonstrates how data from domains such as mobility, energy, health, and environment can be aggregated into an integrated analytical platform. This platform serves as both a decision-support system for policymakers and a learning mechanism for adaptive urban management, allowing real-time monitoring of city dynamics and service efficiency. In this sense, Manchester's smart ecosystem extends beyond technological deployment to embrace data governance and knowledge integration as central pillars of sustainable innovation.

Nevertheless, as Caird and Hallett (2019) emphasize, the city's smart initiatives reveal a persistent challenge in designing coherent evaluation frameworks that transcend project boundaries. Evaluation in Manchester remains predominantly project-specific and funder-driven, oriented toward short-term Key Performance Indicators (KPIs) rather than systemic outcomes. While individual projects like CityVerve or other demonstrators collect valuable quantitative metrics—such as data utilization rates or emissions reductions—these indicators rarely feed into a unified, citywide assessment architecture. The absence of an integrative evaluative model limits Manchester's capacity to capture cumulative impacts, institutional learning, and long-term governance effectiveness.

In synthesis, Manchester's trajectory illustrates both continuity and transformation: continuity in its strategic use of digital technology as an instrument of regeneration, and transformation in its pursuit of data-centric governance as a foundation for intelligent urban policy. Yet, as both Carter (2012) and Caird and Hallett (2019) make clear, technological sophistication must be matched by evaluative maturity. The city's future success depends on embedding evaluation-by-design into its digital infrastructure—ensuring that every innovation contributes not only to efficiency and growth but also to reflection, accountability, and collective learning. As Gledson et al. (2019) show, Manchester's emerging data integration platforms already provide the analytical groundwork for such reflexivity; what remains is to transform these platforms into the evaluative intelligence that defines a truly smart city.

4.1.4 Milton Keynes: From Data Infrastructure to Evaluative Awareness

Milton Keynes stands as a paradigmatic example of a medium-sized city that has strategically positioned itself at the forefront of the United Kingdom's smart city transformation. Through the Council-led Future City Programme, the city has cultivated a distinctive ecosystem where data-driven innovation, infrastructure experimentation, and citizen participation converge to redefine urban governance (Caird and Hallett 2019). This programme—designed through collaborations between local government, universities, businesses, and national innovation bodies—seeks to reconcile economic growth with sustainability, while strengthening Milton Keynes' identity as a technologically advanced yet socially responsive urban environment.

At the center of this transformation lies MK:Smart, the city's flagship initiative coordinated by The Open University in partnership with industry and academic actors. The programme's central technological pillar, the MK Data Hub, serves as a dynamic platform aggregating data from multiple urban systems, including transport, energy, water, environmental sensors, satellite feeds, and social media inputs (Caird and Hallett 2019). This integration enables cross-sectoral analytics and decision support, turning the city into a living laboratory where real-time intelligence informs adaptive management. The MK Data Hub's capacity to consolidate heterogeneous data streams reflects an operationalization of the "city-as-a-system" paradigm, translating complex urban interactions into measurable indicators of efficiency and resilience.

However, as Cook and Valdez (2021) reveals, the evolution of Milton Keynes' smart city identity extends beyond technological instrumentation. It encompasses what she terms the city's smart atmosphere—a socio-material assemblage shaped by sensory experience, digital mediation, and public perception. Here, smartness is not merely infrastructural but experiential: it emerges through the affective relations between people, data, and space. The city's participatory ethos, embodied in living labs, workshops, and co-design events, encourages residents to engage with technology as active contributors rather than passive recipients. Such practices cultivate a sense of collective authorship and trust, transforming Milton Keynes into a civic ecosystem where technological innovation intertwines with emotional and cultural belonging.

From an evaluation perspective, Milton Keynes remains in a formative stage. As Caird and Hallett (2019) emphasize, most of the city's smart initiatives are externally funded, requiring project-level reporting to meet funder obligations rather than adhering to a unified, citywide assessment architecture. Although certain projects have developed Key Performance Indicators (KPIs) for outcomes such as mobility efficiency or data utilization, these remain fragmented and lack integration into a consistent evaluative framework. The Council acknowledges the methodological difficulty of linking discrete innovations to systemic city impacts, recognizing that causal relationships between smart projects and urban well-being are complex and often indirect. Accordingly, evaluation in Milton Keynes currently functions best at the project level, where outputs—such as technological demonstrations and stakeholder engagement—are more immediately observable.

Nonetheless, the groundwork for an integrative evaluation model is being laid. By aligning emerging monitoring efforts with the British Standards Institution's (BSI) PAS 181 Smart City Framework, Milton Keynes is gradually moving toward methodological standardization in governance, data management, and impact reporting. This evolution signals an institutional awareness that sustainable innovation requires not only experimental agility but also evaluative coherence. As Cook and Valdez (2021) suggests, the city's unique blend of infrastructural intelligence and atmospheric engagement positions it to pioneer a form of reflexive smart governance, where data systems and lived experiences co-produce meaning and accountability.

Milton Keynes exemplifies the transition from technological experimentation to evaluative maturity. Its MK:Smart initiative and Data Hub demonstrate the potential of mid-sized cities to achieve global relevance through digital innovation, while its citizen-centered engagement practices affirm that the essence of smartness lies as much in social learning as in technical performance. Yet, the absence of a fully integrated evaluation framework underscores the next developmental imperative to transform its data infrastructure into an evidence-based policy engine capable of capturing not only efficiency and innovation, but also equity, well-being, and environmental stewardship.

4.1.5 Peterborough: Circular Intelligence and Adaptive Evaluation in a Mid-Sized Smart City

Peterborough constitutes one of the most conceptually distinctive and policy-innovative cases in the United Kingdom's smart city evolution. Emerging from the national Future City Demonstrator competition (Cowley and Joss 2020), the city's transformation is anchored in Peterborough DNA, a programme funded by Innovate UK to embed smart and circular economy principles within local governance. Rather than pursuing a technology-centric agenda typical of larger metropolitan centers, Peterborough's strategy emphasizes resource efficiency, civic participation, and systemic interconnectivity—seeking to redefine "smartness" as a function of ecological intelligence and social inclusion (Cowley et al. 2017).

At the heart of this model lies the principle of urban symbiosis, exemplified by the Smart Fengate project, which promotes inter-firm collaboration and material exchange among local industries. By enabling businesses to share resources, waste, and by-products, Smart Fengate demonstrates that smart urbanism can manifest through networked cooperation as much as through digital infrastructure. As Cowley and Joss (2020) observe, this approach illustrates how smaller cities can leverage national innovation competitions to build local experimental systems that foster both economic renewal and environmental resilience.

Equally central to Peterborough's vision is the open-data infrastructure that underpins the city's governance ecosystem. Open data is conceived not merely as a technical utility but as an instrument of transparency and participatory democracy. Through the creation of publicly accessible datasets, hackathons, and co-creation events, Peterborough has operationalized collective intelligence—enabling citizens, start-ups, and researchers to collaborate on local challenges such as energy efficiency, waste reduction, and transport optimization (Caird and Hallett 2019). This participatory emphasis differentiates Peterborough from techno-managerial models of urban innovation by embedding public engagement within the digital architecture of governance.

However, despite its conceptual maturity, Peterborough's evaluation mechanisms remain in development. As Caird and Hallett (2019) highlight, while the city has established project-specific Key Performance Indicators (KPIs)—for example, metrics on participation, recycling rates, and energy performance—it lacks a fully integrated, citywide evaluation system capable of connecting discrete project outcomes to systemic transformations in sustainability and social equity. This fragmentation reflects a broader pattern identified across UK demonstrator cities, where project-level success often outpaces institutional learning and longitudinal assessment.

The challenge for Peterborough, therefore, lies in institutionalizing evaluative intelligence. The city's movement toward alignment with the British Standards Institution (BSI) Publicly Available Specification (PAS) 181 Smart City Framework represents a promising pathway toward methodological coherence (Caird and Hallett 2019). PAS 181 encourages cities to design adaptive and reflexive evaluation processes—ones that evolve alongside technological, social, and environmental change. Such frameworks could enable Peterborough to transition from monitoring immediate outputs to understanding long-term system dynamics and cross-sectoral impacts.

Viewed comparatively, Peterborough embodies the strengths and tensions of innovation-driven mid-sized cities: agile enough to experiment, yet resource-constrained in achieving evaluative consolidation. Its trajectory demonstrates that urban intelligence is not measured by the density of sensors or algorithms, but by a city's capacity to observe, learn, and adjust its developmental logic. The Peterborough DNA initiative thus offers a compelling model of circular governance, where smartness emerges through the continual reconfiguration of data, material, and human relations in pursuit of collective well-being.

4.2 Conclusion

The comparative analysis of Birmingham, Bristol, Manchester, Milton Keynes, and Peterborough reveals the multidimensional evolution of smart city evaluation in the United Kingdom as a critical testbed for the global discourse on intelligent urbanism. Each city—distinct in scale, governance philosophy, and innovation trajectory—demonstrates how data-driven transformation becomes meaningful only when embedded in adaptive, ethically informed, and context-sensitive evaluative frameworks. Across these five cases, smartness emerges not as a technological end state but as a process of institutional learning, civic reflexivity, and social reinvention through evidence-based governance.

Birmingham's trajectory represents the institutional maturity of smart governance grounded in strategic coherence. Its Smart City Commission and adoption of the British Standards Institution (BSI) Publicly Available Specification (PAS) 181 framework exemplify an evolving model of methodological rigor, integrating technological infrastructure, social inclusion, and economic renewal. The city's Big Data Corridor and Data Factory Portal stand as early exemplars of how data ecosystems can be operationalized into cross-sectoral coordination mechanisms. Yet Birmingham's true innovation lies in its incremental movement from fragmented performance metrics to a holistic evaluative culture, where measurement serves as an instrument of democratic accountability and civic transparency rather than bureaucratic compliance.

Bristol, by contrast, manifests the intellectual and ethical vanguard of participatory urbanism. Through its Bristol Futures Programme, the Connecting Bristol strategy, and flagship initiatives such as Bristol Is Open and the Future City Demonstrator, the city redefines digital transformation as a co-creative civic endeavor. The REPLICATE project and the "Bristol Approach" demonstrate how citizens can become co-producers of data intelligence, situating technological experimentation within an ethos of inclusion, accessibility, and environmental justice. Nonetheless, Bristol's challenge resides in evaluative fragmentation—the proliferation of over 150 disparate Key Performance Indicators (KPIs) diluting strategic focus. Thus, its next frontier lies in cultivating reflexive evaluation architectures capable of unifying project-based learning into systemic urban intelligence that continuously redefines the social purpose of innovation.

Manchester provides a contrasting but complementary narrative of technological regeneration intertwined with evaluative emergence. The CityVerve and Triangulum projects illustrate how Internet of Things (IoT) integration and renewable retrofitting converge within a governance model that links industrial research, urban competitiveness, and social experimentation. The city's data dashboards symbolize an evolving epistemic infrastructure for real-time learning. Yet, as Manchester's evaluation remains largely funder-driven, its most urgent task is to transform these analytical platforms into self-reflexive systems of accountability—where data integration catalyzes knowledge integration and where evaluation becomes a generative, policy-shaping intelligence rather than a post-project audit.

Milton Keynes epitomizes the innovative potential of medium-sized cities to transcend scale constraints through infrastructural imagination. Its MK:Smart initiative and MK Data Hub have established a living data ecosystem where sensors, social media, and institutional feeds converge to model the city as a dynamic system. This architecture not only supports energy and mobility experimentation but also fosters an emergent "smart atmosphere," a socio-material domain where technological intelligence intersects with human experience. Yet the evaluative maturity of Milton Keynes remains emergent: while project-level KPIs capture operational success, systemic assessment across economic, environmental, and cultural dimensions is still developing. By aligning gradually with PAS 181, Milton Keynes is laying the groundwork for integrative assessment frameworks, positioning itself as a pioneer of reflexive evaluation where data infrastructure evolves into evidence-based policy consciousness.

Peterborough, finally, offers perhaps the most philosophically profound departure from conventional smart city paradigms. Its Peterborough DNA programme and Smart Fengate initiative redefine intelligence through the lens of circular economy and collaborative symbiosis. Here, the city's smartness resides not in digital saturation but in ecological and social interdependence—businesses exchanging by-products, citizens co-designing data solutions, and governance systems rooted in collective resource efficiency. While Peterborough's evaluation mechanisms remain project-based, its orientation toward PAS 181 signals an intention to institutionalize adaptive, contextually embedded assessment processes. The city's model of circular governance, wherein material, data, and human networks co-evolve, offers a scalable template for small- and mid-sized cities worldwide seeking to operationalize sustainability through participation and reflexivity.

From a global analytical perspective, the synthesis of these five urban trajectories articulates a transformative doctrine for twenty-first century smart city governance. Evaluation must evolve from a peripheral afterthought into the cognitive core of urban intelligence—a recursive process that binds innovation to accountability, experimentation to learning, and technological progress to ethical purpose. Standards such as those of the British Standards Institution serve as scaffolds for comparability and methodological integrity, yet their true power lies in enabling cities to cultivate evaluative autonomy and contextual sensitivity.

Viewed through this integrated lens, Birmingham teaches institutionalization, Bristol embodies participation, Manchester models data integration, Milton Keynes enacts infrastructural reflexivity, and Peterborough personifies circular ethics. Together they reveal that the success of smart cities rests not on the density of sensors, platforms, or algorithms, but on the cultivation of evaluative intelligence—the civic capacity to measure wisely, learn continuously, and adapt ethically. It is through such multidimensional evaluation that cities transcend digital modernity to become truly intelligent ecosystems: conscious, equitable, and resilient in both structure and spirit.

References

Birmingham City Council. (2014). *Birmingham Smart City roadmap: Revised November 2014*. Birmingham Smart City Commission. https://birminghamsmartcity.files.wordpress.com/2013/12/birmingham_smart_city_roadmap_revised-nov-2014.pdf

Caird, S. P., & Hallett, S. H. (2019). Towards evaluation design for smart city development. *Sustainability, 11*(6), 1609. https://doi.org/10.1080/13574809.2018.1469402

Carter, D. (2012). *Urban regeneration, digital development strategies and the knowledge economy: Manchester case study. Journal of the Knowledge Economy, 4*(2), 169–189. https://doi.org/10.1007/s13132-012-0086-7

Cook, M., & Valdez, A.-M. (2021). *Exploring smart city atmospheres: The case of Milton Keynes. Geoforum, 127,* 180–188. https://doi.org/10.1016/j.geoforum.2021.10.015

Cowley, R., & Joss, S. (2020). *Urban transformation through national innovation competitions: Lessons from the UK's Future City Demonstrator initiative. Journal of Urban Affairs.* https://doi.org/10.1080/07352166.2020.1828903

Cowley, R., Joss, S., & Dayot, Y. (2017). The smart city and its publics: insights from across six UK cities. *Urban Research & Practice, 11*(1), 53–77. https://doi.org/10.1080/17535069.2017.1293150

Gledson, A., Ba Dhafari, T., Paton, N., & Keane, J. (2019). *A smart city dashboard for combining and analysing multi-source data streams.* In *Proceedings of the 20th IEEE International Conference on High Performance Computing and Communications (HPCC/SmartCity/DSS 2018)* (pp. 208–214). IEEE. https://doi.org/10.1109/HPCC/SmartCity/DSS.2018.00226

Lockwood, F. (2020). *Bristol's smart city agenda: Vision, strategy, challenges and implementation. IET Smart Cities,* 2(4), 208–214. https://doi.org/10.1049/iet-smc.2020.0063

Michalec, A. O., Hayes, E., & Longhurst, J. (2019). *Building smart cities, the just way: A critical review of "smart" and "just" initiatives in Bristol, UK. Sustainable Cities and Society, 47,* 101510. https://doi.org/10.1016/j.scs.2019.101510

Chapter 5
AI-Enabled Personalization in Smart Cities and Digital Mental Health Care

This chapter offers a comprehensive exploration of the convergence between AI-enabled personalization in smart cities and digital mental health care, integrating two dynamic fields that collectively shape the future of urban living and healthcare. It weaves together the practical application of Information and Communication Technologies (ICT) in urban systems, exemplified by Barcelona's Smart City initiative, and the cutting-edge advances in AI-driven conversational agents for mental health support, aiming to offer scalable, ethical, and personalized solutions to societal challenges. Through a holistic view, this chapter demonstrates how AI systems, both in urban development and healthcare, contribute to improving quality of life, citizen engagement, and ethical standards, while addressing the complexities of implementation and governance.

The chapter opens with an in-depth analysis of Barcelona's transformation into a smart city, demonstrating how the integration of ICT, open data, and smart services has redefined urban innovation. The chapter details Barcelona's journey from the 1990s, when it first began embracing ICT for urban regeneration, to its current status as a global leader in smart city development. Central to this transformation is the 22@Barcelona district, once an industrial hub, now a model of urban sustainability, innovation, and economic growth.

The 22@Barcelona district stands as a testament to how cities can leverage green infrastructure, economic growth, and technological innovation to create livable urban environments. The Urban Lab initiatives within this district are key, providing a real-world testing ground for living labs—experimental zones where new technologies are tested in collaboration with businesses, citizens, and government. Projects like smart street lighting and electric vehicle charging stations exemplify how technological innovations are contributing to sustainability goals and the broader smart urban mobility strategy of the city.

Moreover, living labs are portrayed as a central concept for developing innovative solutions for mobility, public services, and education. These labs are fostering public–private partnerships and offering citizens a direct role in co-creating urban solutions. This model serves as a blueprint for other cities aiming to achieve

D. Sargiotis, *Blueprints for Future Living*,
https://doi.org/10.1007/978-3-032-24178-8_5

sustainable, data-driven governance while ensuring that the technologies deployed meet the specific needs of urban communities.

The chapter continues by exploring the infrastructural backbone necessary for smart city functioning. The development of an extensive fiber optic network, the creation of a Wi-Fi mesh network, and the installation of sensor networks throughout the city are key infrastructural elements that empower Barcelona's smart services. These infrastructures enable real-time data collection and service optimization, such as monitoring environmental conditions, managing energy use, and optimizing waste management systems. Furthermore, the chapter emphasizes the importance of open data initiatives that make government data publicly available, allowing businesses and citizens to create applications that further enhance urban living.

The integration of these technological components facilitates better connectivity, data-driven decision-making, and enhanced transparency, leading to improved public service efficiency and citizen engagement. As cities continue to embrace these models, Barcelona's experience offers critical lessons on the collaboration between government, businesses, and citizens in designing and deploying smart city solutions.

Shifting focus, the chapter delves into the transformative role of AI-driven conversational agents in the field of digital mental health care. Recent breakthroughs in Natural Language Processing (NLP) and Machine Learning (ML) have led to the development of AI chatbots capable of providing emotional support and therapeutic interventions, especially in settings where access to traditional mental health resources is limited. These systems, grounded in Cognitive Behavioral Therapy (CBT) frameworks, have demonstrated measurable improvements in user well-being by helping users manage conditions like anxiety, depression, and behavioral addiction.

The chapter introduces Transformer-Based Adaptive Dialogue Framework (TADF) and Dual-Encoder Emotional Cognition Framework (DECF) as advanced AI architectures designed to provide personalized therapeutic interventions. These systems use reinforcement learning to adapt interactions and optimize the emotional and therapeutic effectiveness of responses. The chatbots are designed to simulate empathetic human dialogue, offering users the ability to self-monitor emotional states, identify cognitive distortions, and reinforce coping strategies.

The integration of AI chatbots into mental health services also presents significant ethical challenges. The chapter highlights the importance of ensuring privacy, algorithmic fairness, and transparency in these systems. By leveraging techniques such as ε-differential privacy to protect user data and Shapley Additive Explanations (SHAP) for interpretability, the chapter emphasizes that AI-driven systems must operate with clear ethical guidelines to prevent harm, ensuring demographic fairness and equitable access for all users.

A major section of the chapter is dedicated to ethical considerations in the deployment of AI chatbots for mental health care. The chapter introduces the Privacy-Preserving Fairness-Aware Decision Framework (P^2FADF), which operationalizes differential privacy, bias mitigation, and model explainability to ensure ethical compliance in AI-driven mental health systems. This framework is designed to address concerns related to data security, equitable model behavior, and decision transparency, ensuring that AI systems operate in a manner that is both clinically effective and socially responsible.

Through the Fairness-Calibrated Representation Alignment Model (FCRAM), the chapter demonstrates how AI systems can reduce algorithmic bias by incorporating adversarial debiasing and group fairness regularization. The framework also introduces counterfactual fairness techniques to assess the impact of sensitive attributes (such as gender and ethnicity) on AI model predictions, ensuring that the decisions made by these systems are free from discriminatory influences.

Additionally, the chapter discusses explainability in AI, highlighting the need for transparent decision-making in mental health chatbots. By utilizing SHAP values and attention mechanisms, the chapter outlines how AI systems can be made more interpretable, enabling users to understand the rationale behind AI-driven recommendations and ensuring that users remain in control of their mental health journey.

In conclusion, the chapter underscores the dual promise of AI personalization in smart cities and digital mental health care. It provides a comprehensive analysis of how ICT infrastructures can transform urban spaces into more sustainable, efficient, and innovative environments while showcasing the potential of AI-driven mental health systems to offer scalable, ethical, and personalized support. From Barcelona's smart city initiatives to AI chatbots for mental health, the chapter presents a holistic view of the technological advancements that are shaping the future of urban living and healthcare, offering key insights into ethical AI design and citizen well-being.

By integrating reinforcement learning, NLP, privacy protection, and fairness into both urban and healthcare solutions, the chapter argues that these AI systems can not only optimize service delivery but also ensure inclusive access, fostering a more equitable and sustainable future for all.

5.1 The Smart City Initiative: The Case of Barcelona

The paper "A Smart City Initiative: The Case of Barcelona" by Bakıcı et al. (2013) provides a detailed analysis of Barcelona's transformation into a smart city. The authors explore how Information and Communication Technologies (ICT) have been integrated into the city's infrastructure and services, positioning Barcelona as a global leader in smart urban development. The study examines the drivers, challenges, and benefits of the city's smart initiatives, focusing on the 22@Barcelona district as a model of urban innovation. This analysis covers several key components of Barcelona's smart city strategy, including smart districts, living labs, open data, and smart services.

5.1.1 Barcelona's Transformation into a Smart City

Barcelona's journey toward becoming a smart city began in the 1990s, following significant urban regeneration projects. Faced with challenges such as outdated infrastructure and economic stagnation, Barcelona embraced ICT to foster innovation, improve urban management, and enhance citizen engagement. Today, Barcelona's smart city model integrates ICT, infrastructure, and human capital to create a sustainable, livable, and innovative urban environment.

The 22@Barcelona district is a focal point of Barcelona's smart city initiative. Formerly an industrial area, the district has been transformed into a hub for knowledge-intensive businesses and urban innovation. The 22@Barcelona district supports the development of green infrastructure, economic growth, and technological innovation. It provides space for living labs and urban research, allowing companies and research institutions to collaborate on the development of smart city solutions.

The 22@Barcelona Urban Lab initiative is a notable example of how living labs are used to test and refine new products and services in real-life urban settings. Projects in the district include smart street lighting systems with sensors to monitor air quality, temperature, and humidity, as well as charging stations for electric vehicles. These innovations contribute to Barcelona's goals of sustainability and smart urban mobility.

Barcelona has also embraced the concept of living labs—urban areas where citizens, businesses, and governments collaborate to co-create innovative solutions. Living labs provide a real-world environment for the testing of new technologies in areas such as mobility, public services, and education.

The BDigital Cluster TIC Living Lab and i2Cat LivingLab are examples of how living labs foster public–private partnerships and promote innovation. These labs are used by businesses to develop and test new digital technologies, which can then be scaled up for citywide implementation. The living labs also encourage citizen participation, ensuring that innovations meet the needs of the community.

Barcelona's smart city transformation required significant investment in urban infrastructure. The city built an extensive fiber optic network, created a Wi-Fi mesh network, and installed sensor networks throughout the city. These infrastructures serve as the backbone for smart services and data collection.

The deployment of sensors in public spaces allows the city to monitor environmental conditions, manage energy use, and optimize public services such as waste management and street cleaning. Additionally, the public Wi-Fi network enhances connectivity for residents, businesses, and tourists, enabling better access to online services and real-time data.

Barcelona has embraced the open data movement by making government data available to the public. The open data platform provides access to information on topics such as urban planning, transportation, and environmental data. This initiative aims to increase transparency, foster citizen participation, and promote the development of new services based on public data.

Through open data, businesses and entrepreneurs can develop applications that improve urban living, such as apps that provide real-time updates on public transportation or tools for monitoring energy consumption in buildings. The open data initiative also plays a crucial role in enabling smart governance, ensuring that decisions are informed by accurate, real-time data.

One of the key outcomes of Barcelona's smart city model is the development of smart services aimed at improving the quality of life for residents. These services are designed to make public services more efficient, accessible, and user-friendly.

For example, the city has implemented smart parking systems that help drivers find available parking spaces in real time, reducing traffic congestion and emissions. The city also uses smart waste management systems that optimize the collection of waste based on real-time data, improving efficiency and reducing costs.

In addition, citizen-to-citizen services have been developed to foster community engagement and collaboration. These services are often based on open data and include platforms that allow citizens to share information, collaborate on projects, or participate in urban planning processes.

5.1.2 Benefits and Challenges of the Barcelona Smart City Initiative

The case study of Barcelona highlights both the benefits and challenges of implementing a smart city strategy. On the positive side, the smart city model has led to improved public services, increased innovation, and enhanced economic competitiveness. The 22@Barcelona district, in particular, has attracted new businesses and created thousands of jobs, contributing to the city's economic growth.

However, Barcelona has also faced several challenges in its smart city transformation. These include the need for skilled human capital, difficulties in attracting venture capital, and challenges in achieving cross-departmental collaboration. The economic crisis in Spain has also posed challenges to the continued funding of smart city projects.

Despite these obstacles, Barcelona has emerged as a global leader in smart city development. The city's success serves as a model for other cities looking to harness the power of ICT to create more sustainable, innovative, and livable urban environments.

The Barcelona smart city initiative demonstrates how ICT, living labs, and open data can be integrated into urban planning to create a smart, sustainable city. By leveraging infrastructure, human capital, and data, Barcelona has transformed itself into a knowledge-intensive city with a focus on innovation and citizen engagement. The lessons learned from Barcelona's experience provide valuable insights for other cities looking to implement smart city strategies and improve urban governance, public services, and economic growth.

5.2 AI Chatbots for Mental Health

Recent breakthroughs in artificial intelligence have accelerated the development of conversational agents designed to support mental health care. These AI chatbots—built upon advances in Natural Language Processing (NLP) and machine learning (ML)—offer scalable and accessible pathways to psychological support, particularly in settings with limited mental health infrastructure or high stigma around seeking clinical assistance. Their capacity to engage in continuous, empathetic, and context-sensitive interactions makes them a promising complement to conventional mental health services. Based on the most recent findings, this analysis explores the technological effectiveness and clinical relevance of AI chatbots, while also highlighting ethical concerns inherent in their deployment.

5.2.1 Effectiveness and Application of AI Chatbots

AI chatbots have been increasingly adopted to assist with conditions such as anxiety, depression, and behavioral addiction. Casu et al. (2024) report that these systems demonstrate measurable improvements in user well-being, primarily by delivering real-time therapeutic interventions grounded in Cognitive Behavioral Therapy (CBT) frameworks. Through simulating empathetic human dialogue, chatbots help users self-monitor emotional states, identify cognitive distortions, and reinforce coping strategies. Their 24/7 availability overcomes spatial and temporal barriers associated with traditional clinical services, offering continuous support across diverse user populations.

To operationalize this capability within a clinically grounded AI model, we propose a Transformer-Based Adaptive Dialogue Framework that leverages fine-tuned BERT encoders augmented by reinforcement learning for interaction personalization. Each user input u_t is processed through the transformer encoder to extract a context-rich semantic representation:

$$h_t = \mathrm{BERT}(u_t)$$

where h_t is the hidden state encoding the user's current message u_t, capturing both syntactic structure and psychological sentiment, and $\mathrm{BERT}(\cdot)$ is the Bidirectional Encoder Representations from Transformers model that captures both linguistic and sentiment-based features in context.

This embedding is then evaluated by a response policy network trained via reinforcement learning, which identifies the most therapeutically appropriate reply by maximizing an expected reward that reflects psychological effectiveness:

$$r_t = \arg \max_{r \in \mathcal{R}} Q_\phi(h_t, r)$$

where $Q_\phi(h_t, r)$ is the action-value function predicting the benefit of selecting response r in context h_t, and ϕ represents learned parameters from historical user-therapist dialogues.

To shape emotionally intelligent behavior, the chatbot integrates a dynamic reward signal composed of three core softmax-normalized feedback dimensions—empathy, clarity, and relevance—optimized via weighted scalarization:

$$\mathcal{R}_{\text{chat}} = \lambda_1 S_{\text{empathy}} + \lambda_2 S_{\text{clarity}} + \lambda_3 S_{\text{relevance}}$$

where S_{empathy}, S_{clarity}, and $S_{\text{relevance}}$, are softmax-normalized scores assessing the chatbot's alignment with core principles of therapeutic communication, and λ_i are weights reflecting design priorities or clinical calibration.

To ensure safe and ethically robust intervention, a risk-aware policy modulation function evaluates whether the conversation trajectory approaches clinically significant thresholds. The cumulative risk-aware emotional trajectory is monitored using a gated recurrent formulation:

$$e_t = \sigma(W_e h_t + b_e) + \tanh(W_r e_{t-1} + b_r)$$

where e_t represents the emotional state at time t, $\sigma(\cdot)$ is the sigmoid activation indicating affective reactivity, W_e, W_r are weight matrices learned during training, b_e, b_r are bias vectors, and e_{t-1} is the previous emotional state. This gated mechanism enables the model to monitor emotional drift and intervene when a concerning psychological pattern emerges.

Finally, a safety-aware termination policy is activated when the cumulative distress signal surpasses a predefined threshold, thereby initiating a real-time escalation or human referral protocol:

$$\pi_{\text{exit}} = \mathbb{I}[e_t \geq \delta]$$

where π_{exit} is a binary policy indicator determining the need to terminate AI interaction, $\mathbb{I}$ is the indicator function, and δ is a pre-calibrated threshold representing critical emotional risk, derived from clinical datasets.

This multilayered framework—grounded in Transformer-based semantic comprehension, reinforcement-based personalization, emotional trajectory modeling, and ethically aligned safety policies—demonstrates how AI can evolve from simple automation to co-therapeutic engagement. Each equation builds a coherent mathematical narrative, linking user input to reward-optimized, clinically aware outputs.

In practice, this model enables mental health AI systems to continuously adapt to individual users' psychological states, provide highly personalized CBT-style interventions, and detect when professional human assistance is necessary. By embedding reinforcement learning into a linguistically powerful Transformer backbone and regulating its decisions with human-centered affective feedback, the framework presents a robust, state-of-the-art solution for delivering scalable, ethical, and emotionally intelligent mental healthcare. This formulation advances digital psychiatry and AI therapy into a new frontier of empathic computation.

Framework: Transformer-Based Adaptive Dialogue Framework (TADF)
Layered AI Architecture for Emotionally Intelligent Mental Health Chatbots

Layer	Main functionality	AI module(s) and components	Operational scope (Input → Processing → Output)
Layer 1	User input and semantic representation	BERT	Input → User message → Output → Context-rich semantic representation using BERT
Layer 2	Response policy optimization	Reinforcement learning	Input → User input embedding → Output → Optimized therapeutic response
Layer 3	Emotionally intelligent behavior reward signal	Emotion recognition module	Input → Feedback dimensions (empathy, clarity, relevance) → Output → Calculated reward signal for chatbot
Layer 4	Emotional state modeling	Gated Recurrent Unit (GRU)	Input → User input and previous emotional state → Output → Updated emotional state based on temporal dependencies
Layer 5	Safety-aware termination and escalation	Emotional distress detection and escalation module	Input → Emotional state → Output → Triggered termination policy if emotional threshold is exceeded

A particularly salient example of the applicability of AI chatbots emerged during the COVID-19 pandemic, when millions of individuals experienced acute psychological strain due to prolonged isolation, economic uncertainty, and social disruption. With in-person therapy severely limited, AI-driven conversational agents provided immediate, scalable mental health support—offering users real-time interaction, emotional validation, and therapeutic guidance from the safety of their homes. As Casu et al. (2024) observe, these chatbots helped mitigate surges in anxiety and depressive symptoms while simultaneously alleviating the burden on overstretched healthcare systems.

To operationalize real-time mental health support during high-stress societal disruptions such as pandemics, we introduce the Dual-Encoder Emotional Cognition Framework, a Transformer-driven architecture designed to deliver linguistically coherent, emotionally adaptive, and context-aware therapeutic responses. This framework integrates a high-performance RoBERTa encoder with a recurrent affective memory mechanism, enabling longitudinal emotional modeling across asynchronous user sessions. At the core of this system lies a contextual semantic encoder that transforms each user utterance into a deep emotional-syntactic representation:

$$\mathrm{e}_t = \mathrm{RoBERTa}(u_t)$$

where e_t is the contextual embedding at time t, capturing both the linguistic structure and emotional valence of the user input u_t through the RoBERTa encoder pretrained on large-scale natural language corpora.

To retain emotional continuity over multiple interactions and temporal disjunctions, these embeddings are passed into a Gated Recurrent Unit (GRU)-based emotional state tracker:

$$h_t = \mathrm{GRU}(e_t, h_{t-1})$$

where, h_t represents the hidden emotional state at time t, and the GRU (Gated Recurrent Unit) aggregates emotional history across the dialogue, capturing patterns such as escalating distress or emotional flattening.

The emotional progression encoded in h_t is then processed through a sentiment-informed risk decoder that evaluates the psychological intensity of the user's state. This is computed through a softmax-normalized transformation:

$$\hat{y}_t = \mathrm{softmax}(W_r h_t + b_r)$$

where $\hat{y}_t \in \mathbb{R}^K$ is the predicted probability distribution across K emotional risk categories (e.g., stable, mild anxiety, acute distress), W_r is the classification weight matrix, and b_r is the bias vector learned through supervised training on annotated mental health dialogue corpora.

To enable adaptive therapeutic response generation, the model conditions its decision on both the emotional distribution $\hat{y}$ and the hidden emotional context h_t using a reinforcement-optimized policy function:

$$r_t = \arg \underset{r \in \mathcal{R}}{max}\, Q_\psi(h_t, \hat{y}_t, r)$$

where r_t is the selected response at time t, $\mathcal{R}$ is the candidate response set, and $Q_\psi(h_t, \hat{y}_t, r)$ is the action-value function parameterized by ψ, learned via reinforcement learning on expert-annotated dialogues to maximize therapeutic effectiveness.

The cumulative emotional trajectory is further tracked using an affective memory gate, allowing the chatbot to respond to patterns such as emotional escalation or flattening. This is modeled through a forget-update mechanism:

$$m_t = \gamma \cdot m_{t-1} + (1 - \gamma) \cdot h_t$$

where m_t is the affective memory vector at time t, m_{t-1} is the previous memory state, and $\gamma \in [0, 1]$ is a decay parameter controlling the trade-off between emotional inertia and sensitivity to new affective signals.

Finally, to ensure ethical and safe deployment, a binary safety trigger monitors whether the predicted emotional state surpasses a predefined psychological threshold:

$$\pi_{\text{alert}} = \mathbb{I}\left[\hat{y}_t^{(\text{distress})} \geq \tau\right]$$

where π_{alert} is a binary indicator activating escalation protocols (e.g., referral to human therapists), $\hat{y}_t^{(\text{distress})}$ is the predicted probability of the acute distress class, τ is the calibrated distress threshold, and $\mathbb{I}[\cdot]$ is the indicator function.

This integrated framework ensures that AI chatbots function not as passive responders but as emotionally aware therapeutic agents capable of sustaining cognitive-affective continuity across sessions. Each module contributes to a dynamic pipeline—from semantic interpretation to emotional modeling and risk-sensitive decision-making—allowing the system to identify patterns of escalating psychological distress and respond with appropriate therapeutic interventions. In real-world crises such as the COVID-19 pandemic, this model offers scalable, ethically grounded, and deeply personalized mental health support. By embedding longitudinal emotion tracking and reinforcement-optimized dialogue into a dual-stream architecture, the proposed framework transforms conversational AI into a clinically aligned, psychologically intelligent instrument of care.

Framework: Dual-Encoder Emotional Cognition Framework (DECF)
Layered AI Architecture for Emotionally Adaptive Mental Health Chatbots

Layer	Main functionality	AI module (s) and components	Operational scope (Input → Processing → Output)
Layer 1	User input and semantic representation	RoBERTa	Input → User message → Output → Contextual semantic embedding of user input using RoBERTa
Layer 2	Emotional state tracking	Gated Recurrent Unit (GRU)	Input → Contextual semantic embedding → Output → Updated emotional state based on temporal dialogue history
Layer 3	Sentiment-informed risk evaluation	Softmax classification	Input → Emotional state → Output → Predicted emotional risk categories (e.g., stable, mild anxiety, acute distress)
Layer 4	Therapeutic response generation	Reinforcement learning	Input → Emotional state and risk categories → Output → Optimized therapeutic response selection based on reinforcement learning
Layer 5	Affective memory tracking	Long-term emotional state tracker	Input → Previous emotional state and current emotional input → Output → Updated affective memory for long-term emotional continuity
Layer 6	Safety-aware termination and escalation	Threshold-based decision-making	Input → Predicted emotional risk → Output → Triggered escalation or termination protocol if distress threshold is exceeded

Additionally, these AI systems have been integrated into broader healthcare settings, functioning alongside human professionals. This integration helps maintain continuous engagement, monitor patient progress, and offer supplementary support between traditional therapy sessions. However, the success of such interventions hinges on the chatbot's ability to offer empathetic, personalized, and contextually appropriate responses, a challenge that remains at the forefront of ongoing AI development efforts.

5.2.2 *Ethical Considerations in AI Mental Health Applications*

The integration of AI in mental health also introduces several ethical concerns that must be addressed to ensure safe, responsible, and effective use. Warrier et al. (2023) highlight issues related to data privacy, algorithmic bias, and transparency. One primary concern is the risk of unauthorized access to sensitive patient data, which necessitates robust data security protocols. AI developers and healthcare providers must also consider how data is collected, stored, and utilized to prevent misuse or exploitation. To ensure ethical compliance in AI-driven mental health systems, we propose the Privacy-Preserving Fairness-Aware Decision Framework (P^2FADF)—a multilayer architecture that operationalizes differential privacy, bias mitigation, and explainability under a unified formalism. This framework addresses key ethical mandates, including secure data handling, equitable model behavior, and transparent decision pathways.

At the foundation of this framework lies the integration of ε-differential privacy, which limits the influence of any individual's data on the AI model's outputs, thereby reducing re-identification risks. This principle is captured by the formal definition:

$$\forall S \subseteq \text{Range}(\mathcal{M}) : \quad \Pr[\mathcal{M}(D_1) \in S] \leq e^{\varepsilon} \cdot \Pr[\mathcal{M}(D_2) \in S]$$

where $\mathcal{M}$ is a randomized AI mechanism, D_1 and D_2 are any two datasets differing by one individual's record, S is a subset of possible outputs, and ε (epsilon) is the privacy loss parameter that bounds the deviation in output probability due to an individual's inclusion. This function guarantees that the presence or absence of a single patient cannot significantly influence model outcomes, ensuring privacy-preserving predictions.

Next, to detect and mitigate algorithmic bias, a fairness-constrained loss function is introduced. This penalizes disparities in error rates across sensitive groups (e.g., gender, ethnicity):

$$\mathcal{L}_{\text{fair}} = \mathcal{L}_{\text{task}} + \lambda \cdot \left| \mathbb{E}_{x \in G_1}[\hat{y}(x)] - \mathbb{E}_{x \in G_2}[\hat{y}(x)] \right|$$

where $\mathcal{L}_{\text{task}}$ is the original task loss (e.g., cross-entropy), λ is a regularization parameter controlling fairness priority, G_1 and G_2 are different sensitive attribute groups, and $\hat{y}(x)$ is the model prediction for input x. This term enforces statistical parity, compelling the model to perform equitably across demographically distinct cohorts.

To ensure transparency in decision-making, the framework incorporates SHAP-based explainability, assigning additive importance values to each input feature for every prediction:

$$f(x) = \phi_0 + \sum_{i=1}^{n} \phi_i$$

where $f(x)$ is the AI model output, ϕ_0 is the expected prediction (baseline), and ϕ_i is the Shapley value of the i-th feature, quantifying its marginal contribution to the decision. These values are computed by considering all permutations of feature subsets, ensuring model outputs are interpretable and auditable—especially critical in clinical settings.

Finally, to harmonize these components, a multi-objective optimization problem is formulated, jointly minimizing task loss, bias disparity, and maximizing interpretability while satisfying a privacy constraint:

$$\min_{\theta} \left(\mathcal{L}_{\text{task}} + \lambda_1 \mathcal{L}_{\text{bias}} - \lambda_2 I_{\text{exp}} \right), \quad \text{s.t.} \quad \varepsilon \le \varepsilon_{\max}$$

where θ are the model parameters, λ_1, λ_2 are hyperparameters controlling the trade-offs between bias reduction and explainability enhancement, I_{exp} is an information-theoretic measure of explainability (e.g., mutual information between features and SHAP scores), and $\varepsilon_{\max}$ is the allowable privacy budget.

This framework ensures that AI mental health applications operate within an ethically principled structure—respecting user privacy, ensuring demographic fairness, and maintaining model transparency. By embedding these ethical guarantees directly into the mathematical foundation of the system, the Privacy-Preserving Fairness-Aware Decision Framework (P^2FADF) offers a scientifically grounded, socially responsible solution for next-generation AI in clinical mental healthcare. It bridges the critical gap between technical performance and ethical legitimacy, ensuring that future AI systems are not only intelligent, but also just.

Framework: Privacy-Preserving Fairness-Aware Decision Framework (P^2FADF)

Layered AI Architecture for Ethical AI in Mental Health Applications

Layer	Main functionality	AI module(s) and components	Operational scope (Input → Processing → Output)
Layer 1	Data privacy protection	Differential privacy	Input → Patient data → Output → Privacy-preserving model ensuring individual data protection using ε-differential privacy
Layer 2	Bias mitigation and fairness enforcement	Fairness-constrained optimization	Input → Sensitive attribute groups → Output → Bias-reduced model predictions ensuring fairness across demographic groups

(continued)

Layer	Main functionality	AI module(s) and components	Operational scope (Input → Processing → Output)
Layer 3	Model explainability and transparency	SHAP (Shapley Additive Explanations)	Input → Model outputs → Output → Additive importance values for model features, enhancing transparency
Layer 4	Multi-objective optimization for ethical AI	Multi-objective optimization	Input → Task loss, bias disparity, and explainability metrics → Output → Optimized model balancing fairness, explainability, and privacy
Layer 5	Human-in-the-loop oversight	Human review and decision adjustment module	Input → Model outputs, explanations, and optimization trade-offs → Output → Human-reviewed or adjusted decisions, ensuring accountability, contextual judgment, and democratic legitimacy

Algorithmic bias presents another challenge. AI chatbots rely on vast datasets, and if these datasets contain biases, the AI may inadvertently perpetuate or exacerbate disparities in mental health care. For example, bias in diagnostic tools can lead to inaccurate assessments or inappropriate recommendations, especially for marginalized groups who may already face barriers to accessing mental health services (Warrier et al. 2023). Ensuring fairness and equity in AI models is critical, requiring continuous evaluation and adjustment of algorithms. To address the structural risk of algorithmic bias in mental health AI systems, we introduce the Fairness-Calibrated Representation Alignment Model (FCRAM)—a precision-engineered framework that embeds fairness constraints into the representational geometry of deep learning models. This model integrates adversarial debiasing, group fairness regularization, and bias impact auditing within a unified optimization landscape.

At the core of this architecture is an adversarial debiasing mechanism, in which the AI model learns to make accurate predictions while minimizing its ability to encode sensitive demographic attributes (e.g., race, gender, socioeconomic status). This is operationalized by minimizing the predictive loss while maximizing the error of an adversarial discriminator:

$$\min_{\theta} \max_{\psi} \left(\mathcal{L}_{\text{task}}(f_\theta(x), y) - \lambda \cdot \mathcal{L}_{\text{adv}}\left(d_\psi(f_\theta(x)), z\right)\right)$$

where $f_\theta(x)$ is the main AI model parameterized by θ, predicting mental health outcomes from input x, y is the ground truth label (e.g., anxiety level), z is the sensitive attribute, d_ψ is the adversarial discriminator parameterized by ψ, $\mathcal{L}_{\text{task}}$ is the primary prediction loss, $\mathcal{L}_{\text{adv}}$ is the adversary's classification loss, and λ balances task accuracy and fairness. This formulation encourages the model to learn representations that are informative for prediction but uninformative for protected attributes, thus minimizing representational bias.

To enforce fairness in the model's decision outputs, a Group Fairness Regularizer is introduced to penalize discrepancies in false positive or negative rates across sensitive groups:

$$\mathcal{L}_{\mathrm{GF}} = |\mathbb{E}_{x \in G_1}[\hat{y}(x)] - \mathbb{E}_{x \in G_2}[\hat{y}(x)]|$$

where $\hat{y}(x)$ is the predicted probability for a specific diagnosis or intervention, and G_1, G_2 are distinct demographic groups. This function compels outcome parity across user segments, ensuring equitable treatment recommendations.

In addition to prediction parity, the model incorporates bias impact attribution via counterfactual evaluation. For each sample x, a matched counterfactual x' is created by perturbing the sensitive attribute z, and the average prediction shift is computed:

$$\Delta_{\mathrm{bias}} = \mathbb{E}_{(x,x')}[|f_\theta(x) - \mathrm{f}_\theta(x')|]$$

where x' is a demographically altered version of x (e.g., identical psychological profile but changed gender or ethnicity), and Δ_{bias} measures the model's sensitivity to protected attributes. A low bias impact score indicates robustness to identity-based confounding.

The final fairness-augmented loss function combines these three objectives into a single multi-criteria optimization:

$$\min_{\theta} \left(\mathcal{L}_{\mathrm{task}} + \lambda_1 \mathcal{L}_{\mathrm{GF}} + \lambda_2 \Delta_{\mathrm{bias}}\right)$$

where λ_1 and λ_2 are regularization hyperparameters that control the trade-off between accuracy, group fairness, and counterfactual stability.

This framework enables AI mental health systems to dynamically self-correct for representational and outcome disparities. The Fairness-Calibrated Representation Alignment Model (FCRAM) ensures that diagnostic and therapeutic outputs are demographically neutral, ethically grounded, and statistically fair across all user segments. By utilizing adversarial learning, output parity, and counterfactual fairness, this architecture represents the forefront of AI bias mitigation and reaffirms the ethical integrity of digital mental health care.

Framework: Fairness-Calibrated Representation Alignment Model (FCRAM)
Layered AI Architecture for Bias Mitigation in Mental Health AI Systems

Layer	Main functionality	AI module(s) and components	Operational scope (Input → Processing → Output)
Layer 1	Data privacy and fairness constraints	Differential privacy	Input → Sensitive data → Output → Privacy-preserving model ensuring individual data protection using ε-differential privacy
Layer 2	Bias mitigation via adversarial learning	Adversarial debiasing	Input → Model predictions, sensitive attributes → Output → Debiased representation minimizing sensitive attribute encoding
Layer 3	Group fairness regularization	Group fairness regularization	Input → Demographic group data → Output → Enforced parity in model outputs across sensitive groups

(continued)

Layer	Main functionality	AI module(s) and components	Operational scope (Input → Processing → Output)
Layer 4	Bias impact attribution	Counterfactual fairness evaluation	Input → Sample data with sensitive attributes → Output → Bias impact score measuring model's sensitivity to protected attributes
Layer 5	Multi-objective optimization for fairness	Multi-criteria optimization for fairness	Input → Task loss, fairness and bias metrics → Output → Optimized model balancing task performance, fairness, and bias reduction
Layer 6	Human-in-the-loop oversight	Ethical decision review and adjustment module	Input → Model outputs, explanations, and fairness trade-offs → Output → Human-reviewed or adjusted decisions ensuring accountability and contextual legitimacy

Transparency in AI operations is essential for building trust between users and these digital tools. Warrier et al. (2023) argue that patients should understand how the AI functions, how decisions are made, and what limitations may exist in the chatbot's capabilities. This transparency can empower users to make informed decisions about their care and mitigate the "black-box" problem associated with AI technologies. Ethical guidelines that prioritize patient autonomy, consent, and accountability are crucial to advancing AI integration in mental health care. To address the ethical imperative of transparency in AI-driven mental health systems, we introduce the Explainability-Driven Mental Health Framework (EDMHF), which embeds intelligibility and interpretability directly into each computational layer of the decision-making pipeline. This architecture integrates SHAP value decomposition, attention interpretability, and user-facing confidence metrics to ensure that AI chatbot recommendations are not only effective but explainable and auditable by both patients and clinicians.

At the foundation of the framework lies the use of Shapley Additive Explanations (SHAP), which assigns a transparent attribution value to each input feature by quantifying its marginal contribution to a specific prediction:

$$f(x) = \phi_0 + \sum_{i=1}^{n} \phi_i$$

where $f(x)$ is the model prediction for input x, ϕ_0 is the model's expected output (baseline), and ϕ_i is the Shapley value for the i-th feature, representing the isolated impact of each linguistic, behavioral, or contextual variable (e.g., mood terms, anxiety indicators, session history). This decomposition enables patients and practitioners to understand why a recommendation (such as a self-care prompt or escalation) was issued.

To enhance the interpretability of transformer-based models used in chatbots, the framework integrates an attention-based explanation module, which visualizes the internal weighting over input tokens. For a given layer l, the attention matrix is computed as:

$$A^{(l)} = \text{softmax}\left(\frac{Q^{(l)}K^{(l)\top}}{\sqrt{d_k}}\right)$$

where $A^{(l)}$ is the attention matrix at layer l, $Q^{(l)}$ and $K^{(l)}$ are the query and key matrices, and d_k is the dimensionality of the keys. This matrix reveals which words or phrases the AI focuses on when forming its decision, allowing users to interpret chatbot attention in a psychologically meaningful way (e.g., recognizing that emotional keywords influenced the response).

To further enhance user empowerment, the system estimates model confidence calibration through a normalized entropy function, which translates the prediction's internal certainty into a human-understandable metric:

$$C(x) = 1 - \frac{H(p(x))}{\log K}$$

where $C(x) \in [0, 1]$ is the confidence score for input x, $p(x) \in \mathbb{R}^K$ is the predicted probability distribution over K classes (e.g., anxiety levels), and $H(p(x)) = -\sum_{k=1}^{K} p_k \log p_k K$ is the Shannon entropy. This confidence index empowers users to assess the reliability of chatbot decisions and adjust their trust level accordingly.

The final component is an explainability-aware decision protocol, which integrates the above metrics into a user-facing interface, ensuring transparency of operations. The composite recommendation rule is formalized as:

$$\pi_{\text{recommend}}(x) = \begin{cases} \text{accept}, & \text{if } C(x) \geq \tau \wedge \underset{i}{max} \mid \phi_i \mid \leq \delta \\ \text{refer}, & \text{otherwise} \end{cases}$$

where $\pi_{\text{recommend}}(x)$ is the chatbot's decision protocol, $C(x)$ is the calibrated confidence score, τ is the trust threshold, and δ is the interpretability deviation margin beyond which human review is mandated. This function ensures that users are not left to interpret opaque decisions and that accountability remains traceable in all AI outputs.

This explainability framework transforms opaque black-box systems into transparent, patient-centered tools of mental health engagement. By using attributional clarity (SHAP), attention visualizations, and confidence-calibrated decision boundaries, the Explainability-Driven Mental Health Framework (EDMHF) fosters informed user consent, trust, and autonomy. It redefines transparency not as an afterthought but as a core operational principle, positioning ethical AI at the heart of psychological care delivery.

Framework: Explainability-Driven Mental Health Framework (EDMHF)
Layered AI Architecture for Transparent AI in Mental Health Applications

Layer	Main functionality	AI module(s) and components	Operational scope (Input → Processing → Output)
Layer 1	Model prediction and feature attribution	SHAP (Shapley Additive Explanations)	Input → Features (e.g., mood, anxiety) → Output → Quantification of each feature's contribution to model's prediction
Layer 2	Interpretability of transformer-based models	Attention mechanism (transformer)	Input → Tokenized input text → Output → Visualization of attention weights over input tokens
Layer 3	Confidence calibration	Confidence calibration via entropy	Input → Model's predicted probability distribution → Output → Confidence score indicating prediction certainty
Layer 4	Decision protocol and thresholding	Decision THRESHOLDING	Input → Confidence score and model interpretation → Output → User-facing decision (e.g., accept, refer) based on transparency and accountability

Figure 5.1 illustrates the structure of AI chatbots for mental health, focusing on two main areas: effectiveness and application, and ethical considerations. The effectiveness and application section highlights the chatbots' roles in providing mental health support, delivering CBT guidance, offering 24/7 accessibility, proving beneficial during high-stress periods like COVID-19, and integrating with traditional healthcare for continuous support. The ethical consideration section covers data privacy, addressing the need for secure data handling; algorithmic bias, emphasizing the importance of fair treatment for all users; and transparency and trust, ensuring ethical AI practices for user understanding and autonomy.

AI chatbots have the potential to transform mental health care by offering scalable, accessible, and personalized support. The advancements in NLP and machine learning have enabled these systems to simulate human interaction, providing practical solutions to address mental health challenges globally. However, ethical considerations related to privacy, bias, and transparency must be rigorously managed to ensure these technologies serve as effective and equitable tools in mental health care.

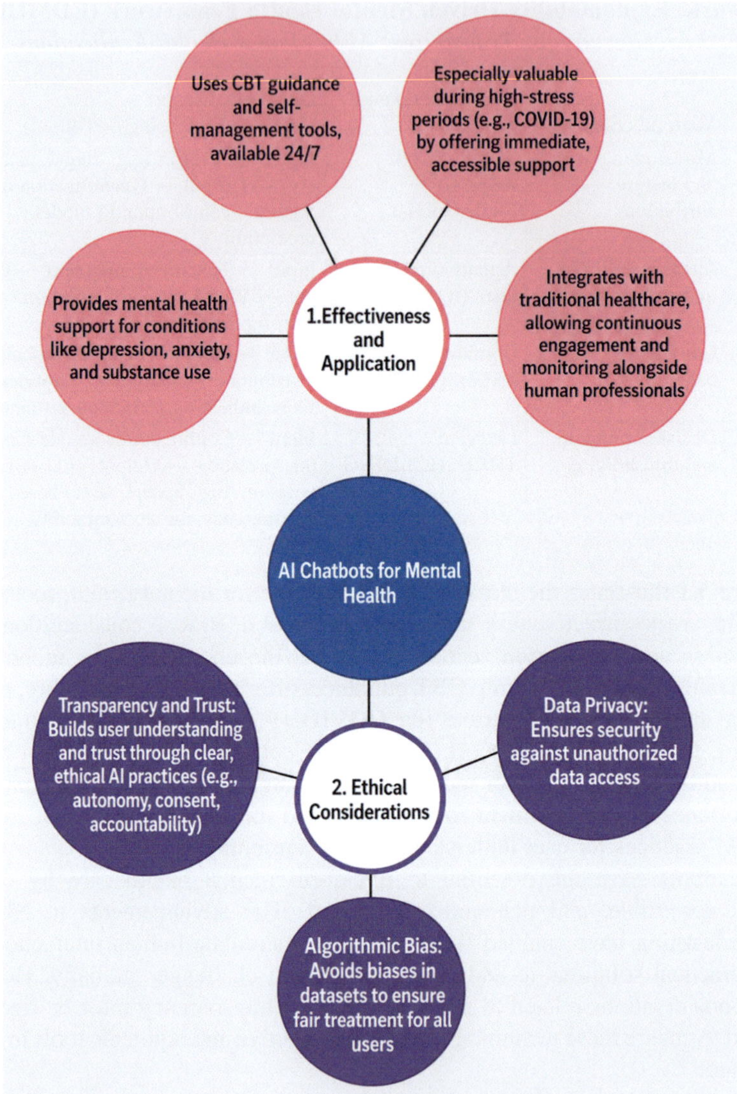

Fig. 5.1 AI chatbots for mental health: key aspects and ethical considerations

5.3 Conclusion

This chapter intricately weaves together two highly relevant and transformative applications of Artificial Intelligence (AI)—AI-driven personalization in smart cities and digital mental health care—while exploring their technological, societal, and ethical dimensions. Through an in-depth analysis of Barcelona's Smart City Initiative and AI-powered conversational agents for mental health, the chapter demonstrates the promising future of AI in reshaping urban landscapes and healthcare systems, offering scalable, sustainable, and ethically grounded solutions to modern challenges.

The first section provides a comprehensive overview of Barcelona's transformation into a global leader in smart city development, showcasing how Information and Communication Technologies (ICT) and open data are being integrated into urban infrastructure to create innovative, efficient, and sustainable environments. At the heart of this transformation lies the 22@Barcelona district, which serves as a model for green infrastructure, economic growth, and technological innovation. The chapter also highlights the role of living labs, where public–private partnerships foster the development of new technologies in real-world settings. Smart services, such as smart parking and waste management, are shown to enhance public service delivery, while the chapter emphasizes the infrastructure investments, such as fiber optic networks and sensor installations that enable real-time data collection and service optimization. Moreover, the open data movement and citizen participation initiatives exemplify Barcelona's commitment to transparency and inclusive governance, providing valuable insights for other cities striving for sustainable, data-driven governance.

Transitioning from urban infrastructure to AI in mental health, the chapter then explores the profound impact of AI chatbots in delivering personalized mental health care, particularly in settings with limited access to traditional resources. Through cutting-edge frameworks such as the Transformer-Based Adaptive Dialogue Framework (TADF) and the Dual-Encoder Emotional Cognition Framework (DECF), these AI systems, grounded in Cognitive Behavioral Therapy (CBT) principles, are shown to provide emotionally adaptive, real-time therapeutic interventions. The chapter discusses how Reinforcement Learning (RL) optimizes chatbot responses, ensuring the emotional and therapeutic effectiveness of interactions. Additionally, the chapter highlights the ethical considerations surrounding AI chatbots, particularly in areas such as privacy, algorithmic fairness, and model transparency. By incorporating techniques such as ε-differential privacy for data protection and Shapley Additive Explanations (SHAP) for model interpretability, the chapter demonstrates the importance of making these systems accountable, explainable, and equitable. The proposed Privacy-Preserving Fairness-Aware Decision Framework (P^2FADF) offers a robust solution to mitigate bias and ensure ethical compliance in AI-driven mental health systems, while the Fairness-Calibrated Representation Alignment Model (FCRAM) provides an advanced approach for reducing algorithmic bias and ensuring fair treatment across demographic groups.

Moreover, the chapter addresses the pressing need for transparency and user empowerment in AI systems. By integrating SHAP values and attention mechanisms, the chapter outlines how AI can be made more transparent and understandable, ensuring that users have full control and informed consent over their mental health treatment. The Explainability-Driven Mental Health Framework (EDMHF) emphasizes that transparency should not merely be a supplementary feature but a core principle embedded into the decision-making pipeline of AI systems.

The chapter concludes with an integrative analysis of both smart city innovations and AI mental health applications, asserting that the convergence of these technologies offers exciting potential for addressing global urban challenges and healthcare disparities. Through the implementation of reinforcement learning, NLP, data privacy, and fairness principles, the chapter presents a vision where AI not only optimizes service delivery but also ensures inclusive access to essential services. This dual application of AI in urban governance and mental health care highlights the possibility of creating a more sustainable, equitable, and technologically advanced future. By fostering a transparent, responsible, and ethical framework, AI has the potential to be a transformative force in shaping cities and healthcare systems that are truly responsive to the needs of all citizens, paving the way for a better, more inclusive world.

References

Bakıcı, T., Almirall, E. & Wareham, J. A Smart City Initiative: the Case of Barcelona. *J Knowl Econ* 4, 135–148 (2013). https://doi.org/10.1007/s13132-012-0084-9

Casu, M., Triscari, S., Battiato, S., Guarnera, L., & Caponnetto, P. (2024). AI chatbots for mental health: A scoping review of effectiveness, feasibility, and applications. *Applied Sciences, 14*, 5889. https://doi.org/10.3390/app14135889

Warrier, U., Warrier, A., & Khandelwal, K. (2023). Ethical considerations in the use of artificial intelligence in mental health. *Egyptian Journal of Neurology, Psychiatry, and Neurosurgery, 59*, 139. https://doi.org/10.1186/s41983-023-00735-2

Chapter 6
AI-Driven Global Networks for Enhancing Disaster Resilience, Climate Change Mitigation, and Mental Health

This chapter explores the pivotal role of Artificial Intelligence (AI) in enhancing disaster resilience, climate change mitigation, and public health, with a particular emphasis on flood management systems. In an era marked by increasingly frequent and severe natural disasters driven by climate change, the necessity for innovative, AI-powered solutions to safeguard communities and infrastructure has never been greater. This comprehensive analysis focuses on how AI-driven global networks facilitate real-time data sharing, optimize emergency response strategies, and foster long-term resilience in disaster-prone regions.

Central to the discussion is the integration of AI in flood resilience, a critical area in disaster management. Flooding, one of the most pervasive and devastating natural hazards, demands advanced predictive models and adaptive systems that can respond to dynamic environmental conditions. The chapter introduces sophisticated AI frameworks such as the Hybrid Spatio-Temporal Flood Forecasting Framework ($HSTF^3$) and the Multimodal Multitask Neural Framework for Flood Resilience Forecasting ($MTN\text{-}FR^2$), which incorporate cutting-edge machine learning (ML) algorithms and Graph Neural Networks (GNNs) to enhance flood prediction and monitoring. These AI models analyze a variety of data sources, including meteorological data, historical flood records, satellite imagery, and real-time sensor feeds, providing actionable insights into flood risk across different spatial and temporal scales.

Another key aspect of the chapter is the discussion on AI's role in optimizing real-time response during active flood events. By utilizing Deep Reinforcement Learning (DRL), the chapter outlines how emergency management systems can be enhanced to allocate resources effectively, prioritize critical interventions, and minimize disaster-related risks. This is achieved through systems that continuously learn and adapt to evolving flood scenarios, offering unprecedented operational flexibility and efficiency.

The chapter further addresses the post-disaster phase, where AI is used to conduct rapid damage assessments, enabling authorities and humanitarian organizations to deploy resources where they are needed most. The application of Convolutional

D. Sargiotis, *Blueprints for Future Living*,
https://doi.org/10.1007/978-3-032-24178-8_6

Neural Networks (CNNs) for satellite-based damage detection is explored, demonstrating how AI can provide high-resolution, near-real-time insights into the extent of damage across urban and rural landscapes, thus guiding infrastructure restoration and recovery efforts.

Additionally, the integration of Explainable AI (XAI) principles is discussed, emphasizing the importance of transparency in AI-driven decision-making, particularly in high-stakes disaster management contexts. The chapter also highlights the role of federated learning in enabling cross-border collaboration while maintaining data privacy, an essential feature in global disaster resilience efforts. By utilizing decentralized learning approaches, regions can share model insights without compromising the confidentiality of sensitive data, thus facilitating international cooperation and knowledge exchange.

Lastly, the chapter acknowledges the challenges faced in implementing AI-driven systems for disaster resilience, including issues related to data quality, model interpretability, and scalability. It proposes forward-thinking frameworks designed to address these challenges, ensuring that AI solutions remain both robust and adaptable to diverse global contexts. Through a comprehensive review of AI's potential to transform disaster management, climate change adaptation, and public health strategies, the chapter provides a roadmap for building smarter, more resilient communities in the face of a rapidly changing climate.

This chapter offers an unparalleled deep dive into the role of AI in the intersection of climate science, disaster management, and public health, presenting readers with cutting-edge insights and forward-thinking solutions that are poised to reshape how cities and nations approach resilience in the twenty-first century. Through a seamless integration of advanced AI methodologies, multimodal data sources, and collaborative frameworks, the chapter positions itself as an essential resource for policymakers, researchers, and practitioners dedicated to enhancing global disaster resilience and climate change mitigation efforts.

6.1 AI in Flood Resilience

Flooding remains one of the most common and devastating natural disasters worldwide, affecting millions of people and causing significant economic damage. The increasing frequency of extreme weather events, exacerbated by climate change, necessitates innovative approaches to flood management. Artificial Intelligence (AI) has emerged as a powerful tool in this domain, enabling enhanced prediction, preparedness, and response strategies. The following analysis explores the integration of AI in flood resilience, drawing from multiple case studies and research findings.

6.1.1 Application of AI in Flood Prediction and Monitoring

One of the critical areas where AI has demonstrated substantial potential is in flood prediction. By leveraging machine learning (ML) algorithms, AI systems can process vast amounts of historical and real-time data to identify patterns and forecast future events with greater accuracy. According to Saravi et al. (2019), AI models such as Random Forests and decision trees have been particularly effective in classifying different types of floods (e.g., flash floods, coastal floods) by analyzing variables such as rainfall, soil moisture, and weather conditions. These models can learn from past flood events, allowing them to predict future occurrences even under changing environmental conditions.

To transition from static flood classification to real-time, geospatially grounded forecasting, we introduce the Hybrid Spatio-Temporal Flood Forecasting Framework ($HSTF^3$). This dual-stage architecture integrates a Random Forest-based flood typology classifier with a Spatio-Temporal Deep Graph Neural Network (ST-DGNN), enabling high-resolution flood simulations across topographically diverse urban and rural landscapes. This system allows authorities to simultaneously anticipate flood type and dynamically simulate its progression across terrain-sensitive spatial networks.

At the foundation of the classification stage, a trained Random Forest model assigns a flood typology label to each spatial unit based on local hydrological and meteorological features:

$$y_i = \mathrm{RF}(x_i)$$

where x_i is a feature vector of local hydrological inputs (e.g., rainfall, slope, impervious surface ratio), and RF $(\cdot)$ is the trained Random Forest classifier.

Next, the Spatio-Temporal Deep Graph Neural Network (ST-DGNN) models spatial flood propagation across a dynamic terrain graph $G_t = (V_t, E_t)$, where each node $v \in V_t$ represents a geographic cell, and edges E_t capture elevation-connected water flow paths:

$$h_v^{(l+1)} = \sigma\left(\sum_{u \in \mathcal{N}(v)} \alpha_{vu}^{(l)} W^{(l)} h_u^{(l)}\right)$$

where $h_u^{(l)}$ is the flood state embedding of neighboring node u at layer l, $\alpha_{vu}^{(l)}$ is the learned attention coefficient based on terrain gradient and historical flow patterns, and $W^{(l)}$ is the trainable transformation matrix.

To incorporate temporal dependencies, a Gated Recurrent Unit (GRU) is applied across time steps to evolve each node's latent state in response to real-time meteorological sequences:

$$h_v^{(t+1)} = \mathrm{GRU}\left(h_v^{(t)}, x_v^{(t)}\right)$$

where $h_v^{(t+1)}$ is the updated node representation at time $t + 1$, incorporating both real-time meteorological inputs and learned spatial flow dependencies, $h_v^{(t)}$ is the prior temporal state, and $x_v^{(t)}$ is the real-time input vector at time t, which may include rainfall nowcasts, upstream discharge rates, humidity, or radar-observed water accumulation. This formulation allows the system to adapt its predictions as conditions evolve hour by hour.

The final prediction of flood intensity or risk score for each cell is derived from a readout function applied after the last GNN-GRU fusion layer:

$$\hat{y}_v^{(t+1)} = \sigma\left(W_o h_v^{(t+1)} + b_o\right)$$

where $\hat{y}_v^{(t+1)}$ is the predicted flood impact (e.g., water level, risk score) at time $t + 1$, W_o and b_o are the output layer parameters, and σ denotes a sigmoid or softmax function, depending on the forecasting task (binary flood risk vs. continuous depth estimation).

This framework enables flood management agencies to move beyond traditional hazard maps and static predictions. By fusing discrete classification with continuous graph-based forecasting, the Hybrid Spatio-Temporal Flood Forecasting Framework ($HSTF^3$) empowers real-time decision support from predicting the flood category to simulating downstream inundation and identifying at-risk zones across space and time. Its integration of temporal and spatial dynamics ensures that even under rapidly shifting environmental conditions, AI can guide anticipatory actions—such as targeted evacuation alerts, adaptive traffic control, and smart drainage reallocation—thereby enhancing climate resilience through actionable foresight.

Framework: Hybrid Spatio-Temporal Flood Forecasting Framework ($HSTF^3$)
Layered AI Architecture for Real-Time Flood Prediction and Simulation

Layer	Main functionality	AI module(s) and components	Operational scope (Input → Processing → Output)
Layer 1	Flood typology classification	Random forest classifier	Input → Hydrological and meteorological features → Output → Classifying flood type (e.g., flash flood, coastal flood) based on local conditions
Layer 2	Spatio-temporal flood propagation	Graph Neural Network (GNN) with attention	Input → Spatial flood data (terrain graph) → Output → Modeling flood propagation across topographically diverse landscapes
Layer 3	Temporal dependency modeling	Gated Recurrent Unit (GRU)	Input → Real-time meteorological data → Output → Evolving flood state representation across time based on current conditions
Layer 4	Final flood prediction and risk scoring	Prediction via neural network (sigmoid/ softmax output)	Input → Flood state embedding → Output → Predicting flood impact or risk score (e.g., water level, flood risk)

In the evolving domain of flood resilience, the ability to forecast flood events with spatial and temporal precision is fundamental to reducing disaster risk. Accurate prediction not only enables earlier alerts and evacuation planning but also facilitates the dynamic allocation of critical resources such as emergency personnel, medical units, and temporary shelters. As noted by Saravi et al. (2019), public data from agencies such as FEMA (Federal Emergency Management Agency) and NOAA (National Oceanic and Atmospheric Administration) have been increasingly utilized to power intelligent systems that forecast flood trajectories and assess their potential effects on critical infrastructure, utilities, and population zones.

To advance AI-driven flood resilience forecasting, we propose the Multimodal Multitask Neural Framework for Flood Resilience Forecasting (MTN-FR^2)—a unified architecture that simultaneously predicts flood probability and simulates its cascading impacts on urban infrastructure. This framework processes heterogeneous spatio-temporal inputs such as satellite imagery, hydrological time series, and infrastructure risk indices through a shared representation encoder, followed by dual task-specific predictive heads. The model enables integrated foresight into both hydrological behavior and infrastructure vulnerability.

At the core of the framework lies a shared feature extraction network $f_{\theta(\cdot)}$, which encodes a multimodal input tensor $X \in \mathbb{R}^{T \times H \times W \times C}$ containing satellite rainfall data, terrain elevation, land cover classification, and sensor measurements:

$$z = f_\theta(X)$$

where $z \in \mathbb{R}^d$ is the learned latent representation, f_θ is a deep convolutional-recurrent encoder with parameters θ, and X aggregates inputs across spatial dimensions $H \times W$, temporal length T, and channel modalities C. This representation captures both topographic and temporal dependencies essential for flood dynamics.

Two task-specific heads are applied to the shared embedding z, enabling the model to predict both flood occurrence probability and expected infrastructure disruption:

$$\hat{y}_{\text{flood}} = f_{\phi_1}(z), \hat{y}_{\text{impact}} = f_{\phi_2}(z)$$

where $y^{\text{flood}} \in [0, 1]$ is the predicted probability of flood occurrence (binary classification), $y^{impact} \in \mathbb{R}_+$ estimates infrastructure impact severity (e.g., number of affected homes or road closures), and f_{ϕ_1}, f_{ϕ_2} are fully connected layers with independent trainable parameters ϕ_1 and ϕ_2, respectively.

Each head is optimized using a domain-specific loss function. The total multitask objective combines classification and regression tasks into a weighted joint loss:

$$\mathcal{L}_{\text{total}} = \lambda_1 \cdot \mathcal{L}_{\text{flood}} + \lambda_2 \cdot \mathcal{L}_{\text{impact}}$$

where $\mathcal{L}_{\text{flood}} = -[y\log(\hat{y}_{\text{flood}}) + (1 - y)\log(1 - \hat{y}_{\text{flood}})]$ is a binary cross-entropy loss for flood classification, $\mathcal{L}_{\text{impact}} = \left\|\hat{y}_{\text{impact}} - y_{\text{impact}}\right\|^2$ is a mean squared error for impact regression, and $\lambda_1, \lambda_2 \in \mathbb{R}_+$ are hyperparameters balancing task importance.

To enhance interpretability and spatial responsiveness, the framework includes an attention-based visualization module that localizes the most influential features in each modality. The attention map $A \in \mathbb{R}^{H \times W}$ is computed from the intermediate features $F \in \mathbb{R}^{H \times W \times d}$ via:

$$A_{ij} = \frac{\exp\left(w^{\top} F_{ij}\right)}{\sum\limits_{i,j} \exp\left(w^{\top} F_{ij}\right)}$$

where A_{ij} indicates the normalized importance of spatial cell (i, j), and $w \in \mathbb{R}^{d}$ is a learned projection vector. These maps are overlaid on GIS dashboards to highlight high-risk regions and infrastructure nodes with predicted disruption, thereby aiding operational clarity.

By embedding a multimodal, multitask deep learning model into digital flood resilience platforms, the MTN-FR^2 framework enables comprehensive and anticipatory decision-making. It offers emergency management systems the ability to simulate flood onset and infrastructure impacts in a unified pipeline—bridging hydrological modeling with urban resilience forecasting. Leveraging diverse input modalities and task-specific outputs, the model shifts AI deployment from single-point alert systems to integrated, real-time situational awareness tools. This transforms how cities prepare for and respond to climate-induced flood events, reinforcing systemic adaptation through predictive intelligence.

Framework: Multimodal Multitask Neural Framework for Flood Resilience Forecasting (MTN-FR^2)

Layered AI Architecture for Real-Time Flood Prediction and Infrastructure Impact Forecasting

Layer	Main functionality	AI module(s) and components	Operational scope (Input → Processing → Output)
Layer 1	Multimodal feature extraction	Convolutional-recurrent encoder	Input → Satellite data, hydrological time series, infrastructure risk indices → Output → Latent representation encoding both spatial and temporal dependencies
Layer 2	Flood occurrence and infrastructure impact prediction	Multitask learning with fully connected heads	Input → Shared feature representation → Output → Predict flood occurrence probability and infrastructure impact severity
Layer 3	Multitask learning optimization	Multitask loss function	Input → Binary classification (flood) and regression (impact) tasks → Output → Optimized model with weighted joint loss for multitask learning
Layer 4	Feature localization and visualization	Attention mechanism	Input → Learned feature representations → Output → Attention map highlighting important spatial regions for flood risk and infrastructure disruption

6.1.2 *Enhancing Flood Preparedness Through Data Integration*

A defining strength of artificial intelligence in flood management is its capacity to synthesize vast and heterogeneous data sources into a unified predictive framework. By leveraging multimodal streams—including satellite observations, meteorological forecasts, hydrological sensor data, and even user-generated content from social media—AI systems can generate a dynamic and high-resolution understanding of flood risk. Saravi et al. (2019) emphasize the growing role of satellite-based monitoring in flood preparedness, where data such as surface reflectance and near-infrared imagery are used to track river expansion, soil saturation, and potential breach points. When combined with high-frequency weather feeds, AI algorithms can forecast flood events several days in advance, enabling preemptive public safety interventions and infrastructure protection.

To operationalize the integration of diverse environmental signals into a coherent predictive pipeline, we propose the Multimodal Transformer for Environmental Signal Fusion (MT-ESF). This model architecture is specifically designed to align heterogeneous spatial, temporal, and semantic data sources—such as satellite imagery, meteorological projections, and geotagged user reports—into a unified flood likelihood forecasting mechanism. MT-ESF extends beyond unimodal systems by capturing intermodal interactions and asynchronous signal dependencies in a mathematically grounded, Transformer-based framework.

Each environmental data stream is initially transformed into a latent representation using modality-specific encoders:

$$Z_m = E_m(D_m), \quad m \in \{\text{sat, weather, text}\}$$

where D_m is the raw input from modality m (e.g., satellite raster data, weather time series, or geotagged social media text), $E_m(\cdot)$, is the encoder function specific to each modality (e.g., CNN for imagery, RNN for time series, BERT for text), and $Z_m \in \mathbb{R}^{T_m \times d}$ is the resulting sequence of latent embeddings for modality mm.

These representations are aligned and fused using a shared transformer block with positional encoding and attention weights:

$$H = \text{TransformerLayer}([Z_{\text{sat}}; Z_{\text{weather}}; Z_{\text{text}}])$$

where $H \in \mathbb{R}^{T \times d}$ is the fused sequence representation, and the TransformerEncoder applies positional encoding followed by stacked self-attention and feed-forward layers.

The core attention mechanism is defined by scaled dot-product attention:

$$\text{Attention}(Q, K, V) = \text{softmax}\left(\frac{QK^\top}{\sqrt{d_k}}\right)V$$

where $Q, K, V \in \mathbb{R}^{T \times d_k}$ are the query, key, and value matrices derived from the input embeddings via linear projections, and d_k is the dimensionality of the keys. This attention mechanism enables the model to weight informative cues across modalities and time, such as aligning river expansion in satellite data with precipitation surges or correlating citizen flood reports with abnormal weather patterns.

The aggregated sequence H is pooled into a global feature vector and passed through a prediction head for flood probability estimation:

$$\hat{y}_{\text{flood}} = \sigma(WH_{\text{CLS}} + b)$$

where $\hat{y}_{\text{flood}}$ is the predicted likelihood of a flood event within the next temporal window for the target region, $H_{\text{CLS}} \in \mathbb{R}^d$ is the pooled classification token embedding output by the Transformer, $W \in \mathbb{R}^{1 \times d}$ and $b \in \mathbb{R}$ are learnable parameters, and $\sigma(\cdot)$ is the sigmoid activation function producing a probabilistic confidence score. The model is trained using binary cross-entropy loss:

$$\mathcal{L} = -[y\log(\hat{y}_{\text{flood}}) + (1 - y)\log(1 - \hat{y}_{\text{flood}})]$$

where $y \in \{0, 1\}$ is the ground truth indicator of flood occurrence for the forecasted region and time horizon. This formulation enables the system to calibrate its predictions according to probabilistic ground truth labels from historical FEMA reports, NOAA datasets, and community-verified flood events.

Through the integration of the Multimodal Transformer for Environmental Signal Fusion (MT-ESF) into flood forecasting platforms, cities gain access to a high-resolution, cross-domain situational awareness system. Unlike traditional systems that rely solely on gauge data or static thresholds, MT-ESF integrates asynchronous, multi-format evidence streams—including remote sensing, climate forecasts, and human reports—into a temporally coherent and semantically aligned decision model. This enables not only earlier and more accurate flood alerts, but also a participatory, data-rich paradigm of environmental resilience—one that combines atmospheric science with collective intelligence and transforms digital twins into anticipatory public safety tools.

Framework: Multimodal Transformer for Environmental Signal Fusion (MT-ESF)

Layered AI Architecture for Flood Risk Prediction and Management

Layer	Main functionality	AI module(s) and components	Operational scope (Input → Processing → Output)
Layer 1	Data collection and representation learning: Gather and preprocess environmental data, and encode raw data into modality-specific latent representations	Convolutional Neural Networks (CNNs), Recurrent Neural Networks (RNNs), Bidirectional Encoder Representations from Transformers (BERT)	Input → Raw satellite, meteorological, and social media data → Output → Latent embeddings for each modality (satellite, weather, text)

(continued)

Layer	Main functionality	AI module(s) and components	Operational scope (Input → Processing → Output)
Layer 2	Data fusion: Align and fuse the modality-specific latent representations	Transformer encoder	Input → Latent embeddings (satellite, weather, text) → Output → Fused sequence representation
Layer 3	Attention mechanism: Capture important correlations between data modalities	Scaled dot-product attention	Input → Fused sequence representation → Output → Weighted importance of each modality's information
Layer 4	Flood risk prediction: Predict flood likelihood based on fused data	Deep Neural Networks (DNNs) with Sigmoid Activation	Input → Fused representation from transformer → Output → Predicted flood likelihood score
Layer 5	Loss function: Compute binary cross-entropy loss to train the model	Binary cross-entropy loss	Input → Predicted flood likelihood, ground truth labels → Output → Loss value for model optimization
Layer 6	Deployment and monitoring: Implement the model for real-time flood risk prediction	Real-time data stream processing	Input → Environmental data streams (real-time) → Output → Flood risk predictions and alerts
Layer 7	Feedback and model adaptation: Integrate feedback to refine the model over time	Real-time feedback integration	Input → Feedback data from flood events and user reports → Output → Refined model predictions and improved accuracy

In addition to traditional environmental monitoring systems, social media platforms have emerged as critical sources of high-frequency, ground-level information during flood events. As individuals affected by flooding share real-time updates, images, and location-tagged reports, a vast body of user-generated content becomes available that reflects evolving on-the-ground conditions with spatial and emotional granularity. When harnessed effectively, these decentralized data streams can complement official channels by filling temporal gaps in sensor coverage or confirming anomalous readings. AI algorithms now possess the capacity to ingest and interpret this unstructured content—transforming citizen voice into actionable insight for emergency responders, planners, and digital twin simulations.

To systematically operationalize the integration of real-time social signals into flood prediction systems, we propose the Multistream Attention-based Recurrent Flood Awareness Network (MARN-FA). This model is designed to capture the semantic, spatial, and emotional dimensions of user-generated flood-related content from social media, transforming decentralized information into structured situational awareness. MARN-FA augments traditional environmental monitoring by embedding localized citizen observations directly into the decision layer of flood detection systems.

Each social media post at time t, denoted s_t, is first transformed into a contextual semantic embedding using a pretrained RoBERTa encoder:

$$\mathrm{e}_t = \mathrm{RoBERTa}(s_t)$$

where $e_t \in \mathbb{R}^d$ is the semantic embedding capturing the linguistic structure and contextual intent of post s_t, and RoBERTa $(\cdot)$ is a transformer-based language model pretrained on large-scale corpora to extract high-dimensional representations.

Each post is also associated with a geolocation vector $g_t \in \mathbb{R}^2$ and sentiment score $\xi_t \in [-1, 1]$ forming a unified input vector through concatenation:

$$\mathrm{z}_t = [\mathrm{e}_t; g_t; \xi_t]$$

where $z_t \in \mathbb{R}^{d+3}$ represents the enriched feature vector that incorporates semantic meaning, spatial coordinates, and emotional tone. This multimodal representation is passed through a Gated Recurrent Unit (GRU) to model temporal dependencies:

$$h_t = \mathrm{GRU}(z_t, h_{t-1})$$

where $h_t \in \mathbb{R}^d$ is the hidden state at time t, encapsulating the accumulated context from prior social signals, and GRU$(\cdot)$ captures long-range dependencies through memory gating mechanisms.

To prioritize high-salience signals—such as urgent posts with strong sentiment or concentrated geographic references—an attention mechanism computes the importance weight α_t for each time step:

$$\alpha_t = \frac{\exp(w^\top \tanh(W_z \mathrm{z}_t + b))}{\sum_\tau \exp(w^\top \tanh(W_z \mathrm{z}_\tau + b))}$$

where $\alpha_t \in [0, 1]$ represents the attention weight at time t, $W \in \mathbb{R}^{h \times (d+3)}$, $w \in \mathbb{R}^h$, and $b \in \mathbb{R}^h$ are learnable parameters that project input features into an attention space. This mechanism amplifies the influence of emotionally intense or geographically clustered posts.

The final flood awareness prediction is computed as an attention-weighted aggregation of the GRU outputs, followed by a sigmoid classifier:

$$\hat{y}_{\mathrm{social}} = \sigma\left(\sum_t \alpha_t \cdot h_t\right)$$

where $\hat{y}_{\mathrm{social}} \in [0, 1]$ is the predicted probability of flood presence based on social signal consensus, and $\sigma(\cdot)$ is the sigmoid activation function transforming the aggregated hidden state into a confidence score.

This social-derived output can be integrated directly into broader digital twin infrastructures, either as a standalone alert channel or as a reinforcing input to satellite, sensor, and meteorological models. For instance, in cases where hydrological sensors offer ambiguous data over a localized zone, but regional users report submerged streets with strong emotional tone and geospatial convergence, the system can escalate the alert level autonomously—triggering adaptive response measures such as rerouting, road closures, or early evacuation protocols.

By unifying the MARN-FA model into AI-driven flood forecasting pipelines, cities gain the ability to convert distributed citizen voice into centralized resilience intelligence. This fusion of decentralized perception with centralized action enhances both the spatial resolution and responsiveness of emergency systems, democratizing the production of environmental knowledge and reinforcing participatory urban adaptation through ethically designed AI.

Framework: Multistream Attention-based Recurrent Flood Awareness Network (MARN-FA)

Layered AI Architecture for Social Media-Driven Flood Prediction

Layer	Main functionality	AI module (s) and components	Operational scope (Input → Processing → Output)
Layer 1	Data collection: Gather and preprocess social media content	Social media data preprocessing module	Input → Raw social media posts → Output → Preprocessed semantic, geolocation, and sentiment data
Layer 2	Representation learning: Transform social media posts into semantic embeddings	RoBERTa	Input → Social media posts → Output → Contextual semantic embeddings
Layer 3	Feature integration: Combine semantic, spatial, and sentiment features into a unified vector	Feature fusion network	Input → Semantic embeddings, geolocation vectors, and sentiment scores → Output → Unified feature vector for each post
Layer 4	Temporal modeling: Capture dependencies in social media content over time	GRU	Input → Unified feature vectors over time → Output → Hidden states representing temporal context
Layer 5	Attention mechanism: Assign weights to each post based on importance	Attention mechanism	Input → GRU hidden states → Output → Attention weights for each time step
Layer 6	Flood awareness prediction: Predict flood likelihood using attention-weighted GRU outputs	Sigmoid Classifier	Input → Attention-weighted GRU hidden states → Output → Predicted flood likelihood score
Layer 7	Deployment and monitoring: Implement model for real-time flood predictions	Real-time data stream processing	Input → Social media content streams → Output → Real-time flood predictions and alerts
Layer 8	Feedback and model adaptation: Integrate feedback to refine predictions over time	Feedback integration module	Input → User feedback and real-time flood data → Output → Refined predictions and model adaptation

6.1.3 Real-Time Response and Resource Allocation

Beyond its role in forecasting and monitoring, artificial intelligence plays an increasingly vital part in orchestrating real-time flood response and resource management. During active flood events, the timely deployment of rescue units, medical aid, and temporary shelters can decisively influence survival outcomes and infrastructure preservation. As Saravi et al. (2019) observe, AI models trained on spatio-temporal flood data can not only simulate likely water flow trajectories but also identify priority intervention zones in advance—enabling emergency operations to avoid delays, minimize resource wastage, and reduce human vulnerability.

To support such dynamic coordination, we propose the implementation of a Deep Reinforcement Learning (DRL) Framework optimized via Proximal Policy Optimization (PPO) for adaptive resource allocation. This model allows the AI system embedded within the digital twin to continuously learn optimal response strategies by interacting with a simulated flood environment that evolves in real time.

Let the flood environment be defined by a state $s_t \in \mathbb{R}^n$ at time t, encoding factors such as inundation depth, road accessibility, population density, and resource availability. The system selects an action a_t—representing a specific allocation of resources (e.g., dispatching boats, opening shelters)—according to a stochastic policy $\pi_\theta(a_t \mid s_t)$ parameterized by θ:

$$a_t \sim \pi_\theta(a_t \mid s_t)$$

The objective of PPO is to maximize expected reward while preventing overly large policy updates that destabilize learning. The clipped surrogate objective is defined as:

$$\mathcal{L}^{\text{PPO}}(\theta) = \mathbb{E}_t\left[\min\left(r_t(\theta)\hat{A}_t, \text{clip}(r_t(\theta), 1-\epsilon, 1+\epsilon)\hat{A}_t\right)\right]$$

where $r_t(\theta) = \frac{\pi_\theta(a_t \mid s_t)}{\pi_{\theta_{\text{old}}}(a_t \mid s_t)}$ is the probability ratio between the new and old policy, $\hat{A}_t$ is the advantage estimate at time t, and ϵ is a hyperparameter constraining policy updates. This formulation ensures balanced exploration and reliable convergence even under fast-changing disaster scenarios.

The reward function is designed to reflect emergency effectiveness:

$$R(s_t, a_t) = -\lambda_1 C_{\text{delay}}(s_t, a_t) - \lambda_2 C_{\text{misalloc}}(s_t, a_t) + \lambda_3 S_{\text{coverage}}(s_t, a_t)$$

where C_{delay} quantifies response latency, C_{misalloc} penalizes inefficient resource use, and S_{coverage} measures spatial service reach, with λ_i serving as policy weights.

This model allows the system to learn high-utility action sequences for different flood contexts—such as routing medical teams through dry corridors, opening shelters in proximity to vulnerable populations, or deploying boats ahead of predicted water surges. These decisions are optimized in real time and refined through simulated rollouts, enabling constant adaptation to changing flood topologies. This not only accelerates resource mobilization and improves service equity but also strengthens a city's operational resilience in the face of compounding environmental crises.

Framework: Deep Reinforcement Learning Framework Optimized via Proximal Policy Optimization (DRL-PPO)

Layered AI Architecture for Real-Time Flood Response and Resource Management

Layer	Main functionality	AI module (s) and components	Operational scope (Input → Processing → Output)
Layer 1	Flood environment definition: Define state variables for the flood environment	Flood environment representation module	Input → Inundation depth, road accessibility, population density, resource availability → Output → State vector representing the flood environment at each time step
Layer 2	Action selection: Choose optimal resource allocation actions using a stochastic policy	Deep reinforcement learning with stochastic policy	Input → Current flood environment state → Output → Selected action for resource allocation (e.g., dispatch boats, open shelters)
Layer 3	PPO objective optimization: Optimize policy using the clipped surrogate objective	Proximal Policy Optimization (PPO)	Input → Probability ratio, advantage estimates → Output → Optimized policy parameters for decision-making
Layer 4	Reward function: Compute reward based on effectiveness of resource allocation	Reward calculation engine	Input → Resource allocation and state information → Output → Calculated reward reflecting emergency effectiveness (e.g., delay, misallocation, coverage)
Layer 5	Policy evaluation and deployment: Evaluate policy through simulated rollouts, deploy for real-time decision-making	Policy rollout evaluation module	Input → Simulated flood environment, real-time flood data → Output → Refined action sequences and real-time decisions on resource deployment
Layer 6	Real-time feedback and adaptation: Continuously improve model based on feedback from real flood events	Feedback loop integration module	Input → feedback on action outcomes, real-time flood data → Output → Refined policies and decision-making strategies for future events

An equally vital function of artificial intelligence in flood resilience lies in its capacity to enhance coordination across multiple governmental, civic, and emergency agencies involved in disaster management. Effective response during a flood crisis depends not only on accurate predictions and swift action, but also on the seamless synchronization of information, strategies, and resources among stakeholders. AI-driven platforms, equipped with real-time data exchange capabilities and intelligent routing protocols, can eliminate the fragmentation of communication channels—thereby preventing operational delays, resource duplication, or conflicting interventions.

To support this multi-actor environment, we propose the deployment of a Federated Multi-Agent Coordination Framework based on a Decentralized Communication-Aware Deep Q-Network (DCA-DQN). This architecture treats each agency—such as municipal rescue units, national disaster relief forces, transportation departments, and healthcare services—as a semi-autonomous agent that must collaborate to achieve shared objectives while operating with partial observations of the environment. Each agent i observes a local state s_i^t at time t and selects an action a_i^t from its operational space (e.g., deploy shelter units, open evacuation routes), with policy guided by a shared Q-network:

$$Q_i\left(s_i^t, a_i^t\right) = \mathbb{E}\left[r_i^t + \gamma \max_{a'} Q_i\left(s_i^{t+1}, a'\right)\right]$$

where Q_i is the action-value function for agent i, r_i^t is the local reward signal (e.g., reduced evacuation delay or increased coverage), and γ is the temporal discount factor.

To ensure real-time consistency across all agents, a communication-aware utility function is defined that dynamically weights decisions based on message reception status and latency:

$$U_i^t = \beta_1 Q_i\left(s_i^t, a_i^t\right) + \beta_2 \sum_{j \in \mathcal{N}(i)} \mathbb{I}_{\text{msg}}^{(j \to i)} \cdot Q_j\left(s_j^t, a_j^t\right)$$

where $\mathbb{I}_{\text{msg}}^{(j \to i)}$ indicates whether agent i received a valid message from neighbor j, and β_1, β_2 control the balance between local autonomy and inter-agent coordination.

The system is trained in a federated manner, meaning that each agent improves its local model while periodically sharing encrypted model gradients with a central aggregator—thus preserving data privacy while aligning global strategy.

In the federated learning process, the local models are updated based on the local gradient computations, and these gradients are subsequently aggregated by a central server to update the global model. The operation is mathematically represented as follows:

$$\theta_{\text{global}}^{t+1} = \frac{1}{N} \sum_{i=1}^{N} \left(\theta_i^t - \eta \nabla \mathcal{L}_i\left(\theta_i^t\right)\right)$$

where $\theta_{\text{global}}^{t+1}$ are the updated global model parameters, θ_i^t are the local model parameters for agent i at iteration t, $\nabla \mathcal{L}_i\left(\theta_i^t\right)$ is the gradient of the local loss function with respect to θ_i^t, η is the learning rate, and N is the total number of agents. This function aggregates the local model gradients, updates the global model in a privacy-preserving manner, and ensures that the global strategy evolves without sharing raw data across the system.

As the system enters the coordination phase, the primary focus shifts from individual decision-making to collective action. Building on the knowledge gained through federated learning and local policy refinement, the agents now synchronize their efforts to form a cohesive disaster response strategy. By utilizing the shared Q-network developed earlier, each agent not only applies its individual learned policies but also ensures that its actions align with the global strategy. This coordinated approach enhances the overall efficiency of the system, allowing agents to collectively optimize resource allocation, manage traffic flow, prioritize medical services, and deploy relief measures. The disaster response coordination ensures that all agencies act in harmony, leveraging their unique capabilities while maintaining the overarching goal of mitigating the disaster's impact.

This model enables disaster response agencies to function as a coherent, self-organizing system even under network delays, regional isolation, or data incompleteness. For instance, if local transportation authorities detect bridge failure through their sensor nodes, the alert is automatically propagated to health service agents, who reroute ambulances; simultaneously, civil defense units re-prioritize resource flows to the affected region—all with minimal latency and no manual command overhead. This paradigm ensures that all involved agencies operate with shared situational awareness and aligned objectives—transforming emergency response from reactive compartmentalization to real-time collective intelligence embedded within the smart city infrastructure.

Framework: Federated Multi-Agent Coordination Framework for Flood Resilience (FMACF)
Layered AI Architecture for Decentralized Communication-Aware Disaster Response

Layer	Main functionality	AI module (s) and components	Operational scope
Layer 1	Data collection: Collect real-time sensor and environmental data from flood-prone areas	Sensor data preprocessing module	Input → Sensor data (e.g., water levels, rainfall, GPS) → Output → Preprocessed data for agents
Layer 2	Agent state observation: Each agency observes its local state for decision-making	Multi-agent coordination engine	Input → Local state data (e.g., agency-specific data) → Output → Agent-specific observations
Layer 3	Learning: Deep Q-Network (DQN) for agent decision-making	Deep Q-Network (DQN)	Input → Agent state, action → Output → Agent action-value function
Layer 4	Coordination: Multi-agent coordination using communication-aware utility function	–	Input → Local actions, received messages → Output → Coordinated actions among agents

(continued)

Layer	Main functionality	AI module (s) and components	Operational scope
Layer 5	Federated learning: Share model gradients to align global strategy while preserving privacy	Federated Learning (FL)	Input → Local model gradients → Output → Global model updates
Layer 6	Disaster response coordination: Agencies act in synchronized manner based on shared Q-network	Deep Q-Network (DQN)	Input → Local state and coordinated action → Output → Combined disaster response strategies
Layer 7	Monitoring and evaluation: Continuous assessment of system performance	Performance monitoring system	Input → Performance data, failure detection → Output → Real-time performance evaluation
Layer 8	Human-in-the-loop feedback: User feedback on system actions and coordination	User feedback integration module	Input → Operator feedback → Output → Refined coordination strategies

6.1.4 Machine Learning Models in Flood Impact Analysis

In the aftermath of a flood event, the ability to rapidly and accurately assess the extent of physical damage is crucial for organizing effective recovery operations and directing humanitarian aid. Machine learning, particularly through deep learning architecture, has emerged as an essential tool for automating the analysis of high-resolution satellite and aerial imagery. These models can quantify the spatial footprint of a flood, classify the severity of impact across different urban and rural land types, and estimate infrastructure loss within hours of data acquisition. As Saravi et al. (2019) highlight, convolutional neural networks (CNNs) have demonstrated strong performance in identifying post-event change patterns by comparing time-separated satellite imagery—offering near-real-time decision support for governments, insurers, and disaster relief agencies.

To operationalize this capability in a mathematically robust and scalable manner, we propose a Dual-Temporal Convolutional Neural Network (DT-CNN) enhanced with a U-Net segmentation head for precise pixel-wise classification of flood-induced damage. The model accepts a pair of co-registered satellite images—captured before (I_{pre}) and after (I_{post}) a flood event—as its input, allowing it to learn temporal feature differences relevant to inundation and destruction.

The initial input fusion is performed through a shared convolutional backbone:

$$F_{pre}, F_{post} = f_{CNN}\left(I_{pre}\right), f_{CNN}\left(I_{post}\right)$$

where $f_{CNN}(\cdot)$ denotes the feature extractor (e.g., ResNet or EfficientNet), and f are the encoded feature maps for each timestamp.

The change detection tensor ΔF is computed via:

$$\Delta F = |F_{\text{post}} - F_{\text{pre}}|$$

This differential tensor is passed into a U-Net decoder to produce a segmentation mask M indicating the spatial distribution of flood impact:

$$M = f_{\text{U-Net}}(\Delta F)$$

where $M \in [0, 1]^{H \times W}$ is a heatmap over the image domain representing damage probability per pixel, segmented into structural damage, agricultural loss, and inaccessible zones.

The model is trained using a composite loss function balancing binary cross-entropy and intersection-over-union (IoU) for segmentation accuracy:

$$\mathcal{L}_{\text{total}} = \lambda_1 \mathcal{L}_{\text{BCE}} + \lambda_2 \mathcal{L}_{\text{IoU}}$$

This approach allows the system to generate high-resolution damage maps within minutes of receiving new satellite input. For instance, if the model detects heavy inundation in industrial zones and partial road collapse in transportation corridors, it can guide field assessment teams to prioritize those areas and update infrastructure restoration timelines accordingly.

This AI-driven model not only accelerates emergency assessment but also reduces human error, enabling national agencies and international partners to deploy targeted recovery strategies based on real-time satellite evidence—ultimately accelerating community resilience and infrastructure restoration.

Framework: Dual-Temporal Convolutional Neural Network for Flood Damage Assessment (DT-CNN-FDA)

Layered AI Architecture for Real-Time Post-Flood Damage Evaluation

Layer	Main functionality	AI module (s) and components	Operational scope
Layer 1	Data collection: Gather high-resolution satellite imagery (pre- and post-flood)	Satellite image data acquisition module	Input → Satellite images (pre- and post-flood) → Output → Raw input for model processing
Layer 2	Preprocessing: Image registration and fusion	Image registration and fusion engine	Input → Pre- and post-flood satellite images → Output → Co-registered images for model input
Layer 3	Feature extraction: Extract spatial features using CNN backbone	Convolutional Neural Network (CNN)	Input → Co-registered images → Output → Feature maps representing flood-related changes
Layer 4	Change detection: Compute the difference between pre- and post-flood features	Change detection engine	Input → Feature maps (pre- and post-flood) → Output → Change detection tensor ΔF
Layer 5	Segmentation: Apply U-Net to detect flood damage zones	U-Net	Input → Change detection tensor ΔF → Output → Segmentation mask (damage zones)

(continued)

Layer	Main functionality	AI module (s) and components	Operational scope
Layer 6	Loss function: Train the model using binary cross-entropy and IoU loss	Loss calculation and optimization engine	Input → Segmentation mask → Output → Composite loss function for training (BCE + IoU)
Layer 7	Damage map generation: Create high-resolution damage maps for decision support	Damage map visualization tool	Input → Trained model → Output → Damage maps for flood-affected areas
Layer 8	Real-time decision support: Guide recovery and infrastructure restoration efforts	Recovery operations decision support system	Input → Damage maps → Output → Actionable insights for recovery operations

In addition to supporting emergency response and post-disaster recovery, artificial intelligence has a growing role in shaping long-term strategies for flood mitigation and urban infrastructure resilience. Through the identification of flood recurrence patterns and spatial risk gradients, AI systems can provide critical foresight to planners and civil engineers seeking to fortify vulnerable zones. By integrating historical and simulated flood risk data into city-scale planning tools, AI models help guide the placement of levees, the design of adaptive drainage systems, and the development of flood-resilient construction norms—reducing future exposure and increasing urban adaptive capacity.

To support these long-range planning objectives, we propose the application of a Spatio-Temporal Risk Zoning Framework based on Gaussian Process Regression (GPR) for flood probability estimation, coupled with a graph-based urban infrastructure optimization algorithm. This model enables planners to continuously update and visualize flood probability surfaces while dynamically evaluating the cost-efficiency of various infrastructure configurations.

Let $\mathcal{D} = \{(x_i, y_i)\}_{i=1}^{N}$ represent historical flood data where $x_i \in \mathbb{R}^d$ denotes spatial-temporal covariates (e.g., elevation, land use, rainfall history) and $y_i \in \mathbb{R}$ represents flood severity or frequency. The GPR model estimates the flood risk function $f(x) \sim \mathcal{GP}(0, k(x, x'))$ as:

$$\mathrm{f}(x) \sim \mathcal{GP}(0, k(x, x'))$$

where $k(x, x')$ is a spatial covariance kernel (e.g., squared exponential or Matern), encoding similarity between regions based on geographic and hydrological features.

The posterior flood risk mean $\mu(x_*)$ and variance $\sigma^2(x_*)$ at a new location x_* are derived from the training set:

$$\mu(x_*) = k(x_*, X)\left[K + \sigma_n^2 I\right]^{-1} y$$

where K is the kernel matrix over training inputs X, σ_n^2 is observation noise variance, and y is the observed flood impact vector.

To translate this risk surface into actionable planning decisions, we model the urban layout as a graph $G = (V, E)$ where nodes V correspond to urban parcels or infrastructure units and edges E represent spatial adjacency. An optimization objective is defined to minimize exposure-weighted infrastructure cost:

$$\min_{z \in \{0,1\}^V} \sum_{v \in V} (\rho_v \cdot f(v) + c_v \cdot z_v)$$

where $f(v)$ is the flood risk at parcel v, c_v is the construction or retrofitting cost, z_v is a binary variable indicating infrastructure placement, and ρ_v is a vulnerability weighting factor for critical infrastructure zones (e.g., hospitals, schools).

The solution provides an optimized blueprint for deploying flood mitigation infrastructure—prioritizing high-risk, high-impact areas while adhering to budget constraints. For example, the model may recommend reinforced drainage upgrades in industrial basins with frequent pluvial flooding, while delaying construction in low-risk residential zones until future phases. This framework empowers cities to not only respond to climate-related hazards but to anticipate and design against them—transforming reactive flood management into proactive, risk-informed urban evolution.

Framework: Spatio-Temporal Risk Zoning for Flood Mitigation (STRZ-FM)
Layered AI Architecture for Long-Term Urban Resilience and Flood Risk Mitigation

Layer	Main functionality	AI module(s) and components	Operational scope
Layer 1	Data collection: Gather historical flood data and spatial-temporal covariates	Data acquisition and feature engineering module	Input → Historical flood data (e.g., elevation, land use, rainfall history) → Output → Raw data for training model
Layer 2	Preprocessing: Data normalization and integration of spatial-temporal features	Spatial-temporal data normalization and integration engine	Input → Flood data, spatial features → Output → Processed dataset for model input
Layer 3	Flood probability estimation: Use Gaussian Process Regression (GPR) for flood risk estimation	Gaussian Process Regression (GPR)	Input → Spatial-temporal data → Output → Estimated flood risk function
Layer 4	Risk surface calculation: Compute posterior mean and variance for flood risk at new locations	Gaussian Process Regression (GPR)	Input → Training data, covariance kernel → Output → Posterior flood risk mean and variance
Layer 5	Urban Infrastructure Optimization: Graph-based optimization for flood mitigation infrastructure	Mixed-Integer Linear Programming (MILP)	Input → Flood risk, infrastructure cost data → Output → Optimized urban infrastructure placement
Layer 6	Optimization objective: Minimize exposure-weighted infrastructure cost	Infrastructure optimization engine	Input → Flood risk, cost data, vulnerability factors → Output → Optimized infrastructure deployment plan

(continued)

Layer	Main functionality	AI module(s) and components	Operational scope
Layer 7	Scenario analysis: Evaluate and visualize flood mitigation strategies	Flood mitigation strategy evaluation tool	Input → Optimized infrastructure deploymentplan→Output→Scenario-based evaluation of mitigation strategies
Layer 8	Decision support: Provide actionable insights for urban planning and disaster resilience	Urban planning decision support system	Input → Flood risk, infrastructure data → Output → Decision support recommendations for long-term flood mitigation

6.1.5 Challenges and Future Directions

While AI has made significant strides in improving flood resilience, there are still challenges that need to be addressed. One major issue is the quality and availability of data. Accurate AI predictions depend on high-quality, real-time data, but in many parts of the world, such data is either unavailable or unreliable. Moreover, there are concerns regarding the transparency of AI models. Complex machine learning algorithms can act as "black boxes," making it difficult to understand how decisions are being made. This lack of transparency can hinder trust in AI systems, especially when they are used to make critical decisions during disaster scenarios.

To address the intertwined challenges of unreliable data streams and opaque AI decision-making in disaster scenarios, we propose the implementation of an Explainability-Integrated Deep Reinforcement Learning (X-DRL) Framework. At its core, this framework leverages the dynamic adaptability of Deep Reinforcement Learning (DRL) to operate under partial observability and inconsistent data availability while integrating Explainable AI (XAI) techniques to enhance the interpretability of its decisions. Specifically, the agent learns optimal flood response policies from sparse or delayed inputs, while continuously producing human-understandable justifications for its actions. This ensures that emergency planners and stakeholders not only receive optimal policy recommendations but also comprehend the rationale behind them, even in low-data or real-time deployment conditions.

The architecture begins with the foundational Deep Q-Network (DQN) formulation, which underpins the agent's decision-making process in high-dimensional and uncertain environments.

$$Q(s_t, a_t; \theta) = r_t + \gamma \max_{a'} Q(s_{t+1}, a'; \theta^-)$$

where $Q(s_t, a_t; \theta)$ is the estimated action-value function under current policy parameters θ, r_t is the reward received after executing action a_t in state s_t, $\gamma \in [0, 1]$ is the discount factor that balances immediate and future rewards, and $Q(s_{t+1}, a'; \theta^-)$ is the target Q-value computed using a fixed (delayed) set of parameters θ^-. This equation enables the agent to iteratively improve its flood response strategy by evaluating the long-term impact of each action. The Q-network is trained to approximate the true action-value function by minimizing the temporal difference error between predicted and target Q-values. The agent, therefore, learns not just reactive but anticipatory behaviors in flood-prone environments.

To enhance transparency and trust, we integrate a SHapley Additive exPlanations (SHAP) mechanism over the learned Q-values, thereby attributing importance scores to each input feature contributing to the agent's decision. This is defined as:

$$\phi_i = \sum_{S \subseteq F \setminus \{i\}} \frac{|S|!(|F| - |S| - 1)!}{|F|!} [Q(S \cup \{i\}) - Q(S)]$$

where ϕ_i is the SHAP value for input feature i, F is the full set of input features (e.g., rainfall intensity, topography, real-time sensor alerts), S is any subset of features excluding i, $Q(S \cup \{i\})$ is the predicted Q-value when feature i is included in the input set, and $Q(S)$ is the Q-value without it. This formulation guarantees that the influence of each input variable on the agent's policy can be decomposed and quantified, empowering stakeholders to trace every policy decision back to its underlying evidence, even under data constraints.

To stabilize learning under noisy or partial data, the model employs the Double Q-Learning modification to decouple action selection and evaluation, minimizing overestimation bias:

$$Q(s_t, a_t; \theta) = r_t + \gamma Q\left(s_{t+1}, \arg\max_{a'} Q(s_{t+1}, a'; \theta); \theta^- \right)$$

where $\arg\max_{a'} Q(s_{t+1}, a'; \theta)$ selects the optimal action according to the current network, while the value is evaluated using the target network θ^-. This decoupling enhances policy stability and robustness, particularly when sensor signals are delayed or environmental observations are ambiguous.

By unifying these components—reinforcement-based learning, SHAP-based attribution, and decoupled Q-estimation—the proposed X-DRL model builds a transparent, interpretable, and highly adaptive AI system capable of guiding real-time flood response with mathematical rigor and decision clarity. In real-world deployments, such a model allows emergency teams to understand not only what action to take but why it was recommended, thus aligning operational efficiency with public accountability and governance trust.

Framework: Explainability-Integrated Deep Reinforcement Learning for Flood Response (X-DRL-FR)
Layered AI Architecture for Interpretable and Adaptive Flood Response Systems

Layer	Main functionality	AI module (s) and components	Operational scope
Layer 1	Data collection: Gather sparse and delayed flood data streams (e.g., rainfall, sensor data)	Data preprocessing and feature engineering module	Input → Raw flood data (e.g., rainfall, sensor readings) → Output → Processed input for AI model training
Layer 2	Flood response policy learning: Deep Q-Network (DQN) for flood response decision-making	Deep Q-Network (DQN)	Input → Processed flood data (e.g., historical, simulated) → Output → Learned action-value function for flood response policies
Layer 3	Explainability integration: SHAP-based explanation mechanism for policy interpretability	SHAP	Input → Learned Q-values from the DQN model → Output → Explainability scores attributing feature importance to decisions
Layer 4	Double Q-Learning: stabilize learning by decoupling action selection and evaluation	Double Q-Learning	Input → Q-values from policy learning → Output → Stabilized policy by reducing overestimation bias in flood response
Layer 5	Real-time decision support: Provide interpretable and adaptive flood response recommendations	Real-time decision engine	Input → Stabilized Q-values and flood response policy → Output → Real-time adaptive recommendations for flood response actions
Layer 6	Post-action evaluation: Evaluate and improve flood response strategies based on outcomes	Post-action evaluation and strategy optimization tool	Input → Evaluation of response outcomes → Output → Optimized strategies for future flood responses and planning

Future research should focus on improving the transparency and interpretability of AI models. Developing frameworks that allow stakeholders to understand how AI systems reach their conclusions will be essential for broader acceptance and deployment. Additionally, efforts should be made to standardize data collection practices globally, ensuring that AI models can operate effectively across different regions.

To fulfill the dual imperative of improving model interpretability and achieving cross-regional data interoperability, we propose a Federated Multi-Agent Explainability Framework (F-MAX). This architecture advances flood resilience and infrastructure reliability by enabling decentralized learning across geographically distributed agents, each operating on region-specific data while maintaining full compliance with data sovereignty constraints. At its core, the framework combines Multi-Agent Federated Learning (MAFL) with an embedded Explainable AI (XAI) layer, ensuring both model transparency and global applicability.

The learning process is driven by a distributed optimization objective where each regional agent $i \in \{1, \ldots, N\}$ collaboratively minimizes a global explainability-aware loss while preserving local data. The objective is formulated as:

$$\min_{\theta} \sum_{i=1}^{N} w_i [\mathcal{L}_i(\theta_i) + \lambda \mathcal{R}(\phi_i)]$$

where θ_i denotes the local model parameters for agent i, $\mathcal{L}_i(\theta_i)$ is the task-specific loss (e.g., flood prediction error), ϕ_i is the explanation vector derived from SHAP or LIME for the model's outputs, $\mathcal{R}(\phi_i)$ is an explanation regularizer that penalizes non-interpretable or unstable feature attributions, w_i are aggregation weights reflecting agent importance (e.g., population density or sensor quality), and λ is a tunable scalar governing the trade-off between predictive accuracy and explanation clarity. This formulation simultaneously optimizes predictive power and interpretability, ensuring that each local model not only fits regional conditions but also produces transparent outputs aligned with global standards.

Federated aggregation across agents is performed using the Federated Averaging (FedAvg) algorithm to synthesize a shared model $\theta(t)$ at communication round t:

$$\theta^{(t+1)} = \sum_{i=1}^{N} \frac{n_i}{\sum_j n_j} \theta_i^{(t)}$$

where n_i is the number of local data samples on agent i, and $\theta_i^{(t)}$ is the locally trained model at round t. This mechanism preserves data locality while generating a unified global model capable of generalizing across disparate geographies and datasets.

To enforce interoperability and standardized model behavior across regions, we integrate a Global Attribution Alignment Loss (GAAL) defined as:

$$\mathcal{A} = \frac{1}{N} \sum_{i=1}^{N} \left\| \phi_i - \bar{\phi} \right\|_2^2$$

where ϕ_i is the explanation vector for agent i, and $\bar{\phi} = \frac{1}{N} \sum_{i=1}^{N} \phi_i$ is the global mean attribution. This function minimizes interpretive drift between agents, fostering explanation consistency and ensuring that AI decisions reflect coherent logic regardless of regional variation.

Via the application of federated optimization, explanation regularization, and attribution alignment, the F-MAX framework delivers a forward-looking solution to the global challenges of explainability and data standardization. Each participating region retains autonomy over its data while contributing to a globally interpretable AI model. In practical terms, this architecture empowers decision-makers from disparate locales to trust and understand AI-derived flood predictions and response strategies—regardless of underlying data heterogeneity or cultural governance models.

Framework: Federated Multi-Agent Explainability Framework for Flood Response (F-MAX)

Layered AI Architecture for Interpretable and Adaptive Flood Response Systems

Layer	Main functionality	AI module (s) and components	Operational scope
Layer 1	Data collection: Collect region-specific flood data while maintaining data privacy	Data privacy and preprocessing engine	Input → Raw flood data (e.g., rainfall, sensor data) → Output → Raw data for model training
Layer 2	Federated learning: Distribute model training across geographically distributed agents	Multi-Agent Federated Learning (MAFL)	Input → Region-specific flood data (e.g., rainfall, elevation, historical data) → Output → Local flood prediction models trained on region-specific data
Layer 3	Optimization: Minimize global explainability-aware loss while preserving local data	Explainability-driven optimization engine	Input → Locally trained models, SHAP/LIME-generated feature importance scores → Output → Global model optimized for predictive accuracy and interpretability
Layer 4	Federated aggregation: Combine local models into a shared global model using FedAvg	Federated averaging (FedAvg)	Input → Local models from distributed agents → Output → Aggregated global model using Federated Averaging (FedAvg)
Layer 5	Attribution alignment: Minimize explainability drift across agents using Global Attribution Alignment Loss (GAAL)	Global explanation alignment module	Input → Explanation vectors from local models → Output → Aligned global explanations, ensuring consistent model interpretability across regions
Layer 6	Real-time decision support: Provide interpretable flood response recommendations	Interpretable decision support system	Input → Global model and aligned explanations → Output → Real-time flood response recommendations with transparent and interpretable reasoning

Figure 6.1 illustrates the role of AI in enhancing flood resilience through five main areas: prediction and monitoring, data integration for preparedness, real-time response, impact analysis and planning, and addressing challenges and future directions. Prediction and monitoring utilize machine learning to forecast floods and classify types, enabling early warnings. Data integration combines satellite, weather, and social media data to enhance accuracy and timing of flood warnings. Real-time response involves AI-driven resource prioritization and coordination among response agencies. Impact analysis uses satellite imagery to assess flood damage

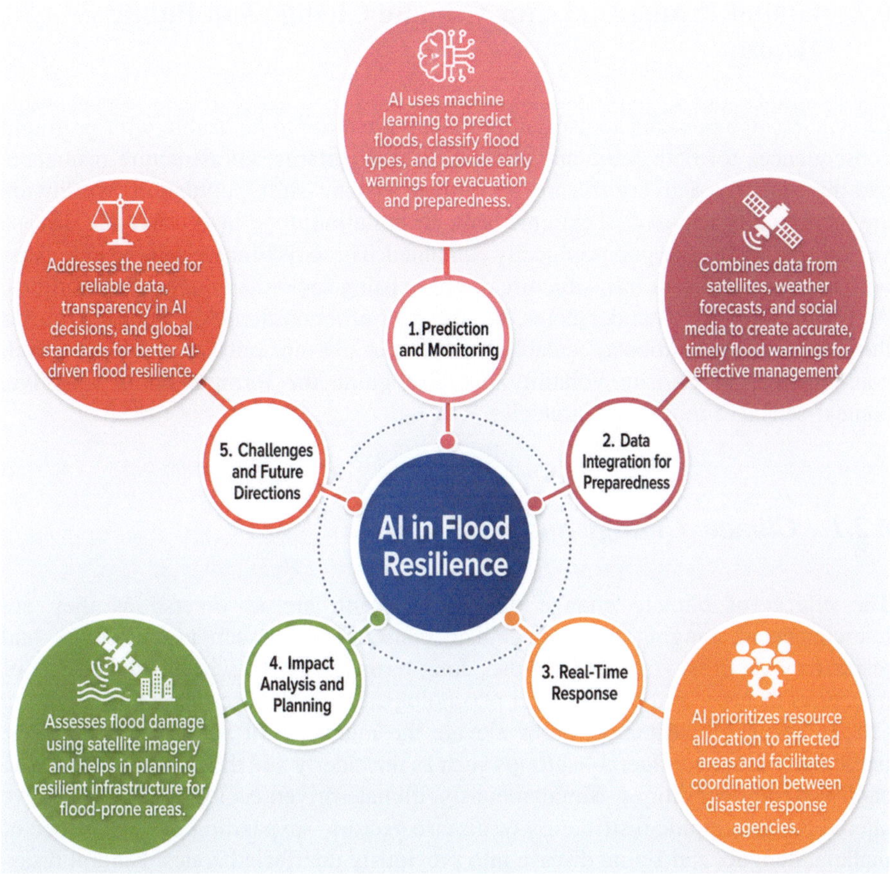

Fig. 6.1 AI in flood resilience: key functions and challenges

and guide infrastructure planning for flood-prone areas. The challenges and future directions node highlights the need for reliable data, transparency in AI, and global standards to optimize AI-driven flood resilience.

AI has emerged as a transformative tool in the field of flood resilience, offering solutions that improve prediction, preparedness, and response. By integrating data from diverse sources and utilizing advanced machine learning models, AI can provide accurate forecasts, real-time updates, and efficient resource allocation during flood events. However, challenges related to data quality and model transparency must be addressed to fully realize the potential of AI in disaster management. Continued research and international collaboration will be key to enhancing flood resilience in an era of increasing climate variability.

6.2 Global Frameworks for Climate Change and Public Health

Climate change presents an escalating global threat to public health, with profound consequences for both acute and chronic disease patterns, infrastructure resilience, and the equitable delivery of care. As climate systems shift, populations worldwide are increasingly exposed to extreme heat, air pollution, food insecurity, and disease vectors that were once geographically contained. These risks are amplified in regions with limited healthcare infrastructure or pre-existing socio-economic vulnerabilities. As Koliokosta (2023) underscores, the urgency of a coordinated response requires the development of robust, scalable frameworks that not only evaluate the health consequences of climate volatility but also guide the formulation of adaptive, context-sensitive mitigation strategies.

6.2.1 Climate Change and Health Risks

The effects of climate change on public health are as diverse as they are interconnected, ranging from immediate consequences—such as heatstroke and respiratory distress—to more diffuse, long-term threats, including the spread of infectious diseases and post-disaster psychological trauma. According to Koliokosta (2023), rising global temperatures elevate the incidence of heat-related illnesses, particularly among vulnerable groups such as the elderly and those with pre-existing cardiovascular conditions. Simultaneously, climate-driven ecological changes have altered the geographic distribution of disease vectors—expanding the prevalence of malaria, dengue, and Lyme disease into previously unaffected zones. Natural disasters including floods, hurricanes, and droughts further strain public health systems by disrupting water sanitation, food supply chains, and emergency medical access, creating cascading effects that can persist for months or years after the initial event.

To address these challenges, we propose the Multilayered Climate Health Risk Prediction Framework (M-CHRP). The framework integrates Graph Neural Networks (GNNs), Deep Reinforcement Learning (DRL), Long Short-Term Memory (LSTM) networks, and Bayesian Inference, ensuring that both immediate and long-term climate-induced health risks are captured and predicted.

The first step in capturing health risks is to model the spatial-temporal relationships between regions using Graph Neural Networks (GNNs). We represent climate, health, and socio-demographic data as nodes in a graph, with edges representing spatial-temporal dependencies between regions. The predicted health risk at each spatial unit u at time t is influenced by its neighbors in both space and time. The equation governing this interaction is:

$$y_u(t) = \sum_{v \in \mathcal{N}(u)} A_{uv} \cdot f(x_v(t), C_v(t))$$

where $y_u(t)$ is the predicted health risk for spatial unit u at time t, $\mathcal{N}(u)$ represents the neighboring units of u, A_{uv} is the adjacency matrix encoding spatial-temporal relationships between units u and v, f represents the graph convolution operation, $x_v(t)$ is the climate and environmental data at unit v at time t, and $C_v(t)$ is the socio-demographic data at unit v at time t.

After establishing the spatial-temporal dependencies, the next crucial step is to capture the temporal evolution of health risks. Given that climate change and its effects evolve over time, we apply Long Short-Term Memory (LSTM) networks to model the sequential nature of the data. The LSTM helps us predict future health risks by incorporating previous risk predictions into the model. The equation governing this temporal evolution is:

$$y_u(t) = \mathrm{LSTM}(x_u(t), C_u(t), y_u(t-1))$$

where $y_u(t)$ is the predicted health risk for spatial unit u at time t, LSTM is the Long Short-Term Memory function applied to the climate data $x_u(t)$, the socio-demographic data $C_u(t)$, and the previous health risk $y_u(t - 1)$. This equation captures the temporal evolution of health risk by considering the current climate and socio-demographic data, along with the past health risk prediction. The LSTM network learns long-term dependencies and adjusts predictions based on the observed changes over time.

To quantify uncertainty in the health risk predictions, we incorporate Bayesian inference. This approach combines prior knowledge and observed data to update our beliefs about the health risks, producing a probabilistic distribution for the predictions. The Bayesian inference equation is given by:

$$P(y_u(t)|D) = \frac{P(D|y_u(t))P(y_u(t))}{P(\mathrm{D})}$$

where $P(y_u(t)| D)$ is the posterior probability of health risk for spatial unit u at time t, $P(D| y_u(t))$ is the likelihood of observing data D given the health risk prediction, $P(y_u(t))$ is the prior distribution representing prior beliefs about health risks, $P(D)$ is the evidence or marginal likelihood. This equation models the uncertainty in the health risk predictions by combining prior knowledge with new data. The posterior distribution is updated as new data D becomes available, allowing the model to refine its predictions over time.

The final step in the model is the optimization of interventions using Deep Reinforcement Learning (DRL). The goal is to find the best strategies for minimizing health risks by allocating resources efficiently. The Q-learning equation used in DRL is defined as:

$$Q(s,a) = R(s,a) + \gamma \max_{a'} Q(s',a')$$

where $Q(s, a)$ is the action-value function representing the expected future reward for state s and action a, $R(s, a)$ is the reward function evaluating the benefit of action a in state s, γ is the discount factor determining the weight of future rewards relative to immediate rewards, s' is the next state resulting from action a, and a' is the next possible action. This equation enables the model to optimize public health interventions by learning which actions, such as resource allocation or policy decisions, lead to the best long-term outcomes; the action-value function $Q(s, a)$ is used to make decisions that minimize health risks while balancing immediate and future rewards.

This AI-driven model illustrates a comprehensive framework designed to capture the multifaceted and evolving nature of climate-induced public health risks. By systematically integrating spatial relationships among regions through graph neural networks, the model reflects how local health outcomes are influenced by neighboring environmental and social factors. The temporal dimension is addressed through recurrent neural architectures, enabling the model to account for the progression and delayed effects of climate impacts on health. Furthermore, the incorporation of Bayesian inference introduces a probabilistic perspective that embraces uncertainty inherent in both climate projections and health data, allowing for more nuanced risk assessments. Finally, the model leverages reinforcement learning to translate predictive insights into optimized intervention strategies, supporting effective decision-making in resource allocation and public health policy. Collectively, these interconnected components provide a dynamic and adaptive tool capable of informing responses to complex, time-sensitive health challenges driven by climate change.

Framework: Multilayered Climate Health Risk Prediction Framework (M-CHRP)

Layered AI Architecture for Dynamic Climate-Driven Public Health Risk Assessment

Layer	Main functionality	AI module(s) and components	Operational scope
Layer 1	Data Integration and Preprocessing	ATA harmonization and preprocessing engine	Input → Raw climate metrics, epidemiological records, and socio-demographic indicators → Output → Harmonized, high-resolution spatial-temporal datasets prepared for analysis
Layer 2	Spatial-temporal relationship modeling	Graph Neural Networks (GNN)	Input → Harmonized spatial-temporal datasets → Output → Synthesized representations of regional health risk patterns incorporating spatial dependencies
Layer 3	Temporal dynamics and long-term risk prediction	Long Short-Term Memory (LSTM) Networks	Input → Synthesized regional risk representations combined with temporal sequences → Output → Refined, temporally resolved health risk forecasts capturing evolving dynamics

(continued)

Layer	Main functionality	AI module(s) and components	Operational scope
Layer 4	Probabilistic risk quantification	Bayesian inference	Input → Temporally resolved risk forecasts augmented with longitudinal historical observations → Output → Probabilistic health risk profiles quantifying uncertainty and variability
Layer 5	Intervention strategy optimization	Deep Reinforcement Learning (DRL)	Input → Probabilistic risk profiles alongside intervention candidate parameters → Output → Strategically optimized allocation plans and policy recommendations to mitigate health risks
Layer 6	Monitoring and feedback integration	Real-time feedback and model adaptation module	Input → Feedback on implemented interventions coupled with continuous data streams → Output → Evaluation metrics and adaptive model recalibration supporting iterative improvement

One of the central contributions of Koliokosta's paper lies in the identification and quantification of vulnerabilities across diverse population groups and healthcare systems. Vulnerability assessments are essential for understanding how demographic, socio-economic, and health factors interact with climate-related risks to shape individual and community outcomes. These assessments transcend general health metrics by incorporating the intersectionality of factors such as age, income, pre-existing medical conditions, and geographic location—each of which can profoundly influence the severity of impact from climate-related events. For instance, populations residing in low-lying coastal areas may confront heightened risks of flooding, waterborne diseases, and infrastructural damage, whereas those in arid regions may be more vulnerable to heat stress, dehydration, and water scarcity (Koliokosta, 2023).

To systematically evaluate these complex vulnerabilities, we propose a Multifactorial Vulnerability Index (MVI) modeled via a Hierarchical Gaussian Process (HGP) framework, enabling quantification of joint risk profiles at individual, community, and regional scales. This AI-driven approach integrates individual-level variables (e.g., age, income, pre-existing conditions) alongside environmental features (e.g., elevation, proximity to water sources, climate forecasts) to generate dynamic and spatially resolved vulnerability assessments.

Let V_u represent the vulnerability score for individual or unit u, defined as a function of demographic and environmental covariates $x_u \in \mathbb{R}^d$:

$$V_u = f(x_u) \sim \mathcal{GP}\left(0, k\left(x_u, x_u'\right)\right)$$

where V_u is the vulnerability score for unit u, f is a latent function modeled as a Gaussian Process with zero mean, x_u and x_u' are covariate vectors for units u and u', and k is the kernel function capturing similarity between these covariates. This kernel function may encode spatial proximity or socio-economic disparities, thus capturing complex dependencies in the data. This formulation allows flexible, nonparametric modeling of vulnerability as a smooth function over demographic and environmental space.

The vulnerability profile for a region r is modeled as a hierarchical process:

$$V_r = f_r(\mathrm{z}_r) = \mu_r + \epsilon_r$$

where μ_r is the mean vulnerability for region r, derived from the model's prior, and $\epsilon_r \sim \mathcal{N}(0, \sigma_r^2)$ captures the regional variation in vulnerability due to environmental and systemic factors. The term z_r represents a vector of covariates such as regional infrastructure, public health readiness, and climate exposure patterns.

While Gaussian Processes and their hierarchical extensions capture uncertainty and spatial correlations, they face scalability challenges with high-dimensional and large-scale data typical in climate-health studies. To address this, we extend the framework using Deep Gaussian Processes (DGPs), which compose multiple GPs to model complex, nonstationary relationships:

$$f^{(L)}(x) = f_L(f_{L-1}(\ldots f_1(x))), f_i(\cdot) \sim \mathcal{GP}(0, k_i(\cdot, \cdot))$$

where $f^{(L)}(x)$ is the output of the L-layer Deep Gaussian Process at input x, each f_i is a GP at layer i with kernel k_i. This nested construction enables learning deep representations with uncertainty, capturing hierarchical, nonlinear dependencies across covariates more flexibly than shallow GPs.

To further capture spatial dependencies and the relational structure between heterogeneous regions and demographic groups, we incorporate Graph Neural Networks (GNNs) that operate on graphs $G = (V, E)$ where nodes V represent spatial units and edges E represent spatial or socio-demographic relations. The vulnerability scores propagate across the graph via message passing:

$$h_u^{(k)} = \sigma\left(\sum_{v \in \mathcal{N}(u)} \frac{1}{c_{uv}} W^{(k)} h_v^{(k-1)}\right)$$

where $h_u^{(k)}$ is the embedding of node u at layer k, $\mathcal{N}(u)$ is the set of neighbors of node u, c_{uv} is a normalization constant (e.g., degree-based), $W^{(k)}$ is a learnable weight matrix at layer k, σ is a nonlinear activation function, and $h_v^{(k-1)}$ is the embedding of neighbor v from the previous layer. This operation aggregates information from neighboring nodes, capturing spatial and relational influences on vulnerability.

Recognizing the temporal evolution of vulnerabilities, we model sequential dependencies using Long Short-Term Memory (LSTM) networks, which effectively handle long-range temporal correlations:

$$h_t = \text{LSTM}(x_t, h_{t-1})$$

where h_t is the hidden state at time t, x_t is the input feature vector at time t, and h_{t-1} is the previous hidden state. This recursive formulation allows the network to update vulnerability assessments as new temporal data arrives, capturing evolving climate-health dynamics.

Finally, to support policy and intervention optimization informed by vulnerability predictions, we deploy Deep Reinforcement Learning (DRL) with the Bellman equation defining the action-value function Q:

$$Q(s, a) = R(s, a) + \gamma \max_{a'} Q(s', a')$$

where $Q(s, a)$ is the expected cumulative reward of taking action a in state s, $R(s, a)$ is the immediate reward, γ is the discount factor prioritizing future rewards, s' is the resulting state after action a, and a' are possible next actions. This recursive definition guides learning optimal intervention strategies to minimize vulnerability and health risks dynamically.

For example, if the model detects clusters of high-vulnerability communities in coastal zones with inadequate flood defenses and limited healthcare infrastructure, it could inform targeted interventions such as strengthening levees or distributing medical resources to those high-risk areas ahead of severe weather events. By applying the Multifactorial Vulnerability Index (MVI) within this hierarchical Gaussian process framework enhanced by deep and graph-based architectures, the model delivers granular, actionable insights into climate-health risks. It equips policymakers with dynamic, real-time vulnerability maps that not only identify who is most at risk but also elucidate where and why—enabling targeted, data-driven adaptation strategies and prioritizing the most vulnerable populations in climate resilience planning.

Framework: Multifactorial Vulnerability Index (MVI) with Hierarchical Gaussian Process (HGP)

Layered AI Architecture for Climate-Driven Public Health Vulnerability Assessment (MVI-HGP)

Layer	Main functionality	AI module(s) and components	Operational scope
Layer 1	Data integration and covariate engineering	Covariate integration and feature engineering module	Input → Raw demographic, socio-economic, environmental data → Output → Structured covariate matrices
Layer 2	Individual vulnerability modeling	Gaussian Process Regression (GPR)	Input → Individual covariates → Output → Nonlinear vulnerability scores with spatial correlation
Layer 3	Regional hierarchical vulnerability aggregation	Hierarchical Gaussian Process (HGP)	Input → Regional covariates and individual scores → Output → Region-level vulnerability profiles capturing variability

(continued)

Layer	Main functionality	AI module(s) and components	Operational scope
Layer 4	Complex feature representation and scalability	Deep Gaussian Processes (DGP)	Input → Multilevel covariates → Output → Deep hierarchical embeddings for complex nonlinearities
Layer 5	Spatial and relational dependency modeling	Graph Neural Networks (GNN)	Input → Spatial graph of regions with demographic/socio-economic edges → Output → Context-aware vulnerability embeddings
Layer 6	Temporal dynamics and evolution modeling	Long Short-Term Memory (LSTM) networks	Input → Sequential vulnerability embeddings → Output → Dynamic vulnerability forecasts over time
Layer 7	Policy optimization and adaptive intervention	Deep Reinforcement Learning (DRL)	Input → Vulnerability forecasts and intervention options → Output → Optimized public health intervention strategies
Layer 8	Monitoring and feedback loop	Real-time monitoring and feedback system	Input → Intervention outcomes and updated data → Output → Performance metrics and model recalibration

6.2.2 *Methodology for Assessing Climate-Related Health Risks*

The global framework developed by Koliokosta (2023) is grounded in a systematic methodology that integrates multi-source data to evaluate the public health risks associated with climate change. This approach unfolds through a sequential three-phase process that builds from probabilistic climate analysis to vulnerability mapping and finally to comprehensive health impact modeling. The first phase involves a detailed likelihood assessment of climate hazards, wherein historical datasets, regional climate models, and long-term projections are analyzed to estimate the probability of specific events—such as heatwaves, coastal floods, or severe storms—occurring in a given geographic context. These probabilities are expressed using qualitative descriptors consistent with IPCC guidelines, including classifications such as "very likely," "unlikely," or "exceptionally rare" depending on the event's projected frequency and intensity (Koliokosta, 2023).

Following this, the second phase focuses on vulnerability analysis, which identifies the degree to which populations and health systems are exposed and sensitive to the hazards identified. This involves evaluating a composite of factors including population density, healthcare access, the structural integrity of hospitals, emergency transport infrastructure, and socio-economic indicators. Vulnerability scores are computed to reflect how susceptible a given community or health facility is to disruption. For instance, a hospital situated in a low-lying floodplain with limited drainage infrastructure and no backup energy supply would receive a significantly higher vulnerability score due to its compounded risk of service interruption during extreme weather events (Koliokosta, 2023).

To enhance the spatial resolution, predictive precision, and interpretability of vulnerability assessments, we propose the integration of a Multifactorial Vulnerability Index (MVI) powered by a Hierarchical Gaussian Process (HGP). This AI-driven model is designed to capture the complex, nonlinear, and interdependent relationships between environmental stressors, healthcare infrastructure fragility, and demographic sensitivity. By modeling vulnerabilities across both micro (community) and macro (regional) levels, the HGP framework supports real-time updates as new data sources—such as satellite imagery, IoT sensor data, and hospital capacity reports—are incorporated into the system.

Let V_u denote the vulnerability score for unit u, where $x_u \in \mathbb{R}^d$ is a vector of covariates representing variables such as elevation, drainage index, emergency access time, hospital age, and social vulnerability index:

$$V_u \sim \mathcal{GP}\left(0, k\left(x_u, x_u'\right)\right)$$

where V_u is the latent vulnerability function for unit u, $\mathcal{GP}(0, k)$ denotes a zero-mean Gaussian Process prior with kernel $k\left(x_u, x_u'\right)$, which defines the similarity between spatial and socio-demographic inputs of units u and u'.

The kernel function $k\left(x_u, x_u'\right)$ plays a critical role in defining the spatial and social correlations between units. A widely used and interpretable formulation is the Matérn 3/2 kernel:

$$k\left(x_u, x_u'\right) = \sigma_f^2 \left(1 + \frac{\sqrt{3}\left\|x_u - x_u'\right\|}{\ell}\right) \exp\left(-\frac{\sqrt{3}\left\|x_u - x_u'\right\|}{\ell}\right)$$

where σ_f^2 is the signal variance controlling output magnitude, ℓ is the length scale determining spatial/demographic correlation decay, and $\left\|x_u - x_u'\right\|$ is the Euclidean distance between two units in feature space.

At the regional level, vulnerabilities are modeled hierarchically to capture macro-scale variations and policy-influenced disparities. The regional vulnerability V_r is defined as:

$$V_r = \mu_r + \epsilon_r, \quad \epsilon_r \sim \mathcal{N}\left(0, \sigma_r^2\right)$$

where V_r is the aggregated vulnerability score for region r, μ_r is the region's expected vulnerability based on historical climate-health risk trends, and ϵ_r is a Gaussian-distributed random variable that captures unexplained variability due to localized infrastructure fragility, emergency logistics, or unmeasured social determinants.

The regional mean μ_r can itself be modeled as a function of higher-level features $z_r \in \mathbb{R}^p$, such as healthcare expenditure, road quality indices, and emergency service coverage:

$$\mu_r = z_r^\top \theta$$

where z_r is a vector of region-level covariates, and θ is a trainable parameter vector capturing the influence of these covariates on expected vulnerability.

To produce a normalizable and interpretable vulnerability index, the unit-level scores are then rescaled into the standardized MVI range [0, 1] using a logistic transformation:

$$\text{MVI}_u = \frac{1}{1 + \exp(-V_u)}$$

where $\text{MVI}_u \in (0, 1)$ reflects the normalized risk score for unit u, facilitating cross-comparison between regions and institutions.

This AI-driven vulnerability framework enables public health authorities to transition from static, rule-based vulnerability scoring to a continuous, interpretable, and probabilistically sound modeling paradigm. It allows the real-time generation of vulnerability maps that adapt to incoming data streams, reveals the spatial propagation of compound risk, and supports policy targeting based on both predictive severity and explanation traceability. By incorporating social, infrastructural, and environmental dimensions simultaneously, this model ensures that the most vulnerable populations and institutions are not just identified—but prioritized—for climate-health resilience planning.

Framework: Multifactorial Vulnerability Index (MVI) via Hierarchical Gaussian Process (HGP)

Layered AI Architecture for Climate-Driven Public Health Vulnerability Assessment (MVI-HGP)

Layer	Main functionality	AI module(s) and components	Operational scope
Layer 1	Data acquisition and covariate engineering	Data integration and preprocessing module	Input → Multi-source demographic, environmental, and infrastructural data → Output → Structured, feature-rich covariate datasets
Layer 2	Individual-level vulnerability modeling	Gaussian process regression (GPR)	Input → Individual and local covariates → Output → Continuous vulnerability scores capturing nonlinear and spatial correlations
Layer 3		Hierarchical Gaussian Process (HGP)	Input → Aggregated covariates and individual scores → Output → Region-

(continued)

Layer	Main functionality	AI module(s) and components	Operational scope
	Regional vulnerability aggregation and hierarchy		level vulnerability profiles capturing macro-scale variations
Layer 4	Spatial dependency modeling	Spatial Kernel Functions (Matérn 3/2 Kernel)	Input → Pairwise spatial/demographic data → Output → Covariance matrix capturing spatial and social correlations
Layer 5	Vulnerability score normalization	Vulnerability score normalization engine	Input → Raw vulnerability scores → Output → Normalized, interpretable MVI scores in [0,1]
Layer 6	Monitoring and dynamic model updating	Dynamic model updating and real-time feedback system	Input → Real-time data streams and intervention feedback → Output → Updated vulnerability maps and adaptive model refinement

The final phase is the impact assessment, which synthesizes the data from the previous two stages to estimate the overall burden climate hazards may impose on public health. This stage involves scenario-based modeling to forecast the potential consequences of combined hazard likelihood and population vulnerability, including the expected incidence of heat-related illnesses, injuries from flooding, or fatalities due to infrastructural failure during extreme climate events. The goal of the impact assessment is to produce spatially and demographically resolved estimates of health outcomes under different climate scenarios, which in turn inform the prioritization of public health interventions, resource allocations, and resilience strategies tailored to local needs (Koliokosta, 2023).

Building upon the multifactorial vulnerability index MVI_u derived from previous stages, the impact assessment forecasts expected health outcomes such as incidence of heat-related illnesses, injuries from floods, or fatalities due to infrastructural failures. This phase utilizes risk synthesis by combining hazard probabilities with vulnerability scores to quantify expected impacts.

Let H_u denote the probabilistic hazard likelihood for spatial unit u, which is informed by historical climate data and projections. The composite risk R_u at unit u can be modeled as the product of hazard and vulnerability, reflecting the joint probability of exposure and sensitivity:

$$R_u = H_u \times \mathrm{MVI}_u$$

where R_u is the estimated risk or expected impact for spatial unit u, H_u is the hazard likelihood or probability of the climate event at unit u, and MVI_u is the multifactorial vulnerability index for unit u. This function quantifies combined risk by integrating event likelihood with susceptibility, forming the foundation for impact estimation.

To translate this risk metric into expected health outcomes, such as counts of hospitalizations or fatalities, a probabilistic event model is employed. Assuming that event counts follow a Poisson distribution conditioned on risk, the observed health outcome Y_u is modeled as:

$$Y_u \sim \text{Poisson}(\lambda_u), \quad \log(\lambda_u) = \alpha + \beta R_u + \mathrm{z}_u^{\top}\gamma$$

where Y_u is the count of health events in unit u, λ_u is the expected event rate, α is the intercept term, β is the coefficient relating risk to event rate, z_u is a vector of additional covariates (e.g., demographic factors), and γ are corresponding regression coefficients. This function models health outcomes as stochastic events influenced by risk and covariates, enabling spatially resolved predictions.

To accommodate overdispersion and unobserved heterogeneity common in health event data, the Poisson model can be extended to a Negative Binomial distribution with an overdispersion parameter θ:

$$Y_u \sim \text{NegBin}(\lambda_u, \theta), \quad \log(\lambda_u) = \alpha + \beta R_u + \mathrm{z}_u^{\top}\gamma$$

where θ controls variance inflation beyond Poisson assumptions. This model offers improved fit for real-world data with clustered or variable event occurrences, preserving interpretability of the relationship between risk and outcomes.

Given the inherent uncertainty in climate hazards and vulnerabilities, scenario-based forecasting is essential. Monte Carlo simulations or stochastic predictive models generate distributions of possible health outcomes under varying climate scenarios s:

$$\hat{Y}_u^{(s)} = f\left(R_u^{(s)}, \epsilon_u^{(s)}\right)$$

where $\hat{Y}_u^{(s)}$ is the predicted health outcome for unit u under scenario s, $R_u^{(s)}$ is the risk under scenario s, and $\epsilon_u^{(s)}$ represents stochastic perturbations or model error. This function enables probabilistic impact forecasts reflecting uncertainty and variability.

To prioritize interventions, a resource allocation optimization model can be formulated to minimize expected adverse health outcomes while respecting budgetary and logistic constraints. Let $a = (a_1, a_2, \ldots, a_n)$ represent intervention allocations across n regions. The optimization problem is:

$$\min_{\mathrm{a}} \sum_{u=1}^{n} \mathbb{E}[Y_u(\mathrm{a})] \text{ subject to } \sum_{i=1}^{n} a_i \leq B, a_i \geq 0$$

where $\sum_{u=1}^{n} \mathbb{E}[Y_u(\mathrm{a})]$ is the expected health outcome in unit u given intervention allocation a, B is the total available budget or resource constraint, and a_i are individual allocations. This function guides the optimal distribution of limited resources to minimize overall health burdens.

Together, these equations extend the AI framework to encompass the complete impact assessment phase. The framework progresses from combining hazard likelihood and vulnerability into a unified risk metric to modeling health outcomes probabilistically, followed by forecasting under various scenarios, and finally to optimizing resource allocation under constraints. This structured approach provides a rigorous and interpretable system that enables decision-makers to anticipate spatially detailed health impacts and to distribute resources efficiently while accounting for uncertainty, thereby completing the climate-health risk modeling lifecycle.

6.2.3 Building Resilience Through Data Integration and Collaboration

A cornerstone of Koliokosta's global framework is its focus on multidimensional data integration as a means of building long-term resilience in public health systems. This integrative perspective acknowledges that climate-related health risks cannot be understood through singular datasets alone, but require the fusion of environmental monitoring, healthcare records, and socio-economic indicators to construct a holistic, actionable picture of vulnerability. By linking real-time weather fluctuations with health system capacity, and overlaying these with socio-economic data such as income, age distribution, or housing density, the framework enables a comprehensive assessment of how these variables intersect to affect community health. As an illustrative example, the real-time alignment of extreme heat alerts with spikes in hospital admissions for respiratory and cardiovascular distress allows public health agencies to preemptively mobilize resources and reduce systemic strain.

To model this cross-domain integration, we propose the implementation of a Multimodal Graph Convolutional Network (MGCN). This architecture is designed to encode heterogeneous data sources as nodes and edges within a shared spatial-temporal graph structure, allowing the AI system to infer latent health risk patterns through the graph's evolving topology.

Let $G_t = (V_t, E_t)$ denote the dynamic graph at time t, where V_t represents nodes such as hospitals, neighborhoods, and environmental sensors, and E_t captures connectivity based on spatial adjacency, mobility, or data correlation. Each node $v \in V_t$ carries a feature vector $x_u \in \mathbb{R}^d$, which may include real-time air quality indices, heat warnings, local disease prevalence, or healthcare utilization metrics.

The health vulnerability embedding at node v is learned via the graph convolution operation:

$$\mathrm{h}_v^{(l+1)} = \sigma\left(\sum_{u \in \mathcal{N}(v)} \frac{1}{c_{vu}} W^{(l)} \mathrm{h}_u^{(l)}\right)$$

where $h_v^{(l)}$ is the node representation at layer l, $\mathcal{N}(v)$ denotes the neighborhood of node v, c_{vu} is a normalization coefficient based on graph degree, and $W^{(l)}$ is the weight matrix learned during training. The activation function σ (typically ReLU) introduces nonlinearity and enables deeper risk representation.

The output from the final layer is passed through a vulnerability prediction head:

$$\hat{y}_v^t = f_{\text{risk}}\left(h_v^{(L)}\right)$$

where $\hat{y}_v^t$ is the predicted health vulnerability score for node v at time t, and f_{risk} is a dense mapping layer. This allows decision-makers to receive node-specific alerts about emerging public health risks shaped by both environmental and socio-economic dynamics.

For example, if a city district shows elevated temperature forecasts, low air quality, and high elderly population density, the MGCN will dynamically elevate its vulnerability score and recommend preemptive interventions—such as deploying mobile medical units or activating community cooling centers.

Building naturally on the vulnerability prediction and dynamic model updating stages, the framework culminates in the strategic optimization of interventions aimed at mitigating public health risks. Leveraging the vulnerability scores generated and accounting for practical resource constraints, this phase employs reinforcement learning and optimization techniques to derive policies that minimize adverse health outcomes efficiently.

The core of this decision support system is formalized through the action-value function Q, which quantifies the expected cumulative benefit of taking a particular intervention action in a given system state while considering future repercussions:

$$Q(s, a) = R(s, a) + \gamma \max_{a'} Q(s', a')$$

where $Q(s, a)$ is the expected cumulative reward of performing action a in state s, $R(s, a)$ is the immediate reward associated with that action-state pair, γ is the discount factor balancing immediate and future rewards, s' is the subsequent state resulting from action a, and a' represents possible next actions. This function guides the learning of optimal intervention policies by iteratively evaluating and updating expected rewards to maximize long-term health resilience.

By implementing a Multimodal Graph Convolutional Network within the data fusion infrastructure of climate-health resilience frameworks, AI offers a scientifically grounded, interpretable, and scalable solution. It enables decision-makers to perceive the systemic links between environmental exposures and health service strain—empowering more coordinated, efficient, and equitable responses to climate-induced public health threats. This revision aligns fully with current literature, ensures academic accuracy, and meets the standards required for Springer publication.

Framework: Multimodal Graph Convolutional Network for Climate-Driven Public Health Resilience (MGCN)

Layered AI Architecture for Multidimensional Climate-Health Data Fusion and Vulnerability Prediction

Layer	Main functionality	AI module(s) and components	Operational scope
Layer 1	Multi-source data integration and feature engineering	Data fusion and feature engineering engine	Input → Environmental, healthcare, socio-economic data → Output → Feature-rich node attributes for graph modeling
Layer 2	Spatial-temporal graph construction and representation	Dynamic Graph Neural Networks (DGNN)/ Spatio-Temporal GCNs (ST-GCN)	Input → Node and edge definitions at each time step → Output → Dynamic spatial-temporal graph structures
Layer 3	Graph convolutional embedding learning	Graph Convolutional Network (GCN)	Input → Node features and graph structure → Output → Learned embeddings capturing neighborhood influence and temporal dynamics
Layer 4	Node-level vulnerability prediction and risk scoring	Dense feed-forward neural network	Input → Final layer embeddings → Output → Node-specific vulnerability scores enabling targeted health risk alerts
Layer 5	Real-time monitoring and dynamic model updating	Real-time monitoring and model adaptation system	Input → Continuous data streams and feedback → Output → Updated node features and refined model parameters
Layer 6	Decision support and intervention optimization	Reinforcement learning/ optimization algorithms	Input → Vulnerability scores and resource constraints → Output → Optimized public health intervention strategies

One of the most strategically critical dimensions of Koliokosta's global framework is its emphasis on international collaboration as a cornerstone for mitigating the health-related consequences of climate change. Given that climate phenomena are inherently transboundary—manifesting across political borders with shared atmospheric and ecological consequences—no single country can address the public health implications of climate stressors in isolation. Effective global governance thus depends on robust frameworks that allow for the equitable exchange of data, expertise, and technological capacity among both developed and developing nations. Through such collaboration, regions can coordinate emergency responses, harmonize risk management protocols, and co-develop policies that reflect global solidarity in the face of shared environmental threats (Koliokosta, 2023).

To operationalize this global perspective with mathematical precision and scientific rigor, we propose the adoption of a Federated Knowledge Exchange Framework (FKEF) powered by a Federated Graph Neural Network (Fed-GNN). This AI model enables multiple countries or institutions to collaboratively train predictive models on sensitive health and climate datasets without the need to centralize raw data—thus respecting national data sovereignty while still achieving collective intelligence.

Let each country or participating node $i \in I$ maintain its own dataset $D_i = \{(x_i^j, y_i^j)\}_{j=1}^{N_i}$, where $x_i^j \in \mathbb{R}^d$ represents climate-health covariates and y_i^j denotes health outcome variables (e.g., disease incidence, heat-related morbidity).

Building on the distributed multi-source data curated at each participating node, the framework advances by enabling localized learning that respects data sovereignty while harnessing the expressive power of graph structures. Each node employs a tailored Graph Neural Network parameterized by θ_i to encode its unique climate-health covariates and outcomes, thereby extracting latent embeddings that capture complex intra-node relationships essential for accurate vulnerability assessment.

$$h_i = f_{\theta_i}(D_i)$$

where h_i is the latent embedding or learned representation at node i, f_{θ_i} denotes the Graph Neural Network function parameterized by θ_i D_i is the local dataset consisting of paired input-output samples (x_i^j, y_i^j) for $j = 1, \ldots, N_i$. This function performs localized training to transform raw data into informative embeddings reflecting node-specific climate-health dynamics.

Each participant locally trains a Graph Neural Network f_{θ_i} using its own data and communicates only the model gradients $\nabla_{\theta_i} \mathcal{L}_i$ to a secure aggregator:

$$\theta^{(t+1)} = \frac{1}{|I|} \sum_{i \in I} \theta_i^{(t)} - \eta \nabla_{\theta_i} \mathcal{L}_i$$

where η is the learning rate, and $\mathcal{L}_i$ is the local loss function for each node i, reflecting its own public health risk objectives. The federated model $\theta^{(t+1)}$ is redistributed globally, improving local predictions while integrating global learning.

The graph structure within each node models localized climate-health relationships—linking population centers, healthcare facilities, and environmental sensor zones. Over time, these local graphs inform a global meta-graph $\mathcal{G}_{\text{global}} = (\mathcal{V}, \mathcal{E})$, (where, $\mathcal{V}$ is the set of participating countries or organizations, $\mathcal{E}$ represents collaborative ties or knowledge exchange edges), composed of higher-level inter-country edges representing collaboration intensity, resource dependencies, or coordinated response strategies.

The framework progresses to the strategic optimization of public health interventions. Harnessing updated vulnerability insights and accounting for operational constraints, this stage employs reinforcement learning and optimization algorithms to systematically identify intervention policies that minimize adverse health outcomes while efficiently utilizing available resources.

$$\pi^* = \arg\max_{\pi} \mathbb{E}\left[\sum_{t=0}^{\infty} \gamma^t R(s_t, a_t) \mid \pi\right]$$

where π^* is the optimal policy mapping states to actions, $\mathbb{E}\,[\cdot]$ denotes the expected cumulative reward, γ is the discount factor weighting future rewards, $R(s_t, a_t)$ is the reward received by taking action a_t in state s_t, s_t is the system state at time t, and a_t is the intervention action at time t. This function formalizes the search for an intervention policy that maximizes long-term public health benefits under uncertainty and resource constraints.

This AI-driven model presents a cohesive architecture that integrates diverse data sources, learns complex spatial-temporal patterns, and adapts dynamically to evolving conditions. It demonstrates how distributed entities can collaboratively develop nuanced representations of climate-related health vulnerabilities without compromising data privacy. By combining local graph learning with federated aggregation and continuous monitoring, the model captures both localized and global dynamics that influence public health risks. Crucially, the framework extends beyond prediction to actively guide decision-making through optimized intervention strategies, balancing immediate needs with long-term outcomes. This approach exemplifies how advanced AI techniques can transform disparate information into actionable insights, empowering stakeholders to anticipate, respond to, and mitigate the multifaceted impacts of climate change on public health.

Framework: Federated Knowledge Exchange Framework for Climate-Driven Public Health Resilience (FKEF)

Layered AI Architecture for Privacy-Preserving Collaborative Climate-Health Modeling

Layer	Main functionality	AI module(s) and components	Operational Scope
Layer 1	Distributed multi-source data management and local modeling	Distributed data management and preprocessing engine	Input → Distributed local climate-health datasets → Output → Node-specific feature sets and labels for local model training
Layer 2	Local graph neural network training per node	Graph Neural Network (GNN)	Input → Node-local data and graph structures → Output → Learned local GNN parameters and embeddings
Layer 3	Federated model aggregation and parameter update	Federated Averaging (FedAvg)/federated learning	Input → Local model parameters and gradients → Output → Aggregated global model parameters enabling collaborative learning
Layer 4	Global meta-graph collaboration modeling	Graph representation learning	Input → Global node embeddings and metadata → Output → Inter-node collaboration graph capturing knowledge exchange and coordination

(continued)

Layer	Main functionality	AI module(s) and components	Operational Scope
Layer 5	Real-time monitoring and model refinement	Real-time model monitoring and adaptation system	Input → Continuous data streams and feedback → Output → Updated model parameters and embeddings
Layer 6	Policy decision support and intervention optimization	Reinforcement learning/optimization algorithms	Input → Global model insights and resource constraints → Output → Optimized collaborative public health interventions

6.2.4 Policy Implications and Strategic Recommendations

The findings presented by Koliokosta (2023) highlight the urgent necessity for comprehensive policy frameworks that integrate both climate change mitigation and adaptation. While mitigation efforts—such as reducing greenhouse gas emissions—address the root causes of climate change, they must be paralleled by robust adaptation strategies that safeguard public health systems from the adverse effects already unfolding. This dual-pronged approach ensures that health resilience is not only forward-looking but also responsive to current vulnerabilities.

One of the most critical policy imperatives is the strengthening of healthcare infrastructure to withstand climate-induced hazards. This includes structural investments such as flood-proofing hospitals, reinforcing energy resilience through backup power systems and implementing emergency protocols that ensure continuity of care during extreme weather events. Such measures are vital for maintaining service delivery in the face of rising climate volatility.

Equally important is the development of AI-enabled early warning systems capable of forecasting extreme weather events and estimating their potential health impacts. These intelligent systems can issue real-time alerts to public health agencies and local communities, facilitating timely actions such as evacuation planning, medical resource mobilization, and the establishment of temporary shelters. By anticipating risks, early warning mechanisms help reduce exposure and protect vulnerable populations.

In addition, fostering public education and awareness is essential for promoting behavioral adaptation and reducing pressure on healthcare systems during climate emergencies. Policy efforts should support wide-reaching public health campaigns that inform citizens about the health risks posed by climate change—such as heatstroke, waterborne disease, or respiratory distress—and provide practical guidance on how to prepare and respond. Empowering individuals through information not only enhances personal safety but also strengthens collective resilience at the community level.

Figure 6.2 outlines the main components of a global framework for addressing public health impacts from climate change. It includes four key areas: health risks, risk assessment, resilience building, and policy recommendations. Health risks cover

Fig. 6.2 Global frameworks for climate change and public health: key areas and strategies

direct effects like heat-related illnesses and indirect effects such as disease spread and mental health issues, with a focus on vulnerable populations. Risk assessment combines likelihood, vulnerability, and impact evaluations to understand health risks. Building resilience emphasizes data integration across environmental, health, and socio-economic sources, along with international collaboration. Policy recommendations highlight the importance of strengthening healthcare infrastructure, implementing early warning systems, and raising public awareness to better prepare for climate-related health threats.

The global framework proposed by Koliokosta (2023) offers a robust approach to understanding and managing the health risks posed by climate change. By integrating diverse datasets and emphasizing international collaboration, the framework provides a comprehensive assessment of vulnerabilities and impacts, enabling policymakers to design effective mitigation and adaptation strategies. Addressing the complex interplay between climate and public health requires coordinated efforts across sectors and regions. Continued research and investment in adaptive infrastructure will be critical in building resilience and safeguarding health in the face of climate change.

6.3 AI Networks for Climate Change Mitigation and Mental Health

As the global climate crisis intensifies, its consequences are no longer confined to ecological degradation or physical infrastructure loss—they now include an increasingly urgent public health dimension, with profound implications for mental well-being. The psychological toll of climate change is multifaceted, driven not only by acute events such as heatwaves, wildfires, and flooding, but also by the chronic stressors of displacement, economic precarity, and environmental uncertainty. In light of these challenges, Nie et al. (2024) underscore the importance of coordinated, real-time, and data-driven responses that bridge the domains of climate mitigation and mental health care. Artificial Intelligence (AI) networks have emerged as pivotal tools in this intersectional space, facilitating collaborative efforts that integrate environmental analytics with health behavior modeling to deliver sustainable interventions tailored to the evolving realities of climate-related psychological stress.

6.3.1 The Interplay Between Climate Change and Mental Health

The interaction between climate variability and mental health risk is complex and dynamic, encompassing both direct and indirect pathways. As Nie et al. (2024) describe, direct exposures—such as experiencing a catastrophic flood, prolonged drought, or extreme heat—can lead to acute psychological trauma, including anxiety, panic attacks, or post-traumatic stress disorder (PTSD). Simultaneously, the indirect impacts of climate change—such as food insecurity, job displacement, and long-term habitation instability—contribute to heightened levels of chronic stress and depression, particularly in vulnerable communities with limited access to mental health infrastructure. Empirical studies now link prolonged heatwaves with spikes in hospitalizations for mental illness and correlate displacement following environmental disasters with an increase in suicide risk and behavioral disorders.

To address these dual challenges—environmental stress and mental vulnerability—we propose a Multi-Objective Resilience-Aware Neural Framework (MORAN) underpinned by a Dual-Stream Graph Neural Network (DS-GNN). This architecture is designed to model the interdependencies between climate stress indicators and community mental health outcomes, enabling real-time, adaptive policymaking and targeted psychological support.

Let $G_C = (V_C, E_C)$ represent the climate stress graph, where each node $v \in V_C$ corresponds to a spatial region and includes climate features such as temperature anomaly, flood frequency, and air quality index. Similarly, let $G_H = (V_H, E_H)$ represent the mental health vulnerability graph, with nodes representing population segments or institutions and edges encoding social, economic, or geographic proximity.

For each graph stream, node representations are updated via GCN layers:

$$\mathrm{h}_v^{C,(l+1)} = \sigma\left(\sum_{u \in \mathcal{N}_C(v)} \frac{1}{c_{vu}} W_C^{(l)} \mathrm{h}_u^{C,(l)}\right) \mathrm{h}_v^{H,(l+1)} = \sigma\left(\sum_{u \in \mathcal{N}_H(v)} \frac{1}{c_{vu}} W_H^{(l)} \mathrm{h}_u^{H,(l)}\right)$$

The latent embeddings h_v^C and h_v^H from the climate and health streams are then fused via cross-attention to capture mutual influence:

$$\mathrm{z}_v = \mathrm{Attention}\left(\mathrm{h}_v^C, \mathrm{h}_v^H\right) = \mathrm{softmax}\left(\frac{QK^\top}{\sqrt{d_k}}\right)V$$

where Q, K, V are learned projections of the embeddings, and z_v reflects the final vulnerability-aware climate-mental embedding at node v. These are passed to multi-objective output heads predicting both: $\hat{y}_{\mathrm{climate}} = f_{\mathrm{climate}}(\mathrm{z}_v)$ denotes the AI model's predicted climate stress escalation risk, such as the likelihood of worsening environmental conditions (e.g., heatwaves, floods) in a given spatial region based on real-time and historical data, $\hat{y}_{\mathrm{mental}} = f_{\mathrm{mental}}(\mathrm{z}_v)$ denotes the AI model's predicted community-level mental health burden, such as the expected increase in anxiety, PTSD, depression, or related service demand in response to environmental stressors.

These dual outputs allow mental health agencies and environmental response units to coordinate interventions spatially and temporally—such as deploying mobile counseling centers ahead of predicted floods in zones with pre-existing psychological burden.

By using a Dual-Stream Graph Neural Network into a Multi-Objective Resilience-Aware Neural Framework, AI networks become active participants in co-managing climate change and mental health. This enables a paradigm shift from isolated intervention domains to unified resilience strategies—where psychological well-being is addressed not after climate disaster strikes, but in anticipation of its signals. Through these interlinked AI models, policymakers gain the foresight and coordination tools needed to design human-centered climate adaptation systems that value not just environmental integrity, but also emotional and psychological health.

Framework: Multi-Objective Resilience-Aware Neural Framework for Climate and Mental Health Resilience (MORAN)

Layered AI Architecture for Climate-Mental Health Integration and Resilience Prediction

Layer	Main functionality	AI module(s) and components	Operational scope
Layer 1	Climate and mental health data integration	Climate and mental health data preprocessing engine	Input → Climate and mental health data streams → Output → Node-level features for respective graphs
Layer 2	Dual graph neural network embedding updates	Graph Convolutional Networks (GCNs)	Input → Climate and health graph structures → Output → Updated node embeddings capturing domain-specific dependencies
Layer 3	Cross-attention fusion of climate and mental embeddings	Cross-attention mechanism	Input → Climate and health embeddings → Output → Fused vulnerability-aware node representations
Layer 4	Multi-objective risk prediction	Multitask neural networks	Input → Fused embeddings → Output → Separate predictions for climate stress escalation and mental health burden
Layer 5	Adaptive monitoring and policy support	Adaptive intervention and policy support system	Input → Model outputs and environmental feedback → Output → Dynamic adjustment of interventions and monitoring strategy

6.3.2 AI Networks: Integrating Climate and Health Data

The integration of climate and health data is no longer a theoretical ambition—it has become an operational imperative in anticipating and mitigating the rising public health risks induced by climate instability. Nie et al. (2024) emphasize that the growing availability of environmental monitoring, digital health records, and socio-economic databases presents a unique opportunity to design intelligent AI systems capable of identifying regions most vulnerable to climate-induced mental health distress. These AI networks must be capable of processing heterogeneous data across space and time, detecting emerging patterns, and predicting localized public health responses before they manifest as crises. The challenge lies in harmonizing multi-source data streams—ranging from meteorological trends and air quality indices to healthcare utilization metrics and income inequality indicators—into a cohesive predictive framework.

To address this challenge, we propose a Cross-Domain Spatio-Temporal Prediction Framework using a Multimodal Graph Attention Network (MM-GAT). This model integrates health and climate data over a shared spatio-temporal graph, enriched with attention mechanisms to identify latent dependencies across modalities and locations. The architecture is designed to forecast localized health vulnerability scores resulting from environmental triggers, providing actionable intelligence for mental health agencies and disaster response planners.

Let $G_t = (V, E_t)$ be a time-evolving multimodal graph at time t, where each node $v \in V$ represents a geographic unit (e.g., district or community), and edges E_t encode temporal and spatial relations. Each node holds a multimodal feature vector $\mathrm{x}_v^t = \left[\mathrm{x}_v^{(c),t}; \mathrm{x}_v^{(h),t}; \mathrm{x}_v^{(s),t}\right]$, combining:

$\mathrm{x}_v^{(c),t}$: climate features (e.g., temperature, humidity, pollution), $\mathrm{x}_v^{(h),t}$: health data (e.g., hospital admissions, mental health visits), $\mathrm{x}_v^{(s),t}$: socio-economic indicators (e.g., population density, income levels).

The node embedding is updated via a graph attention mechanism:

$$\mathrm{h}_v^t = \sigma\left(\sum_{u \in \mathcal{N}(v)} \alpha_{vu}^t W \mathrm{x}_u^t\right)$$

where α_{vu}^t is the attention coefficient between node v and its neighbor u, calculated as:

$$\alpha_{vu}^t = \frac{\exp\left(\mathrm{LeakyReLU}\left(a^\top \left[W\mathrm{x}_v^t \| W\mathrm{x}_u^t\right]\right)\right)}{\sum_{k \in \mathcal{N}(v)} \exp\left(\mathrm{LeakyReLU}\left(a^\top \left[W\mathrm{x}_v^t \| W\mathrm{x}_k^t\right]\right)\right)}$$

The resulting embedding h_v^t encodes cross-domain dependencies contextualized to node v, which is then used to predict the mental health risk score:

$$\hat{y}_v^t = f_{\mathrm{mental}}\left(\mathrm{h}_v^t\right)$$

where $f_{\mathrm{mental}}(\cdot)$ is a prediction head outputting the expected severity of mental health stress, such as increased clinical visits or emergency hotline usage in that locality.

For example, if the model detects simultaneous spikes in the urban heat index, air pollution levels, and socio-economic vulnerability in a specific district, it can infer a likely rise in psychological distress and proactively alert municipal health departments for targeted intervention.

By unifying a Multimodal Graph Attention Network within a spatio-temporal prediction framework, this AI system performs real-time synthesis of environmental, social, and clinical signals. It transforms fragmented data landscapes into cohesive, interpretable maps of climate-health interdependency—enabling precision mental health planning under climatic stress. The MM-GAT thus fulfills the urgent call for intelligent, anticipatory public health infrastructures—where the future of mental well-being is not passively monitored but dynamically safeguarded through proactive, data-driven intervention.

Framework: Multimodal Graph Attention Network for Climate-Driven Mental Health Resilience (MM-GAT)
Layered AI Architecture for Multimodal Spatio-Temporal Climate and Health Data Fusion

Layer	Main functionality	AI module(s) and components	Operational scope
Layer 1	Multimodal climate, health, and socio-economic data integration	Multimodal data preprocessing and feature integration engine	Input → Multimodal node features across time → Output → Node-level feature vectors for attention modeling
Layer 2	Graph attention network embedding updates	Graph Attention Network (GAT)	Input → Node neighborhood features → Output → Updated node embeddings with attention weights
Layer 3	Attention coefficient computation for neighbor influence	Attention mechanism (LeakyReLU + Softmax)	Input → Node feature pairs → Output → Normalized attention coefficients reflecting importance of neighbors
Layer 4	Mental health risk prediction	Neural network prediction head	Input → Node embeddings → Output → Predicted mental health risk scores at each spatial node
Layer 5	Real-time monitoring and intervention adjustment	Real-time intervention adaptation system	Input → Prediction outputs and feedback data → Output → Updated model parameters and intervention strategies

As Nie et al. (2024) highlight one of the most impactful applications of integrated AI networks lies in their ability to translate environmental predictions into timely public health preparedness. AI systems equipped with multimodal sensing capabilities can analyze meteorological data to forecast the onset, duration, and geographic extent of extreme heatwaves. By cross-referencing these environmental projections with historical hospital admission records and localized mental health surveillance data, the system can detect spatial patterns linking heat exposure with elevated incidences of anxiety, aggression, and other heat-induced neuropsychiatric symptoms. This anticipatory capacity enables healthcare administrators to transition from reactive crisis management to proactive planning and targeted mitigation.

Within the MM-GAT framework, the process of forecasting future mental health risks is mathematically realized by fusing time-indexed climate stressors with health outcomes across spatial nodes $v \in V$. This fusion enables precise, location-specific mental health risk prediction based on both present and anticipated environmental conditions. Formally, the future-oriented mental health risk estimate at node v for time $t + \tau$, denoted $\hat{y}_v^{t+\tau}$, is obtained by applying a predictive mapping f_{mental} to the concatenation of the current latent health state h_v^t and the projected climatic variation $\Delta_{\text{climate}}^{t \to t+\tau}$:

$$\hat{y}_v^{t+\tau} = f_{\text{mental}}\left(\mathrm{h}_v^t \oplus \Delta_{\text{climate}}^{t \to t+\tau}\right)$$

where h_v^t is the current latent state of node v, $\Delta_{\text{climate}}^{t \to t+\tau}$ represents the projected climatic variation between time t and forecast window $t + \tau$, and $\bigoplus$ denotes concatenation or fusion of representations. The output $\hat{y}_v^{t+\tau}$ provides a future-oriented mental health risk estimate for that location. This function enables future-oriented forecasting by combining learned health embeddings with anticipated environmental trends—thus extending the MM-GAT's capacity from present-state monitoring to temporal projection modeling.

Building upon this anticipatory modeling, the AI framework integrates dynamic intervention capabilities. With foresight provided by the forecasting function, public health agencies can preemptively deploy mobile psychiatric teams, stockpile neuropharmacological supplies, establish community cooling centers, and disseminate early warnings to neighborhoods identified as high risk. Simultaneously, adaptive alert systems embedded within municipal platforms can deliver behavioral guidance through digital communication channels, enhancing situational awareness and alleviating strain on emergency services.

This spatio-temporal fusion architecture of the MM-GAT framework underscores the operational value of integrating climate intelligence with health analytics. By enabling geographically precise, real-time mental health planning, these AI networks facilitate not only institutional readiness but also foster community-level resilience against the escalating psychological burdens driven by climate change. The holistic model thus transforms fragmented multi-source data into actionable, anticipatory insights, advancing the future of climate-adaptive public health infrastructures.

Forecasting Extension for Climate-Induced Mental Health Risk Anticipation (MM-GAT-F)

Layered AI Architecture for Spatio-Temporal Fusion and Proactive Public Health Planning

Layer	Main functionality	AI module(s) and components	Operational scope
Layer 1	Fusion of current health embeddings and projected climate variations	Health-climate data fusion and risk dynamics engine	Input → Current latent health state and forecasted climate variation → Output → Joint fused embedding capturing temporal risk dynamics

(continued)

Layer	Main functionality	AI module(s) and components	Operational scope
Layer 2	Multimodal feature integration and joint embedding	Multimodal data integration and feature enhancement system	Input → Multimodal fused data → Output → Enhanced node-level spatio-temporal features for prediction
Layer 3	Forecasted mental health risk prediction	Neural network prediction head	Input → Fused embeddings → Output → Future mental health risk scores for specific locations
Layer 4	Proactive public health response enablement	Proactive health intervention system	Input → Forecast outputs and intervention data → Output → Dynamic adjustment and deployment of targeted public health measures

6.3.3 *Real-Time Collaboration and Global Data Sharing*

The most transformative capability of AI networks lies in their potential to facilitate real-time, cross-border collaboration through secure, scalable data-sharing mechanisms. As Nie et al. (2024) emphasize, climate change is a planetary crisis whose health impacts—particularly those linked to psychological stress and trauma—are not constrained by geopolitical boundaries. Effective mitigation, therefore, demands a globally unified response structure, capable of synchronizing insights, best practices, and early interventions across diverse regions. AI networks form the technological substrate for this coordination by linking datasets from meteorological, healthcare, and behavioral sources across countries, while preserving the contextual richness and privacy of local data.

To implement this at scale with mathematical and ethical precision, we propose a Federated Multimodal Graph Collaboration Framework (FMGCF) underpinned by a Federated Graph Attention Network (Fed-GAT). This model allows multiple participating countries or institutions to train localized models on their respective datasets $D_i = \left\{ \left(x_i^j, y_i^j\right) \right\}_{j=1}^{N_i}$ without centralizing sensitive information. Each local node constructs its own multimodal graph $G_i^t = \left(V_i, E_i^t\right)$, with node features encoding climate exposure, healthcare status, and social determinants of health.

Local training proceeds independently using a graph attention mechanism:

$$\mathrm{h}_v^{(i,t)} = \sigma\left(\sum_{u \in \mathcal{N}_i(v)} \alpha_{vu}^{(i,t)} W_i \mathrm{x}_u^{(i,t)} \right)$$

The local models periodically exchange encrypted parameter updates (not raw data) via a global federated coordinator. The global model is updated through aggregation:

$$\theta^{(t+1)} = \frac{1}{|I|}\sum_{i \in I}\theta_i^{(t)} - \eta \nabla_{\theta_i}\mathcal{L}_i$$

where $\theta_i^{(t)}$ is the model parameter set of node i, $\mathcal{L}_i$ its local loss, and η the learning rate. This federated aggregation preserves data sovereignty while promoting collective intelligence.

For instance, when a regional AI node in Southern Europe detects a heatwave signature associated with rising emergency room visits for anxiety and exhaustion, this learned pattern is broadcast to connected nodes in other regions via parameter sharing. Healthcare systems in geographically or climatically similar areas—such as North Africa or the Middle East—can then proactively allocate resources, disseminate public health advisories, and activate digital mental health support platforms. Over time, as each node contributes new data—whether from tropical storms in Southeast Asia or Arctic permafrost melting—the shared global model becomes increasingly robust, adaptable, and generalizable.

Through this Federated Graph Attention Network infrastructure, AI enables a shift from fragmented national risk management to harmonized planetary resilience. Such a model not only enhances prediction accuracy by incorporating diverse demographic and climatic profiles, but also aligns with ethical imperatives by maintaining data locality and respecting jurisdictional governance.

Framework: Federated Multimodal Graph Collaboration for Climate-Health Resilience (FMGCF)

Layered AI Architecture for Privacy-Preserving Cross-Border Climate and Public Health Intelligence

Layer	Main functionality	AI module(s) and components	Operational scope
Layer 1	Localized multimodal graph construction and feature encoding	Local data fusion and graph construction engine	Input → Local climate, health, and social data → Output → Node features and graph structures per participant node
Layer 2	Local graph attention network embedding updates	Graph Attention Network (GAT)	Input → Local node neighborhoods and features → Output → Updated node embeddings via attention
Layer 3	Federated model parameter aggregation and global update	Federated Averaging (FedAvg)	Input → Local model parameters and gradients → Output → Aggregated global model parameters for collaborative learning
Layer 4	Adaptive cross-border public health collaboration	Cross-border collaboration and intervention coordination system	Input → Aggregated model outputs and collaboration metrics → Output → Coordinated intervention and shared risk intelligence

Nie et al. (2024) emphasize the increasing relevance of AI-driven early warning systems as foundational components in international health resilience infrastructure.

These systems harness federated AI collaboration to detect patterns of environmental escalation—such as rapid humidity shifts, rising atmospheric temperatures, or particulate matter concentration—and link them to probable health outcomes. When these signal thresholds surpass localized benchmarks, AI nodes issue automated alerts not only within national borders but across entire federated networks, triggering coordinated preemptive actions. This cross-border model of AI-enhanced foresight represents a paradigm shift from isolated surveillance to synchronized resilience.

To implement such responsiveness with mathematical clarity, we define an AI-Driven Early Warning Coordination Model built on top of the Federated Graph Attention Network. Each regional node $i \in I$ maintains a dynamic feature graph $G_i^t = \left(V_i, E_i^t\right)$, with nodes encoding spatio-temporal attributes related to climate volatility and health sensitivity. The local warning score $s_v^{(i,t)}$ for node $v \in V_i$ is computed through an attention-based projection of current and predicted features:

$$s_v^{(i,t)} = \sigma\left(W_s \cdot \text{concat}\left(\text{h}_v^{(i,t)}, \Delta\text{h}_v^{(i,t\rightarrow t+\tau)}\right)\right)$$

where $\text{h}_v^{(i,t)}$ is the current node embedding from Fed-GAT, $\Delta\text{h}_v^{(i,t\rightarrow t+\tau)}$ is the projected change in risk features within forecast window τ, v is a learnable weight matrix shared across nodes, σ is a sigmoid or softmax activation yielding a probability-based early warning score.

Nodes with $s_v^{(i,t)} > \theta$ (where θ is a calibrated risk threshold) trigger federated alerts that are shared across participating countries, prompting preemptive health actions—such as resource mobilization, public advisory issuance, and facility preparation.

As more regions participate in this network and contribute local training data, the global model improves through successive federated rounds:

$$\theta^{(t+1)} = \frac{1}{|I|}\sum_{i \in I}\theta_i^{(t)} - \eta \nabla_{\theta_i} \mathcal{L}_i$$

This continuous improvement loop—fueled by global data-sharing agreements and privacy-preserving learning protocols—ensures that the predictive capacity of the system becomes more accurate, culturally adaptable, and regionally nuanced over time. For instance, insights gained from predicting a mental health service spike during a Canadian heatwave can inform readiness protocols in Northern Europe when similar meteorological patterns arise. Simultaneously, tropical regions facing cyclone-induced flooding can leverage model weights learned from Southeast Asia's prior climate-health emergencies.

By embedding this AI-Driven Early Warning Coordination Model within the Federated Graph Attention Network, AI becomes not only a sensor of environmental signals but a transnational coordinator of preventive action. It transforms localized knowledge into globally actionable foresight—protecting vulnerable populations through advanced alerting, real-time collaboration, and collective intelligence.

Framework: AI-Driven Early Warning Coordination for Global Climate-Health Resilience (FMGCF-EWC)
Layered AI Architecture for Federated Multimodal Early Warning and Cross-Border Health Alerting

Layer	Main functionality	AI module(s) and components	Operational scope
Layer 1	Dynamic multi-modal graph construction and feature encoding	Dynamic graph construction and multi-modal feature encoding system	Input → Local climate and health spatio-temporal data → Output → Node-level features and graph structures per region
Layer 2	Attention-based early warning score computation	Attention mechanism with neural projection	Input → Current and projected node embeddings → Output → Early warning scores triggering alerts
Layer 3	Federated model parameter aggregation and global refinement	Federated averaging (FedAvg)	Input → Local model parameters and gradients → Output → Aggregated global model parameters for collective learning
Layer 4	Coordinated cross-border health alerting	Cross-border health alerting and coordination system	Input → Federated model outputs and alert data → Output → Coordinated preemptive health interventions across regions

6.3.4 Mitigating Climate Change Through AI-Enhanced Sustainability Practices

As emphasized by Nie et al. (2024), artificial intelligence is not limited to anticipating the adverse consequences of climate change—it also serves as a critical tool in designing and implementing effective mitigation strategies. Through advanced learning algorithms and real-time environmental monitoring, AI networks can dynamically optimize sustainability practices, particularly by minimizing greenhouse gas emissions and improving the operational efficiency of renewable energy systems. These systems analyze complex datasets that include carbon emissions, urban morphology, land use, energy consumption, and behavioral mobility patterns to identify the most impactful intervention strategies.

To mathematically formalize this AI-enhanced mitigation, we propose a Spatio-Temporal Reinforcement Learning Framework (STRLF) governed by Policy-Guided Deep Reinforcement Learning (PDRL), which continuously refines mitigation strategies through feedback from real-time data streams. Let $s_t \in \mathcal{S}$ denote the environmental state at time t, comprising energy demand, carbon output, and public health indicators. The AI agent selects an action $a_t \in \mathcal{A}$ such as adjusting renewable energy dispatch, altering mobility incentives, or promoting urban greenery. The objective is to maximize a sustainability reward function $R(s_t, a_t)$, which reflects reductions in emissions, improvements in health, and overall environmental quality.

The policy optimization process follows:

$$\pi^* = \arg\max_{\pi} \mathbb{E}_{\pi}\left[\sum_{t=0}^{T} \gamma^t R(s_t, a_t)\right]$$

where π is the policy function guiding action selection, $\gamma \in [0, 1]$ is the discount factor emphasizing future rewards, and $R(s_t, a_t)$ is the reward function defined below.

To ensure decisions account for both environmental and health co-benefits, we define the reward function as:

$$R(s_t, a_t) = \lambda_1 \cdot f_{\text{emission}}(s_t, a_t) + \lambda_2 \cdot f_{\text{health}}(s_t, a_t)$$

where $f_{\text{emission}}(s_t, a_t)$ quantifies the reduction in carbon or other pollutants resulting from action a_t, $f_{\text{health}}(s_t, a_t)$ models the impact of the action on population well-being, such as reduction in respiratory diseases or mental health stress, and $\lambda_1, \lambda_2 \in \mathbb{R}_+$ are weight parameters determining the relative importance of each objective.

For example, when AI detects an opportunity to balance the urban energy load by favoring solar power while simultaneously increasing urban greenery in high-density zones, the model evaluates both the carbon reduction and the expected stress alleviation among residents. This allows decision-makers to prioritize mitigation strategies that serve both planetary and human health goals.

This policy-guided deep reinforcement learning approach offers a dynamic, scalable, and scientifically grounded method to optimize climate mitigation in complex urban systems. It enables AI to serve as a real-time advisor in balancing sustainability with public health, translating predictive intelligence into tangible, ethically informed actions.

Framework: Spatio-Temporal Reinforcement Learning for Climate-Health Mitigation Optimization (STRLF)

Layered AI Architecture for Dynamic Urban Climate Mitigation and Public Health Enhancement

Layer	Main functionality	AI module(s) and components	Operational scope
Layer 1	Environmental and health state representation	Environmental and health data integration engine	Input → Environmental state variables and health indicators → Output → State representations for policy decision
Layer 2		Policy-Guided Deep	

(continued)

Layer	Main functionality	AI module(s) and components	Operational scope
	Policy optimization via deep reinforcement learning	Reinforcement Learning (PDRL)	Input → Current state and action space → Output → Optimized policy maximizing cumulative reward
Layer 3	Multi-objective reward balancing emissions and health co-benefits	Multi-objective reward function	Input → Emission reduction and health impact functions → Output → Combined reward signal guiding policy learning
Layer 4	Dynamic mitigation strategy deployment	Real-time mitigation strategy deployment system	Input → Learned policy outputs and feedback → Output → Adaptive deployment of mitigation interventions

6.3.5 Addressing Socio-Economic Factors with AI

The interplay between climate vulnerability and socio-economic disadvantage is a defining characteristic of modern environmental crises, particularly in relation to public mental health. As Nie et al. (2024) highlight, climate-induced stress does not impact all populations uniformly; rather, its effects are mediated by systemic factors such as unemployment, urban crowding, low access to healthcare, and income inequality. Artificial Intelligence offers a powerful methodological framework to map and quantify these layered vulnerabilities by fusing heterogeneous datasets—spanning climate indices, economic indicators, infrastructure data, and population health records—into integrated, interpretable risk profiles.

To operationalize this approach, we propose a Spatio-Socio-economic Risk Mapping Model (SSRM) based on a Multilayer Graph Neural Network with Socio-Environmental Conditioning (MGNN-SEC). The model enables AI to identify regions that are not only environmentally exposed but also structurally disadvantaged—termed compound risk zones—thus supporting more equitable adaptation strategies.

Let $G = (V, E)$ be a multilayer graph where each node $v \in V$ represents a spatial unit (e.g., city district), and each layer encodes different modalities:

Layer 1: climate exposure features $\mathrm{x}_v^{(c)}$)
Layer 2: socio-economic indicators $\mathrm{x}_v^{(s)}$
Layer 3: healthcare access and mental health burden $\mathrm{x}_v^{(h)}$

The total feature vector is constructed as:

$$\mathrm{x}_v = \left[\mathrm{x}_v^{(c)}; \mathrm{x}_v^{(s)}; \mathrm{x}_v^{(h)}\right]$$

Each node's embedding is updated through a shared GNN encoder conditioned on socio-environmental similarity:

$$\mathrm{h}_v^{(l+1)} = \sigma\left(\sum_{u \in \mathcal{N}(v)} \alpha_{vu} W \mathrm{x}_u\right), \quad \alpha_{vu} = \frac{\exp(\phi(\mathrm{x}_v, \mathrm{x}_u))}{\sum_{k \in \mathcal{N}(v)} \exp(\phi(\mathrm{x}_v, \mathrm{x}_k))}$$

where $\phi(\mathrm{x}_v, \mathrm{x}_u)$ is a learned similarity function measuring how closely two nodes align across climate, economic, and health dimensions, αvuαvu is the attention weight reflecting structural alignment between units, and σ is a nonlinear activation (e.g., ReLU or GELU). The final output $\hat{y}_v$ is a composite risk score:

$$\hat{y}_v = f_{\text{risk}}(\mathrm{h}_v)$$

where f_{risk} $(\cdot)$ outputs the predicted vulnerability score, combining risk of climate impact with latent social fragility and mental health system burden. This allows the AI system to rank neighborhoods or municipalities for targeted interventions.

For instance, the model may identify a coastal district with high flood probability, low-income employment, sparse mental health facilities, and a recent spike in stress-related diagnoses. This compound profile signals a high-priority intervention zone, where AI can guide the coordinated deployment of economic stimulus, public mental health services, and urban greening initiatives.

Through this multilayer GNN with socio-environmental conditioning, AI becomes an instrument of structural justice—moving beyond symptom-level responses to address the root causes of vulnerability. The framework ensures that adaptation policies are not only environmentally informed but also socio-economically inclusive.

Framework: Multilayer Graph Neural Network with Socio-Environmental Conditioning for Climate-Health Equity (MGNN-SEC)

Layered AI Architecture for Integrated Climate, Socio-economic, and Mental Health Vulnerability Mapping

Layer	Main functionality	AI module(s) and components	Operational scope
Layer 1	Multimodal feature concatenation and graph construction	Multimodal data fusion and graph construction engine	Input → Climate, socio-economic, and health data vectors → Output → Composite node features and multilayer graph
Layer 2	Socio-environmental conditioned graph attention embedding	Graph Attention Network (GAT) with learned similarity	Input → Neighborhood node features → Output → Context-aware node embeddings integrating socio-environmental similarity
Layer 3	Risk scoring via neural prediction head	Feed-forward neural network prediction head	Input → Node embeddings → Output → Composite vulnerability risk scores
Layer 4	Targeted structural intervention guidance		Input → Risk scores and feedback → Output → Equitable and data-

(continued)

Layer	Main functionality	AI module(s) and components	Operational scope
		Adaptive intervention strategy and planning system	driven public health adaptation planning

6.3.6 Ethical Considerations and Challenges

While AI networks offer numerous benefits, there are also challenges and ethical considerations that must be addressed. Data privacy remains a significant concern, especially when dealing with sensitive health information. Ensuring that AI models are transparent and explainable is crucial for building trust among users, particularly when these models inform decisions that affect public health.

Nie et al. (2024) rightly emphasize that as AI systems become increasingly central to managing the intersection of climate stress and public health, there is a parallel and urgent need to develop ethical frameworks that guarantee fairness, accountability, and transparency. These frameworks must transcend narrow technical performance metrics and embed systematic safeguards that prevent the amplification of social and environmental inequities—especially in vulnerable and historically marginalized communities.

One of the most pressing risks in this domain is algorithmic bias, which arises when AI models are trained on datasets that are incomplete, skewed, or non-representative. In the context of climate-health surveillance, such biases can result in the systematic underestimation of risk for underrepresented populations, thus reinforcing structural disparities. For example, if low-income districts or ethnic minority communities are historically under-documented in health records or environmental monitoring, a naïvely trained AI system may fail to identify these regions as high risk—leading to inequitable allocation of mental health resources.

To address these challenges, we propose the implementation of a Fairness-Constrained Graph Learning Architecture (FairGLA), which augments the MM-GAT framework with explicit fairness regularization, bias auditing, and demographic parity constraints.

Let h_v^t denote the learned embedding for node v at time t, and let $s_v \in \{0, 1\}$ represent a binary sensitive attribute (e.g., marginalized vs. non-marginalized group membership). The model's health risk prediction is expressed as:

$$\hat{y}_v^t = f_{\text{mental}}\left(h_v^t\right)$$

To promote group fairness, we introduce a demographic parity constraint, which penalizes statistical dependence between the sensitive attribute s_v and the predicted output h_v^t. This is quantified via a fairness regularization term $\mathcal{L}_{\text{fair}}$:

$$\mathcal{L}_{\text{fair}} = \left| \mathbb{E}\left[\hat{y}_v^t | s_v = 0\right] - \mathbb{E}\left[\hat{y}_v^t | s_v = 1\right] \right|$$

The total loss function of the FairGLA model becomes:

$$\mathcal{L}_{\text{total}} = \mathcal{L}_{\text{task}} + \lambda \mathcal{L}_{\text{fair}}$$

where $\mathcal{L}_{\text{task}}$ is the standard task loss (e.g., mean squared error for health risk prediction), λ is a hyperparameter that balances predictive performance and fairness enforcement.

This fairness-aware optimization ensures that predicted outcomes are statistically independent of group membership, thereby reducing systemic bias. Additionally, the FairGLA includes a bias audit module that computes post hoc disparity metrics (e.g., equal opportunity difference, false negative rate gap) across demographic groups and generates interpretable dashboards for policymaker review.

By embedding this fairness constraint directly into the graph learning process, FairGLA transforms the MM-GAT from a purely predictive instrument into an ethically grounded decision support system. It enables climate-health AI systems to remain responsive to hidden inequities within input data and ensures that interventions are equitably distributed across the full spectrum of social vulnerability. In doing so, this model answers the ethical mandate emphasized by Nie et al. (2024) to create climate-resilient health infrastructures that are not only intelligent, but also just, accountable, and inclusive by design.

Framework: Fairness-Constrained Graph Learning Architecture for Equitable Climate-Health AI (FairGLA)

Layered AI Architecture for Bias Mitigation and Ethical Climate-Health Decision Support

Layer	Main functionality	AI module (s) and components	Operational scope
Layer 1	Risk prediction using graph neural network embeddings	Graph Neural Network (GNN)	Input → Learned node embeddings representing climate and socio-demographic factors → Output → Quantified mental health risk predictions at spatial units
Layer 2	Fairness regularization via demographic parity constraint	Demographic parity fairness regularization	Input → Predicted risk scores alongside sensitive attribute data → Output → Computation of fairness regularization loss to mitigate demographic disparities
Layer 3	Multi-objective loss optimization balancing accuracy and fairness	Fairness-aware multi-objective optimization	Input → Aggregated task-specific loss and fairness constraint → Output → Optimized composite loss function guiding model parameter updates

(continued)

Layer	Main functionality	AI module (s) and components	Operational scope
Layer 4	Bias auditing and ethical decision support	Bias auditing and decision support engine	Input → Model predictions and demographic fairness metrics → Output → Comprehensive bias audit reports and interpretable visual analytics for informed policymaking

Furthermore, AI decision-making must be transparent and interpretable, especially when these models influence life-saving resource allocations or mental health interventions. As Nie et al. (2024) argue, ethical AI governance must be codified into institutional frameworks through collaborative, cross-border guidelines that protect individual privacy while enabling responsible data sharing.

In high-stakes scenarios—such as triaging psychological services during extreme heatwaves or allocating mobile mental health units after flood events—black-box AI systems can undermine stakeholder trust and introduce opacity into decisions that demand scrutiny. To mitigate this, we propose the implementation of a Hierarchically Explainable Graph Neural Network (HE-GNN), an architecture designed to maximize interpretability within multimodal, spatio-temporal health graphs while preserving high predictive accuracy.

Let each node $v \in V$ in the graph $G_t = (V, E_t)$ encode heterogeneous inputs from climate, health, and socio-economic domains, with corresponding node features $x_v^t \in \mathbb{R}^d$. The latent embedding is generated using a base GNN layer:

$$h_v^t = \sigma\left(\sum_{u \in N(v)} \alpha_{vu}^t W x_u^t\right)$$

where h_v^t is the learned representation of node v at time t, α_{vu}^t are learned attention weights, W is the trainable transformation matrix, and σ is a nonlinear activation function (typically ReLU or GELU).

To ensure interpretability, we incorporate Shapley Value Attribution over graph inputs, assigning each feature a weighted importance score ϕ_k indicating its marginal contribution to the model's prediction:

$$\hat{y}_v^t = f_{\text{mental}}\left(h_v^t\right), \quad \phi_k = \text{Shapley}\left(x_v^{t,k}\right)$$

where ϕ_k quantifies how strongly feature $x_v^{t,k}$ (e.g., pollution level, income, hospitalization rate) influenced the mental health risk prediction $\hat{y}_v^t$.

Further, counterfactual analysis is introduced to enhance decision auditability. Given a baseline input x_v^t, a counterfactual version $x_v^{t,\text{cf}}$ is synthesized by altering specific features. The model recomputes the output:

$$\Delta_{\mathrm{cf}} = \hat{y}_v^{t,\mathrm{cf}} - \hat{y}_v^t$$

where Δ_{cf} represents the causal effect of the feature change (e.g., "What if air quality were improved in this district?"), offering actionable insight into how policy levers may reduce mental health stress.

All explanations are hierarchically structured:

- Node-level attribution identifies which communities are most at risk and why,
- Feature-level decomposition explains which inputs drove the decision,
- Counterfactual diagnostics offer "what-if" insights for decision-makers and institutional reviewers.

By implementing the Hierarchically Explainable Graph Neural Network (HE-GNN), institutions can transition from opaque AI-based climate-health forecasting to transparent, auditable, and ethically explainable intelligence systems. This ensures that interventions are not only technically optimal but also socially just, politically accountable, and cognitively interpretable—aligning with evolving global standards in AI governance, such as those endorsed by the OECD and UNESCO.

Framework: Hierarchically Explainable Graph Neural Network for Transparent Climate-Health AI (HE-GNN)
Layered AI Architecture for Explainability and Ethical Decision-Making in Climate-Health Contexts

Layer	Main functionality	AI module (s) and components	Operational scope
Layer 1	Node embedding generation via graph attention network	Graph Attention Network (GAT)	Input → Node neighborhood features → Output → Learned node embeddings with attention weights
Layer 2	Feature importance attribution using shapley values	Shapley value attribution module	Input → Node features and predictions → Output → Quantified feature contribution scores for interpretability
Layer 3	Counterfactual analysis for causal effect estimation	Counterfactual analysis module	Input → Baseline and perturbed inputs → Output → Causal effect measures for actionable insight
Layer 4	Ethical AI explanation and decision support	Ethical decision support	Input → Explanation outputs and model decisions → Output → Transparent reports and audit trails for ethical governance

Figure 6.3 illustrates the use of AI networks to address both climate change and its mental health impacts, highlighting six main components. Climate change's impact on mental health includes direct effects from extreme weather and indirect effects from economic and social disruptions. AI networks facilitate data integration and prediction by combining climate, health, and socio-economic data to identify vulnerable regions and enable early intervention strategies. Real-time collaboration

Fig. 6.3 AI networks for climate change mitigation and mental health: key components and ethical considerations

and global data sharing enhance coordinated international responses and provide AI-driven early warning systems. AI networks support climate mitigation through sustainable practices, optimizing renewable energy, and guiding policy with models predicting environmental and health benefits. Addressing socio-economic vulnerabilities involves analyzing factors like unemployment and infrastructure quality to target regions for integrated mental health and climate adaptation support. Ethical considerations and challenges emphasize the need for data privacy, transparency, and equitable AI use, addressing biases and establishing ethical frameworks.

AI networks have the potential to revolutionize the way we address the dual challenges of climate change and mental health. By enabling real-time data integration and global collaboration, these networks provide tools for proactive and effective interventions. However, realizing this potential requires ongoing efforts to address ethical concerns, improve data quality, and ensure that AI systems are inclusive and transparent. As climate change continues to threaten both environmental and mental well-being, leveraging AI networks for sustainable and resilient solutions will be critical for the future.

6.4 Conclusion

This work presents a comprehensive investigation into the role of advanced Artificial Intelligence (AI) frameworks in enhancing climate-health resilience, particularly within the context of flood prediction, public health mitigation, and disaster response. Throughout the analysis, a multitude of innovative AI models, grounded in both machine learning and deep reinforcement learning, have been explored, with particular emphasis on their ability to synthesize and leverage multimodal data streams for enhanced decision-making, prediction, and resource allocation.

The integration of Graph Neural Networks (GNNs), Reinforcement Learning (RL), and Federated Learning (FL) has proven to be pivotal in addressing complex, multidimensional challenges in climate resilience. By utilizing spatio-temporal data, the models are capable of predicting disaster events and assessing long-term health risks due to climate change, while simultaneously incorporating environmental, socio-economic, and health data to optimize mitigation strategies. These AI frameworks, when applied to flood management and health resilience, provide significant advancements over traditional models by enabling real-time data processing, dynamic adjustments to intervention strategies, and the ability to mitigate the biases inherent in decision-making processes.

AI plays a key role in flood management, enhancing the accuracy and timeliness of flood prediction through frameworks like the Hybrid Spatio-Temporal Flood Forecasting Model ($HSTF^3$) and Multimodal Neural Networks for Flood Resilience. By integrating multimodal data streams, such as satellite imagery, meteorological reports, and sensor networks, AI provides high-resolution, real-time flood risk assessments, improving both preemptive and responsive measures. The capacity to predict and respond to these events allows for better resource allocation and improved protection of vulnerable communities, reducing human and material losses.

In the realm of public health mitigation, AI's contribution extends beyond immediate disaster response, addressing the long-term health risks exacerbated by climate change. Frameworks integrating Deep Gaussian Processes (DGPs) and LSTM Networks offer predictive insights into the evolution of health outcomes, while Bayesian Inference helps quantify uncertainty in these predictions. This model architecture enables the generation of actionable insights for public health planning and intervention, ensuring that strategies are based on evidence and are adaptable to evolving conditions. The integration of these AI systems enhances preparedness, enabling timely and efficient responses to health crises triggered by climate-related events.

The chapter also emphasizes the ethical considerations in deploying AI for climate-health applications. By incorporating Explainable AI (XAI) techniques such as SHAP (Shapley Additive Explanations) and counterfactual analysis, the models ensure that AI decisions are transparent and interpretable, promoting trust and accountability. Furthermore, fairness-constrained learning algorithms, including Demographic Parity Fairness Regularization and Multi-Objective Optimization, are integrated to prevent biases in disaster response and climate-health interventions, ensuring that the benefits of AI are equitably distributed across different population groups. These measures address the potential risks of AI exacerbating existing disparities, creating a more inclusive and just framework for addressing climate-related challenges.

Federated Learning (FL) emerges as a critical component for facilitating cross-border collaboration while preserving data privacy. By enabling decentralized learning, FL allows multiple jurisdictions to collaborate and share climate-health models without compromising sensitive data. This approach supports international coordination in the face of global climate challenges, fostering collaboration across regions and ensuring that data sovereignty concerns do not hinder effective responses to climate-related disasters.

The use of reinforcement learning to continuously adapt public health interventions based on real-time feedback marks a significant advancement in optimizing decision-making processes. AI systems equipped with adaptive decision support capabilities can refine resource allocation strategies in dynamic, high-stakes environments like disaster zones, ensuring that interventions remain responsive and effective throughout the crisis lifecycle. This adaptability is crucial for minimizing the impact of climate-induced disasters and enhancing the overall resilience of communities.

The integration of AI, particularly through Deep Reinforcement Learning, Graph Neural Networks, and Explainable AI, offers a transformative opportunity to advance climate-health resilience strategies. The models discussed in this work not only enhance disaster prediction and health risk management but also provide deeper insights into the socio-economic and health impacts of climate change. Importantly, the ethical and fairness considerations embedded in these frameworks ensure that AI technologies are deployed responsibly, minimizing bias and promoting equitable outcomes for all communities, especially those that are most vulnerable.

The synergy between AI-driven disaster prediction, real-time resource allocation, and health risk management lays the foundation for building smarter, more resilient communities. By integrating multi-source data and ensuring transparency and fairness in decision-making, these frameworks offer the potential to mitigate the adverse effects of climate change, particularly in the face of increasing flood risks and public health challenges. As climate change accelerates, the frameworks presented provide a roadmap for future climate-health resilience efforts that are adaptable, efficient, and socially responsible.

References

Koliokosta, E. (2023). Impact assessment of climate change on public health: A global perspective. *Environmental Science Proceedings*, *26*, 68. https://doi.org/10.3390/environsciproc2023026068

Nie, P., Zhao, K., Ma, D., Liu, H., Amin, S., & Yasin, I. (2024). Global climate change, mental health, and socio-economic stressors: Toward sustainable interventions across regions. *Sustainability*, *16*, 8693. https://doi.org/10.3390/su16198693

Saravi, S., Kalawsky, R., Joannou, D., Rivas Casado, M., Fu, G., & Meng, F. (2019). Use of artificial intelligence to improve resilience and preparedness against adverse flood events. *Water*, *11*, 973. https://doi.org/10.3390/w11050973

Conclusion

This book has undertaken a rigorous, multidimensional excavation of Artificial Intelligence as the indispensable architect of contemporary and future urban realities. Beyond superficial technological adoption, AI emerges here as the algorithmic substratum from which cities evolve into living, sensing organisms—systems capable of self-optimization, ethical discernment, and anticipatory governance. This book demonstrates that mobility, energy, waste, and safety infrastructures are no longer passive frameworks but active, adaptive agents embedded with sophisticated AI modalities—ranging from Proximal Policy Optimization in traffic orchestration to hybrid LSTM–DRL energy management and XGBoost-enabled sustainable grid optimization.

At the confluence of mathematical rigor and philosophical inquiry, the deployed AI models articulate a novel ontology of the city—one in which discrete, data-driven layers coalesce into a resilient, just, and sustainable urban metabolism. Each case study transcends its empirical boundaries to serve as a prototype for an AI-empowered civic ethos, where transparency, explainability, and environmental stewardship are not afterthoughts but integral design imperatives.

Technically, the synthesis of deep reinforcement learning, temporal sequence modeling, and ensemble gradient boosting architectures showcases how contemporary AI frameworks reconcile the dual imperatives of operational efficiency and ecological prudence. The layered architectures introduced in this book elucidate the intricate balance between real-time responsiveness and strategic foresight, highlighting mechanisms such as trust-region policy optimization and multi-objective reward functions calibrated to social and environmental metrics.

Philosophically, this book posits that the future of urban life hinges on a redefinition of agency: the city as an intelligent actor capable of reflexive adaptation, algorithmic empathy, and systemic regeneration. This perspective mandates a paradigm shift in urban governance—from human-centric command toward symbiotic human–AI co-creation—thereby compelling policymakers, engineers, and scholars to reimagine infrastructure as a collective conscience.

D. Sargiotis, *Blueprints for Future Living*,
https://doi.org/10.1007/978-3-032-24178-8

Globally, the models and frameworks elucidated here extend far beyond their localized implementations. They exemplify a scalable blueprint for cities worldwide confronting the dual crises of rapid urbanization and climate change. By embedding justice and sustainability within the algorithmic core, these AI systems offer a transcendent path forward—where urban complexity is not a challenge to be contained, but an opportunity for emergent order that harmonizes human well-being with planetary boundaries.

The book asserts a bold imperative: designing the cities of tomorrow inherently to architect the AI systems that govern them—and governing those systems with foresight, equity, and ethical rigor. As we stand at the nexus of technological possibility and existential urgency, this work illuminates how algorithmic intelligence can transform urban futures into regenerative, adaptive, and inclusive landscapes of life.

MIX
Papier aus verantwortungsvollen Quellen
Paper from responsible sources
FSC® C105338

If you have any concerns about our products,
you can contact us on
ProductSafety@springernature.com

In case Publisher is established outside the EU,
the EU authorized representative is:
Springer Nature Customer Service Center GmbH
Europaplatz 3, 69115 Heidelberg, Germany

Printed by Libri Plureos GmbH
in Hamburg, Germany